3 4211 000086534

S0-BJP-155

SYNOPSIS OF SCIENCE-PROCESS CATEGORIES

A OBSERVING

- **Aa** Seeing
- **Ab** Feeling
- **Ac** Hearing
- **Ad** Smelling
- **Ae** Tasting
- **Af** Using Several Senses

B COMMUNICATING

- **Ba** Describing, Speaking, Sounding
- **Bb** Formulating Operational Definitions, Naming
- **Bc** Recording, Tabling, Writing
- **Bd** Researching the Literature, Reading, Referencing
- **Be** Picturing, Drawing, Illustrating
- **Bf** Graphing

C COMPARING

- **Ca** Making General Comparisons or Comparisons from Different Perspectives
- **Cb** Estimating
- **Cc** Making Numerical Comparisons, Counting
- **Cd** Measuring Lengths, Angles
- **Ce** Measuring Temperatures
- **Cf** Weighing
- **Cg** Measuring Areas, Volumes, Pressures
- **Ch** Making Time Comparisons, Measuring Rates

D ORGANIZING

- **Da** Seriating, Sequencing, Ordering
- **Db** Sorting, Matching, Grouping
- **Dc** Classifying

E EXPERIMENTING

- **Ea** Identifying a Problem, Formulating Questions
- **Eb** Hypothesizing
- **Ec** Controlling and Manipulating Variables, Testing

F INFERRING

- **Fa** Generalizing, Synthesizing, Evaluating
- **Fb** Using Indicators, Predicting
- **Fc** Using Explanatory Models, Theorizing

G APPLYING

- **Ga** Using Knowledge or Instruments, Identifying Examples
- **Gb** Inventing, Creating
- **Gc** Constructing
- **Gd** Growing, Raising
- **Ge** Collecting

Lawrence F. Lowery

University of California, Berkeley

Allyn and Bacon, Inc.

Boston · London · Sydney · Toronto

THE EVERYDAY SCIENCE SOURCEBOOK

Ideas for Teaching in the Elementary and Middle School

ABRIDGED EDITION

To Carol, Alison, Diane

Drawings by Scientific Illustrators
Artist: David Lynas

Credits: Table 111.10 and Figures 111.10a-g from *Structure and Properties of Matter: A Teacher's Manual for Chemistry in the Elementary Grades,* Chester T. O'Konski, Department of Chemistry, and D. C. Ipsen, Elizabeth Hansen, and Gretchen Gillfillan, Elementary School Science Project, University of California, Berkeley, 1961. Table 431.02 and Graph 431.02c from *Cartesian Coordinates,* S. P. L. Diliberto and G. E. Housh, Elementary Science Project, University of California, Berkeley. Table 431.04b from *Science and Children,* March 1976. Copyright 1976 by the National Science Teachers Association, Washington, D.C. Used by permission.

Library of Congress Cataloging in Publication Data

Lowery, Lawrence F
The everyday science sourcebook.

Includes index.
1. Science—Study and teaching (Secondary)—Handbooks, manuals, etc. 2. Science—Study and teaching (Elementary)—Handbooks, manuals, etc.
I. Title.
Q181.L872 1978b 500.2'07 77–16181
ISBN 0–205–06050–1

Printing number and year (last digits):
10 9 8 7 6 5 4 3 2 85 84 83 82 81 80

CONTENTS

PREFACE

Publication of this sourcebook comes prior to a major transition to metric measurement in school curricula. Because the transition is just beginning, both standard and metric units are given for about 90 percent of the activities where measurement is needed. The reader, however, should be aware of several aspects related to the metric measures provided in the activities.

First, many activities purposely do not give precise conversions. Often the metric measure is rounded off to an easily measured unit. Sometimes the ratios of sequential units are more important than precise equivalents. For example, if an activity requires the measurement of 5 inches, 10 inches, and 15 inches, the suggested metric measurements might be 10 cm, 20 cm, and 30 cm—the second unit being twice the first and the third unit three times the first.

Second, many industrial conversions are not yet clear. Although liter containers can substitute well for quart containers, gallon equivalents are not established among different industries. Thus a 10-gallon aquarium remains so until an industrial decision is made on container sizes.

Finally, some of the activities have no equivalents because their addition makes the activities awkward and confusing. This is especially apparent on graphs and certain tables.

As time passes, all activities will eventually be presented in the metric system. Until then, as much help toward this end is given to the reader as possible.

Nearly twenty years have passed since the organizational systems worked out for this sourcebook were conceived. Many friends and colleagues contributed throughout these years in various ways to the gradual collection and classification of appropriate activities. Countless teachers and classroom students added to the collection, suggesting activities, testing activities, and providing refinements. A very loud *thank you* goes to those scores of individuals who contributed directly and indirectly in so many different ways.

Other appreciations go to the specific specialists who submitted activities and kept the information scientifically accurate and classroom related. To the science educators, *thanks* for years of fruitful professional interactions: Carl Berger, Professor of Science Education, University of Michigan; Jane Bowyer, Professor of Science Education, Mills College, California; George Castellani, Director of Curriculum, Oakdale Joint Union High School District, California; Diane Conradson, Professor of Science Education, California State University at San Jose; Gene Deady, Professor of Science Education, California State University at Chico; Lou Foltz, Director of Teacher Education, Warner Pacific College, Oregon; Lawrence Hovey, Professor of Science Education, Texas Tech University; Roger Johnson, Professor of Science Education, University of Minnesota; Vince Mahoney, Professor of Science Education, Iowa Wesleyan; Richard Merrill, Director of Science Education, Mt. Diablo Unified School District, California; Rita Peterson, Professor of Science Education, California State University at Hayward; James Shanks, Professor of Science Education, California State University at San Fernando.

And *thanks* to the scientists who read portions of the manuscript, contributed current scientific knowledge, and corrected for accuracy: Stuart Bowyer, Professor of Astronomy, University of California, Berkeley; Gordon Chan, Professor of Marine Ecology, College of Marin, California; Linda Delucchi, Resident Biologist, Lawrence Hall of Science, Berkeley, California; Joseph Hancock Jr., Professor of Plant Pathology and Environmental Studies, University of California, Berkeley; Alan Friedman, Director of Astronomy, Lawrence Hall of Science, Berkeley, California; Watson Laetsch, Professor of Botany and Director of the Lawrence Hall of Science, Berkeley, California; Richard White, Professor of Electrical Engineering and Computer Sciences, University of California, Berkeley.

L.F.L.

THE EVERYDAY SCIENCE SOURCEBOOK

Introduction

The well-prepared and creative educator is always on the alert for fresh activities that will help to improve, enrich, and extend the learning experiences of students. This book, unlike any other that has ever been offered, is a storehouse of science activities suitable for the elementary grades, organized to provide easy access to fresh and supplementary ideas. It is designed to serve as an invaluable resource for activities to enhance any science program at the elementary and intermediate grade levels.

PURPOSE

Each science program must be defined and presented with the objective of a curriculum that is as broadly applicable as possible to all schools, locales, and children. Only rarely does a science program have the flexibility necessary to account for economic and social differences existing in various locales or to provide for differences in individual interests and abilities. To cope with such differences, the teacher must constantly supplement the science program. This book is designed to provide access to appropriate alternatives for supplementary instruction.

A second purpose of this book is to serve as a flexible and useful tool that will be adaptable to needs of the user. Its Index Guide and unique organization are designed to allow individuals with different needs and objectives to locate appropriate activities easily. The guide system facilitates location of experiences pertaining to specific learning behaviors useful in science—generally called *science processes* (e.g., observing, measuring, classifying). It also leads the user to activities relating to specific principles, generalizations, or topic areas of science.

The organizational system is numerical and, within four broad science categories, the entries are grouped by science topics. Each topic is arranged around scientific principles and generalizations. A glance at this organization provides some notion of how scientific ideas are related and how they contribute to the structure of scientific knowledge.

The guide and organizational systems combine to form a reference book that enables the user to locate quickly numerous synonymous experiences relating to particular scientific ideas or behaviors. Readers will be stimulated by fresh ideas and new directions that will make their instruction more valid, powerful, and enjoyable.

WHAT THIS BOOK IS NOT

This book is not a science program, although instructors might develop a new unit or program of instruction from selected entries. Since this is an activity sourcebook, the entries are designed as supplements to science programs to help reinforce student learning and skills.

This is not a methods book. You will not learn from this book how to teach science. As a sourcebook, it will not tell you how to use activities for inquiry or discovery teaching. Also, it will not tell you how to work through activities in a didactic fashion. How you instruct depends upon your own training and abilities. Each activity can be taught didactically or by an inquiry technique, just as any new science program can be taught didactically or by inquiry. Because this is not a methods book, most of the suggestions for activities are intentionally brief. Creative and well-prepared individuals will use the suggestions as springboards for their own ideas and will adapt them in many ways to meet the needs of students.

This book is not a source of scientific knowledge and is not intended to supply all the factual background information and explanations for science topics. However, the entries do supply the contributing idea and generalization inherent in each activity, and each one will contain sufficient information pertaining to the instructional conditions to enable you to prepare and introduce an experience for students.

FORMAT

The book is divided into three parts. The introduction illustrates how to use it effectively. Careful reading of this part will be worthwhile to the user. The second part, the main body of the book, contains the entries—listings of suggested science activities, grouped topically and organized within a structure that indicates interrelationships among ideas. The third part is the Index Guide, the key to efficient use of the book. It lists science topics and science behaviors alphabetically.

ORGANIZATION: SCIENCE CONTENT

The entries in the book are comprised of a number of interlocking science content organizations and a numerical code system. Entries are arranged in broad science categories that are subdivided into topics, subtopics, and specific activities.

Broad Science Content Categories. Four very broad and generally standard content categories are used to provide a framework for the main body of this book.

100–199: Inorganic Matter. Matter, one of two great divisions studied by scientists, makes up all the materials that occupy space in the world around us. The scientist subdivides the materials into two kinds—inorganic and organic. Inorganic matter is the subdivision which comprises all nonliving materials (e.g., the rocks and minerals above and below the surface of the earth). This content category includes such directly observable aspects as the physical and chemical properties of matter and the changes in the states of matter (e.g., solid, liquid, gaseous).

200–299: Organic Matter. The other subdivision of matter studied by scientists is organic matter. Organic matter is found in all living materials (e.g., the various forms of plants and animals). This content category contains entries that pertain to the directly observable physical and chemical properties of living organisms, their growth and response to environmental conditions, and the interrelationships among them.

300–399: Energy. Energy, the second great division studied by scientists, means the ability to do work. This content category includes entries that pertain to the various forms of energy such as light, sound, and heat.

400–499: Instructional Apparatus, Materials, and Systems. This fourth category contains entries dealing with the preparation of various materials useful in teaching science that have wide application throughout the content categories. This category includes such topics as techniques for the cutting and bending of glass to make certain apparatus and construction plans for building measuring devices (e.g., balance and spring scales).

Each of the four broad science content categories is coded by numerals in a decimal system.

100
↳—— *Inorganic Matter*
200
↳—— *Organic Matter*
300
↳—— *Energy*
400
↳—— *Instructional Apparatus, Materials, and Systems*

Topics

Each of the four broad science categories is divided into specialized topics. For example, the category of *100 Inorganic Matter* is subdivided into six topics that are coded by the second numeral in the series of three numerals.

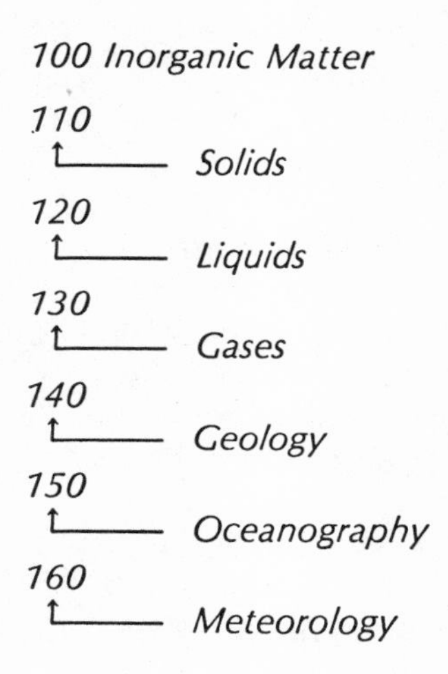

Subtopics

Each of the special topics is further subdivided into finer aspects. For example, topic *140 Geology* is broken up into two subtopics, identified by the third numeral in the series of three.

100 Inorganic Matter

140 Geology

141
↳—— *Characteristics*
142
↳—— *Interactions*

Specific Activities

Each of the subtopics is composed of specific activities. For example, subtopic *142 Interactions* includes activities that are listed sequentially and enumerated as decimals.

100 Inorganic Matter

140 Geology

142 Interactions

142.01 — *Observing wind moving across soil*

142.02 — *Observing how wind carries and deposits materials*

142.03 — *Measuring the distances wind carries and deposits different materials*

142.04 — *Collecting and comparing samples of wind-borne materials*

The table of contents for the book provides a synopsis of the content categories that form the structure for the entries.

ORGANIZATION: SCIENCE ACTIVITIES

The specific activities listed for a subtopic are organized in a classification system of *Generalizations* and *Contributing Ideas* that are logically arranged statements pertaining to order or relationships among phenomena scientists have found, so far as is now known, to be invariable under the given conditions.

A *generalization* in this book might be called a *theory,* a *law,* a *principle,* or a *concept* in elementary science textbooks. The term *generalization* is used to cover broadly the inconsistency of terms used in science education and, in a learning sense, to denote a higher-order idea that is made up of a number of contributing ideas. For example, for a student to understand:

> ***Generalization:*** Adding or removing heat affects materials.

experiences are required that develop knowledge of certain contributing ideas such as:

> *Contributing Idea A:* Adding heat causes nearly all solids to expand; removing heat causes them to contract.
>
> *Contributing Idea B:* Adding heat causes nearly all liquids to expand; removing heat causes them to contract.
>
> *Contributing Idea C:* Adding heat causes all gases to expand; removing heat causes them to contract.

Thus each generalization is subdivided into ideas that, when put together, contribute to the understanding of the generalization. No hierarchical order is suggested by any sequence of ideas, yet each contributes in some way to the larger or more general statement. The term *contributing idea* is used in a comprehensive way to cover the many terms used indiscriminately in science education. *Contributing idea* denotes a description of something actually perceived through the senses or something never perceived directly but deduced from bits of information.

Wherever possible, the decimally sequenced activities contained within a contributing idea are ordered from highly sensory-oriented experiences, through very concrete and manipulative kinds of experiences, to experiences that require more formal thinking by the students.

ORGANIZATION: SCIENCE PROCESS

The nature of science is both content (the scientific ideas and generalizations one should know) and process (the behaviors one uses to come to know information). Since content and process are dependent on each other, they are interrelated at the specific activity level.

The organization uses a letter code system for the processes. Capital letters represent broad superordinate science-process categories. The first four categories tend to encompass science-process activities that are most appropriate to young students (preschool through grade three), while the latter three categories tend to be most appropriate for older students (grade four and above).

A — *Observing*
B — *Communicating*
C — *Comparing*
D — *Organizing*
E — *Experimenting*

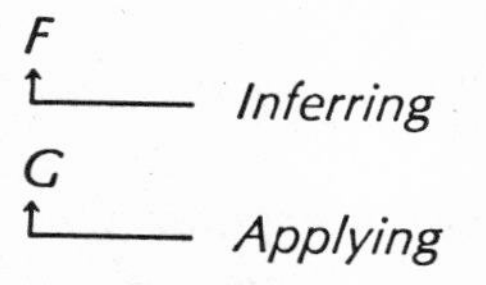

Lower case letters are used to represent more specific behaviors related to the science processes.

A Observing

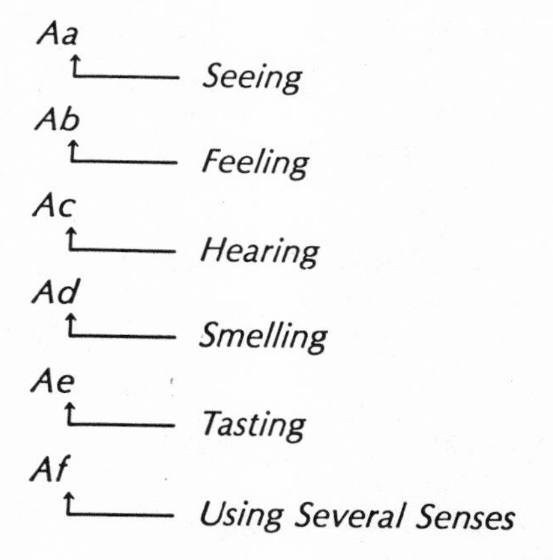

Each activity in the main body of the book is coded by the capital and lowercase letters indicating, respectively, the general science process and the specific behavior expected in the activity. For example, the letter code *Ec* indicates that the activity generally involves experimentation *(E)* and specifically requires the student to control and manipulate variables *(c)*. Naturally, other science processes and behaviors are required in any given activity. This code simply indicates the processes of greatest emphasis.

The charts at the back and front of the book provide a synopsis of the process categories coded in the main body of the book.

HOW TO USE THIS BOOK

This book may be used by current practitioners as well as by the curriculum research groups whose primary concern is developing instructional systems to reach particular objectives. Although individuals may use this book with different purposes in mind, the general purpose for most will be to find the most fitting activity for some specific science objective. The book is structured to accommodate a number of possible objectives.

The following are several examples of how the Index Guide may be used.

To Find the Most Fitting Activity Pertaining to a Science Content Objective or Topic

1. Look up a word related to your objective or topic in the Index Guide.
2. The word will numerically refer to activities or subtopics in the main body of the book. Select the numerical reference most suitable to what you want.
3. Look up the numerical reference in the main body of the book. The pages of the main body are identified by the numerical code.
4. Find the activity most suitable to your objective from the listing of activities located by the reference.

Example 1: Suppose you are looking for an activity that would help students understand a general idea that you have in mind such as "water washes away topsoil." If you look up the word *water* (or *erosion* or *soil*), you will find several possible subtopics listed. The most suitable reference to your idea would be subtopic *erosion 142.17–25.* Look up the numerical reference *(142.17–25)* in the main body of the book and you will find many activities, one or several of which are applicable to your particular situation.

When you locate a numerical reference in the main body of the book, you will find that the scientific content is clearly stated as a generalization and contributing idea. Thus, each activity listed for a generalization or contributing idea has an inherent common scientific finding. In the illustrated example, you would find:

> ***Generalization II:*** Water physically changes the earth's surface by carrying and depositing soil, sand, and other debris.
>
> *Contributing Idea A:* Some water is absorbed into the earth's surface and stored in porous soils and rocks.
>
> *Contributing Idea B:* Water accumulates and flows from higher places to lower places over the earth's surface.
>
> *Contributing Idea C:* Water carries and deposits materials.

Thus each activity listed for *Contributing Idea C* contains the common finding that "water washes away topsoil," yet each activity is different in terms of possible experiences for students. Students may: observe how rain carries away soil *(142.17);* collect samples of water-borne materials *(142.18);* observe how water-carried materials are deposited *(142.19);* and so forth. Note that the activities are generally sequenced from simple to complex or from highly perceptual and sensory to abstract. Such sequencing will help you identify the most appropriate activity for individual students. Also note that a glance at the total structure of this section in terms of other contributing ideas will provide you with related possibilities that might be useful. The other ideas may suggest preliminary experiences of which you had not thought or continuing or branching experiences to follow up the one you referenced.

Example 2: Suppose you have a specific factual content objective in mind such as "heat causes solids to expand." Perhaps your current textbook explains this phenomenon, but you would like to provide some experience to help students understand the idea more clearly. First look up *heat* (or *solids* or *expansion*) in the Index Guide. Next, find the numerical reference most suitable to your objective: "expansion *342.13–17.*" Turn to this numerical reference in the main body of the book to find activities that have inherent in them the specific factual idea that you have as an objective. From the listing of activities, choose the one most appropriate for your students.

The following are several examples of how the process and behavioral aspects of the book can be referenced in the Index Guide.

To Find the Most Fitting Activity Pertaining to a Science Process or Behavioral Objective

1. Look up in the Index Guide an action verb that most closely describes the science process or behavior that you have in mind.
2. The word will numerically refer to activities in the main body of the book. Select the numerical reference most suitable to what you want.
3. Look up the numerical reference in the main body of the book.
4. The activity you find will indicate other science processes and behaviors related to the activity. These are coded and can be identified by looking at the charts at the front and back of this book.

Example 1: Suppose you would like the students to become more skillful in using their sense of touch. Look up *touching* or *feeling* in the Index Guide and select the most suitable science topic area. Look up the numerical reference in the main body of the book to find the activity that involves the sense of touch. Note the letter codes at the side of the activity. The codes indicate other science or behavioral processes that may be emphasized in the same activity. They may introduce some other instructional possibilities for you. To decipher the code, simply look at the chart at the front or back of the book. In the sample shown, communication and comparison processes can be emphasized in addition to the process of touching.

Example 2: Suppose you would like students to be able to classify types of seeds. Look up the appropriate action verb in the Index Guide (classifying). You will find a reference to activities involving the behavior of classifying listed for the content area pertaining to seeds. Turn to the numerical reference in the main body of the book to find the activities.

It should be noted that each activity in the main body of the book begins with a performance or behavioral statement, for example:

221.25 Classifying seeds

and each statement describes a specific behavior with an action verb. As a result,

Seeing and describing some characteristics of some solids. The visual properties of objects that can be selected and controlled in instructional situations are: size, shape, color, and pattern. Although students might begin with grossly different objects, fine discriminations can be later used to sharpen perceptual awareness. Fine discriminations might involve closely matching lengths of sticks, serations or venation of leaves, complex designs, shades or tones of colors.

04
Aa
Ba
Ca

a. *Clue given:* Show students a leaf. Have them find a matching leaf (on the basis of shape) from a collection of differently shaped leaves.

b. *Clue given:* Buttons, coins, or other small objects can be glued to heavy cardboard. Ask students to find the two objects that are alike. Finer discriminations among the objects can be developed by selecting ones that are more closely alike.

c. *Clue given:* String beads (identical in size and color but different in shape) on a string. Every time you string a bead, have a student string a matching one.

d. *Clue not given:* Prepare a collection of polished, colored stones of similar size (include two stones that are identical in color). Ask students to find the two that look the same on the basis of color.

e. *Clue not given:* Prepare a set of objects that vary in size but which are the same in shape and color. Have students match identical objects by size.

Feeling and describing some characteristics of some solids. Tactile properties have the greatest range of instructional possibilities. Materials can be selected to emphasize such aspects as size, shape, or textures (e.g., roughness, stickiness, softness, sharpness, etc.).

05
Ab
Ba
Ca

a. *Clue given:* Prepare about eight squares of sandpaper, identical in size, but different in grade (roughness). Place the squares in a Feely Box. Give students another square that matches, on the basis of roughness, one of those in the box. Ask them to find the matching square. For finer discriminations, you might try using different grades of emery paper.

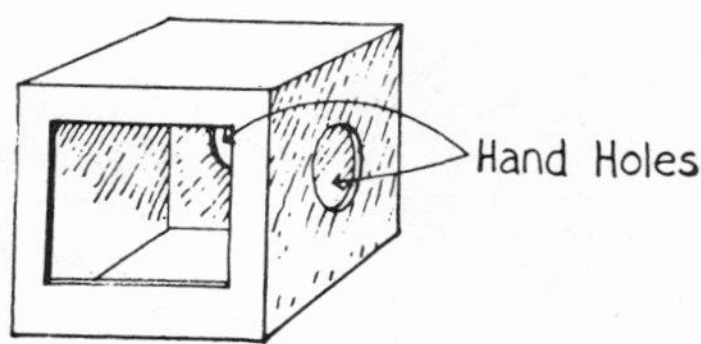

FIGURE 111.05

b. *Clue not given:* Place about six wooden dowels of equal length but of different diameters within a Feely Box. Include two dowels of identical diameters. Have students use their sense of touch to locate the two that feel the most alike.

Hearing some characteristics of some solids. Sounds produced by certain solid objects are sometimes a characteristic of that object. The primary auditory properties are pitch, intensity, and rhythm.

06
Ac
Ba
Ca

a. *Clue given:* Have students listen as you play a single note on a musical instrument. Ask them to identify a matching note from a number of notes played in a series.

b. *Clue given:* Using a pencil, tap simple rhythms on a desk. After students listen to the rhythms with their eyes closed, have them use pencils to repeat the rhythm that you tapped.

represents the number of peas in that pod. Repeat for a second pea pod. After this demonstration, let students come to the board and mark the number of peas found in each of their pods. The result will form a histogram similar to the one shown. From

Tallying the Number of Peas in Pods

3 4 5 6 7 8 9 10 11 12

HISTOGRAM 221.24

the results, questions that will lead to inductive thinking can be asked: "If you were to pull one more pea pod from the bag, what is the most likely number of peas in the pod?"; "Is there a relationship between the length of a pod and the number of peas in the pod?" The activity can be continued with string bean pods or other types of seed pods. If fresh pods are not in season, two #303 cans of whole beans will provide enough bean pods for a class of thirty students. Apples in season can be cut in half between the stem and blossom ends, the seeds can be counted, and a histogram made of the count. *(Note: Seeds from plants in the same family, produce identical numerical seed patterns.)*

Contributing Idea B. Plants can be classified.

Classifying seeds. Let students bring seeds that they find to school. Have them classify the seeds in as many ways as they can. Classifications will depend upon the seeds that are collected and are best if they are invented by the children. They need not agree with scientific classifications; however, students should be able to provide rationales for their groupings. For example, they might group seeds by shapes (triangular, circular, egg-shaped, curved, coiled), by colors (orange, purple, striped, spotted, black, white), by sizes or weights (small or large), or by textures (horns, wings, tails). If fairly large numbers of seeds from the various classifications are obtained, students can set them outside, observe, and keep records concerning the animals that eat the seeds. Judgments about which animals eat which seeds can be made. Students can hypothesize why some seeds are not eaten by any animals.

25 Dc Fa

Classifying seeds by types of dispersion. Divide the class into several small groups. Pass one envelope containing various seeds to each group. Ask students to divide the seeds into sets by the way they think the seeds might travel from one place to another. Encourage group discussion and experimentation to see how seeds might travel (e.g., seeds can be dropped or blown to see what air or wind does to them). Reclassify the seeds on another basis such as texture, shape, size, or color. Reclassify them several times. This will help students realize that there are multiple ways to classify and that some objects can belong to several classifications at one time.

26 Dc

222 INTERACTIONS

Generalization I. Germination is the process by which an embryo starts to grow into a new plant.

Contributing Idea A. Some plants grow from seeds.

any of the statements used in this book may easily be adapted to meet the criteria of behavioral objectives by adding the conditions for the minimal acceptable level of the performance. In a behavioral form, most of the activities in this book can be used for evaluative purposes. For example,

> Given twenty different seeds, students will be able to classify the seeds into groups and describe their own system of classification.

The process and behavioral aspects of the book give it an immediate utility to all concerned—the curriculum developer, the teacher, and the student. It provides an important step toward improvement of science education.

ENTRIES

100-199
Inorganic Matter

110 solids

111 CHARACTERISTICS: SOLID STATE

Generalization I. Solid is one of three states of matter.

> *Contributing Idea.* Solids melt into liquids; liquids freeze into solids.

Observing melting and freezing (solid ⟷ liquid). In small pans, have students heat, in succession, over a hot plate: a piece of paraffin; a cube of sugar; a pat of butter or margarine. Solder or lead (small fishweight) can also be melted in a tin can held by a pair of pliers. After letting students describe what happens in each case, you might tell them that whenever a solid changes to a liquid, it is said to be *melting.* When each pan is removed from the heat and placed in a bowl of ice water, students will see the liquid change back to a solid. Tell them that this change is called *freezing* even though the substance might not feel cold to the touch. **01 Aa Bb**

Determining the melting point of different solids. Students can harden some olive oil in a jar by placing it in a refrigerator for a day. When hard, set the jar of oil in a sunny place. As the oil starts to melt, tip the jar so that the students can take the oil's temperature. Have them record the melting point, then follow the same procedure using margarine, butter, lard, ice, and ice cream. The different materials can be seriated by their melting points. **02 Ce Da**

Observing changes of state: ice (solid ⟷ liquid). Have students watch ice cubes melt in dishes placed in different parts of the classroom. After seeing which melts fastest, let them try to tell why. (It is usually because of heat.) Challenge the students to bring a solid ice cube to school from home. (They will have to be inventive to keep it from melting.) Discuss factors that contributed to those that were brought successfully to school. Next, challenge students to devise ways to make ice cubes that are spherical, hollow, or various other shapes. They might discover that the shape (i.e., amount of surface area) influences the rate at which each melts. For controlled comparisons, differently shaped ice of equal volumes can be made by pouring equal amounts of water into paper cups with flat bottoms and pointed bottoms; various candy cups; cookie molds; round lids or covers; tall cylinders. Set the frozen shapes on dishes in the same locations to keep temperature conditions the same. Students can record and compare the time it takes each to melt. **03 Aa Gb**

Generalization II. Solids have identifiable characteristics.

> *Contributing Idea A.* Our senses can be used to identify solids.

Seeing and describing some characteristics of some solids. The visual properties of objects that can be selected and controlled in instructional situations are: size, shape, color, and pattern. Although students might begin with grossly different objects, fine discriminations can be later used to sharpen perceptual awareness. Fine discriminations might involve closely matching lengths of sticks, serations or venation of leaves, complex designs, shades or tones of colors. **04 Aa Ba Ca**

a. *Clue given:* Show students a leaf. Have them find a matching leaf (on the basis of shape) from a collection of differently shaped leaves.
b. *Clue given:* Buttons, coins, or other small objects can be glued to heavy cardboard. Ask students to find the two objects that are alike. Finer discriminations among the objects can be developed by selecting ones that are more closely alike.
c. *Clue given:* String beads (identical in size and color but different in shape) on a string. Every time you string a bead, have a student string a matching one.
d. *Clue not given:* Prepare a collection of polished, colored stones of similar size (include two stones that are identical in color). Ask students to find the two that look the same on the basis of color.
e. *Clue not given:* Prepare a set of objects that vary in size but which are the same in shape and color. Have students match identical objects by size.

Feeling and describing some characteristics of some solids. Tactile properties have the greatest range of instructional possibilities. Materials can be selected to emphasize such aspects as size, shape, or textures (e.g., roughness, stickiness, softness, sharpness, etc.). **05 Ab Ba Ca**

a. *Clue given:* Prepare about eight squares of sandpaper, identical in size, but different in grade (roughness). Place the squares in a Feely Box. Give students another square that matches, on the basis of roughness, one of those in the box. Ask them to find the matching square. For finer discriminations, you might try using different grades of emery paper.

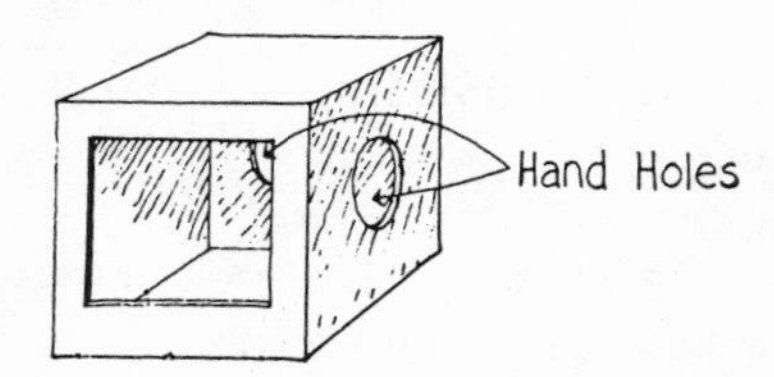

FIGURE 111.05

b. *Clue not given:* Place about six wooden dowels of equal length but of different diameters within a Feely Box. Include two dowels of identical diameters. Have students use their sense of touch to locate the two that feel the most alike.

Hearing some characteristics of some solids. Sounds produced by certain solid objects are sometimes a characteristic of that object. The primary auditory properties are pitch, intensity, and rhythm. **06 Ac Ba Ca**

a. *Clue given:* Have students listen as you play a single note on a musical instrument. Ask them to identify a matching note from a number of notes played in a series.
b. *Clue given:* Using a pencil, tap simple rhythms on a desk. After students listen to the rhythms with their eyes closed, have them use pencils to repeat the rhythm that you tapped.

c. *Clue not given:* Before playing a series of seven notes on a musical instrument, ask students to listen and record the numbers of the two notes among the seven that sound alike. (You might have to repeat the series several times.)

d. *Clue not given:* Tape-record various household sounds. As students listen to the tape, let them describe the characteristics of each sound before identifying the object that makes the sound (e.g., ticking clock, turning egg beater, vacuum cleaner, washing machine).

Smelling some characteristics of some solids. Our sense of smell is not as acute as that of many other animals, and smells are generally described by naming the object that gives off the smell (e.g., "It smells like a skunk"; "It has a mint smell"). **07 Ad**

a. *Clue given:* Place six to eight bars of differently scented soaps into different containers. The bars should be identical in size and shape, and wrapped so that they cannot be distinguished on the basis of color. Place one more bar matching one of the scents in another container and ask students to find, by smelling, the matching container within the set.

b. *Clue not given:* Wrap differently scented candles that are identical in size and shape. Place six to eight of them into boxes or cans. Add one additional box containing a candle that matches one of the others on the basis of its smell. Mix the boxes and ask students to find the two that smell alike.

Tasting some characteristics of some solids. Generally our sense of taste is influenced by our sense of smell. Other than the basic tastes of sweet, sour, bitter, and salty, flavors become complex combinations interwoven with our sense of smell. **(Caution: Tasting unknown materials can be dangerous—warn students against this practice, and let them discuss reasons for the caution.)** **08 Ae Ba Ca**

a. *Clue given:* Give each student his or her own set of four small cups containing salt water, sugar water, quinine water, and water with lemon juice in it. Hand each student a fifth cup containing a liquid that matches in flavor one of those in the other four cups. Have the student find the matching flavor.

b. *Clue not given:* Blindfold students. Let them taste four or five different foods (e.g., slice of lemon, cherry, orange, lime, pineapple). Next, give each a Lifesaver or other candy that matches one of the previous flavors. Ask students to identify which of the previous flavors it matches.

Contributing Idea B. Indicators can be used to identify solids.

Using indicators to identify unknown solid materials. Various tests can be performed on known materials to see what reactions take place. Unknown materials can then be compared to them for similarities. Such tests serve as indicators that show the presence of known materials. To introduce students to some simple indicators, prepare five numbered envelopes for each student or for small groups of students. In the envelopes, place 5 teaspoonsful (five 5 ml spoonsful) each of the following materials: **09 Fb**

1. Granular dextrose or glucose (available at drugstores). *(Note: Ordinary sugar cannot be used. It contains a small*

TIME??

amount of starch to keep it from caking. It will not produce satisfactory results in the tests described here.)

2. Powdered starch (available at grocery stores).
3. Baking soda (available at grocery stores).
4. An equal mixture of dextrose (or glucose) and starch.
5. An equal mixture of dextrose (or glucose) and baking soda.

Have students place 1 teaspoonful (one 5 ml spoonful) each from the numbered envelopes 1, 2, and 3 on a piece of wax or foil paper. Let them describe the materials without naming them—by sight, touch, smell, but *not* by taste. **(Caution: Tasting unknown materials can be dangerous.)** Observations can be recorded on a table, similar to that shown (Table 111.09). Have students discuss limitations of the senses in trying to identify materials. When finished, the materials can be named. Now give students numbered envelopes 4 and 5. Tell them they are mysteries to be solved. Each contains a mixture of two of the known materials. Students can then perform the following tests and record their observations. When finished, they can deduce what materials make up the mystery mixtures.

Test 1—Water as an indicator. Water dissolves some solid materials. Students can half-fill three test tubes with water, put 1 level teaspoonful (one 5 ml spoonful) of the known materials in them, shake the tubes to mix the materials, allow them to stand for a few minutes, and record observations. Students need not agree on descriptive words. It is best to let each record in his or her own terms. Next, test the mystery mixtures and record findings.

Test 2—Heat as an indicator. Heat melts or burns some solid materials sooner than others. Students can put 1 level teaspoonful (one 5 ml spoonful) of each of the known materials into three dry test tubes. Holding them with a test tube holder, heat over a flame by moving the tube in a small circular motion to distribute the heat more evenly. **(Caution: Keep the open end of the test tube pointed away from faces and other people.)** Record observations, then repeat the test using the mixtures.

TABLE 111.09. Identifying the Characteristics of Solid Materials

Material	*Sight*	*Touch*	*Smell*	*Other*
1. sugar				
2. starch				
3. baking soda				
4. unknown				
5. unknown				

Test 3—Iodine as an indicator. Iodine, available at drugstores, indicates the presence of starch by turning a blue-black color. Students can put 1 level teaspoonful (one 5 ml spoonful) of each known material into dry test tubes or place materials in three piles on a piece of waxed or foil paper. Using an eyedropper, put a few drops of iodine solution on each material. Record observations, then test the mystery mixtures. **(Caution: Iodine solution is poisonous and stains are difficult to remove. Have students wash their hands after using it. A dilute solution of sodium thiosulfate, available at drugstores, in water can be used to remove stains.)** Students can use the iodine to test for the presence of starch in various other materials, such as slices of apple, bread, candy, potato, banana, butter, and cheese.

Test 4—Vinegar as an indicator. Vinegar (white or red) indicates the presence of a carbonate by bubbling and fizzing. Students can put 1 level teaspoonful (one 5 ml spoonful) of each of the known materials into three dry test tubes or in three piles on waxed or foil paper. Add a few drops of vinegar to each with an eyedropper. Record observations, then test the mystery materials. Vinegar is acetic acid and reacts in the presence of a carbonate by releasing carbon dioxide gas. Students can use it to test various other materials, such as slices of apple, flour, ground chalk, limestone rock, marble rock, bones, egg shells, or sea shells.

Test 5—Benedict's solution as an indicator. Benedict's solution (or Fehling's solution, both available at drugstores) indicates the presence of simple sugars (sucrose or dextrose). *(Note: It does not react to complex sugars such as table sugar.)* Demonstrate this indicator by placing 1 level teaspoonful (one 5 ml spoonful) of each known material into dry test tubes. Add ½ tube of water and 10 drops of the Benedict's solution. Heat gently, holding the mouth of the tube away from yourself and others. Students will see the color change from blue to green to yellow to orange to brick red depending upon the concentration of sugar present. Next, test the mystery mixtures for the presence of sugar.

Using their own version of Table 111.09 as a guide, students can now deduce the materials that make up materials numbered 4 and 5. Following this experience, other materials can be tested in similar ways: try powdered plaster of paris; granulated sugar; biscuit, pancake, or cake mixes; powdered milk; flour; powdered potatoes. After characteristics are identified, mixtures can provide new mysteries to solve.

Contributing Idea C. Some solids can be identified by their shapes.

Examining the shapes of crystals. Distribute a set of the basic crystalline shapes as shown on the next two pages and/or samples of crystals. The shapes can be constructed from the patterns by cutting them from mimeograph paper and gluing the tabs. Students can sort the shapes by similarities and differences, count the number and kinds of surfaces (sides) and corners on each. A table, such as Table 111.10 on page 24, but without the information, could be used to guide observations. Students will find that each crystal has a definite shape, a certain number of flat surfaces, sharp edges, and pointed corners.

10
Aa
Ba
Ca

Contributing Idea D. Some solids dissolve and some do not.

Testing the solubility of different solids. Obtain a variety of solid materials such as sugar cubes, marbles, salt, nails, soil, coins,

11
Db
Ec

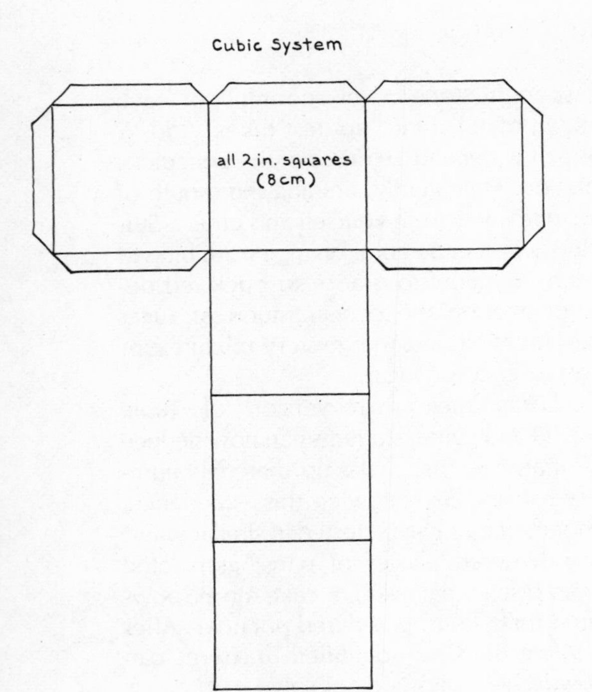

FIGURE 111.10a

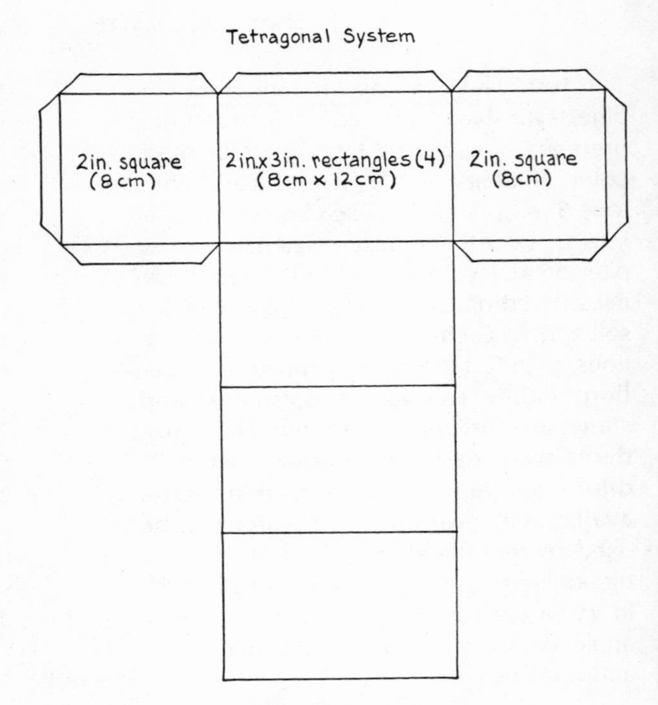

FIGURE 111.10b

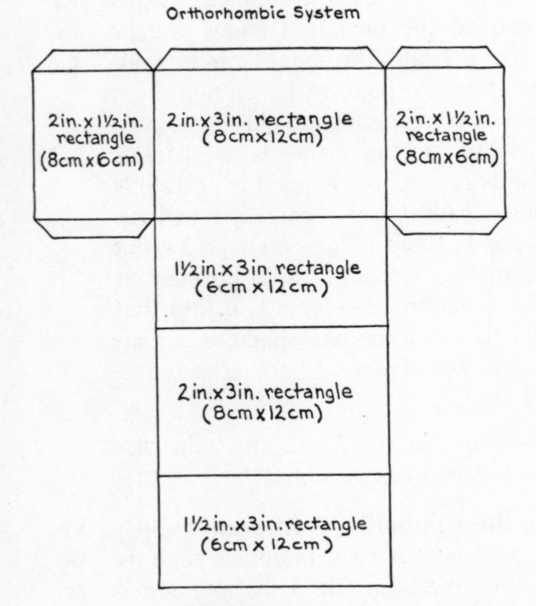

FIGURE 111.10c

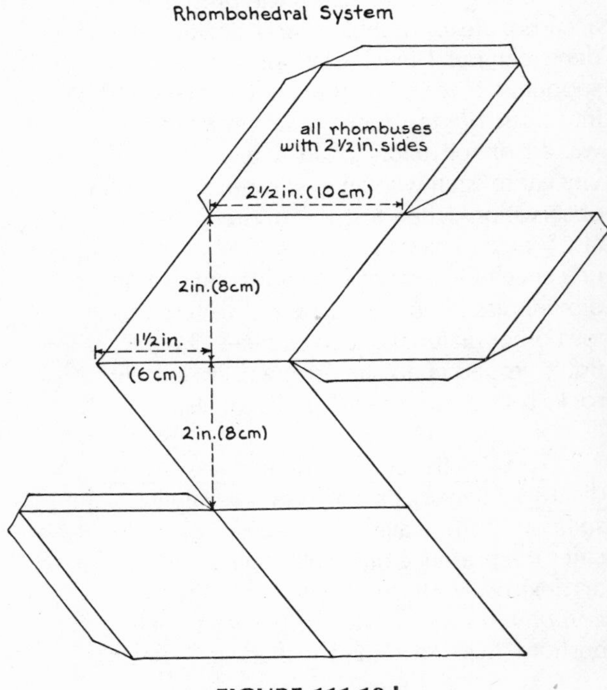

FIGURE 111.10d

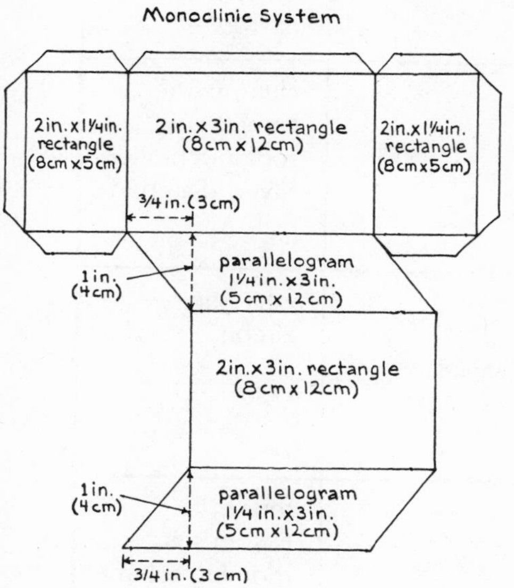

Monoclinic System
2in.x1¼in. rectangle (8cm x 5cm)
2in.x3in. rectangle (8cm x 12cm)
2in.x1¼in. rectangle (8cm x 5cm)
¾in.(3cm)
1in. (4cm)
parallelogram 1¼in.x3in. (5cm x 12cm)
2in.x3in. rectangle (8cm x 12cm)
1in. (4cm)
parallelogram 1¼in.x3in. (5cm x 12cm)
¾in.(3cm)

FIGURE 111.10e

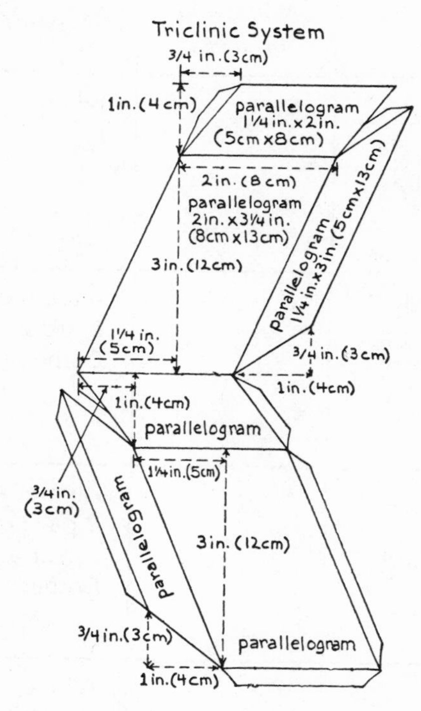

Triclinic System
¾ in.(3cm)
1in.(4cm)
parallelogram 1¼in.x2in. (5cm x 8cm)
2in.(8cm)
parallelogram 2in.x3¼in. (8cm x 13cm)
3in.(12cm)
parallelogram 1¼in.x3in.(5cm x 13cm)
1¼in. (5cm)
¾in.(3cm)
1in.(4cm)
1in.(4cm)
parallelogram
1¼in.(5cm)
¾in. (3cm)
parallelogram
3in.(12cm)
¾in.(3cm)
parallelogram
1in.(4cm)

FIGURE 111.10f

Hexagonal System
2in.x1in.rectangles(6) (8cm x 4cm)
1in.(4cm)
hexagon
2in.(8cm)
hexagon

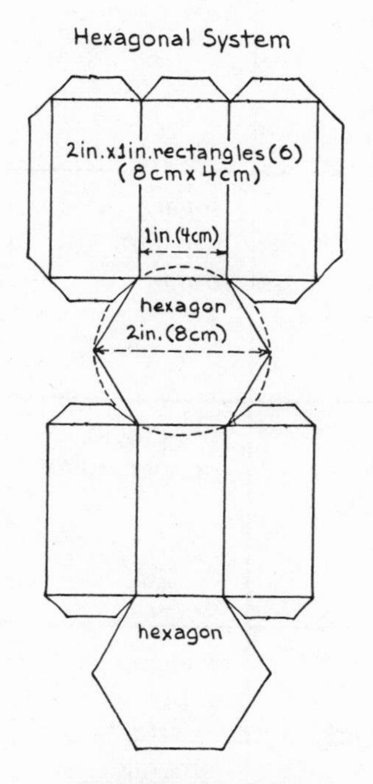

FIGURE 111.10g

soap. Place a small amount of each into glasses or jars of water. Stir or shake the contents, and let stand for several minutes. Students can categorize the materials that dissolve (e.g., sort the materials in terms of those that dissolve, those that do not, those that partially dissolve). Similar tests can be performed using warmer water or different liquids such as alcohol or gasoline. **(Caution: Be sure alcohol or gasoline are used in a well-ventilated area, away from flames or heating units.)**

Contributing Idea E. Solids occupy space and have weight.

Observing that solids occupy space. Half-fill a glass of water and mark the level on the glass. Place a rock, marble, nail, or 12 Aa

TABLE 111.10. Examining the Shapes of Crystals

Type	*Number of Surfaces*	*Shape of Surfaces*	*Crystalline Examples*
Cubic 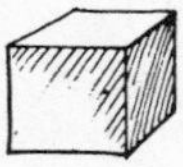	6	All are square	alum, pyrite, garnet, gold, sodium chloride, silver, diamond
Tetragonal 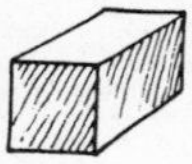	6	4 are rectangles 2 are squares Corner angles are right angles	white tin, zircon
Orthorhombic 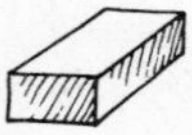	6	All are rectangles 3 pairs of rectangles, each pair a different size Corner angles are right angles	topaz, Epsom salt, rhombic sulfur
Rhombohedral 	6	All are rhombuses (i.e., equilateral parallelograms) No right angles	calcite
Monoclinic 	6	4 are rectangles 2 are parallelograms 16 angles are right angles 8 angles are not right angles	sugar, gypsum, borax
Triclinic 	6	All are parallelograms No right angles at corners	boric acid, copper sulfate
Hexagonal 	8	2 are regular hexagons 6 are rectangles Angles of rectangles are right angles	ice, ruby, sapphire, quartz, emerald, apatite

other object into the water so that it is submerged. Let students note how far the water level rises. They will realize that the solid object has taken up some space in the glass and has pushed some of the water out of the way. To measure solid objects by displacement, see *122.23.*

Measuring solid objects by size and weight. Have students measure the dimensions of a variety of objects such as blocks of wood, desk tops, doors. Be sure they measure the objects in each dimension. After they have practiced somewhat, challenge them to measure such objects as a piece of jig-saw puzzle, a jack, a marble, etc. Ask in what other way objects can be measured (by weight). Students can practice weighing objects on an equal arm balance scale. Let them keep records of the objects measured or weighed and seriate the objects by size or weight. **13 Cd Cf**

Contributing Idea F. Nearly all solids expand when heated and contract when cooled.

Observing that solids expand and contract when heated and cooled. Have students find metal washers made of different materials (e.g., steel, copper, iron) that will just fit through the mouth of a glass bottle. Hold each washer with a pair of pliers and heat it over a flame until it is very hot. Now try to drop the washer through the mouth of the bottle. Students will realize that each washer expands from the heat and will not pass through the mouth. Cool the washers with water or an ice cube to let students see that the washers contract and again fit into the bottle. Students can also note the spaces in concrete walks, bridges, and railroad tracks to see how the structures have been prepared to take care of expansion. **14 Aa**

Contributing Idea G. Some solids conduct heat better than others.

Sensing that solids conduct heat. Have students fill a glass with very hot water, place a silver spoon into it, and hold the end of the spoon for a few minutes to feel it increase in temperature. Students will realize that the heat has been transferred from the water and conducted through the spoon. Let them test other solid objects in the same way and compare how well each conducts the heat. **15 Ab**

Contributing Idea H. Some solids transmit sounds better than others.

Listening to sound through solid materials. Have a student hold a wrist watch in his or her hand at a distance of about 5 feet (2 m) from another student. Let the first student gradually bring it closer until the ticking can be heard by the second student. Measure the distance between the students. Repeat with other individuals. Next, place the face of the watch at one end of a board, and have the students take turns listening. Move the watch toward a student until he or she hears the ticking. Measure this distance and compare it with the earlier measurements. Students can discuss what happens to the volume of sounds when they are transmitted through a solid substance. Try this again using other solid materials. Compare differences. **16 Ac Cd**

Contributing Idea I. Some solids respond to magnetic forces, some do not.

Magnetizing some solids. Obtain some short hard steel rods from a hardware store, and have students stroke one end of each with one end of a magnet (about fifty times) **17 Gc**

from the center to the tip of the steel. Stroke the other end of the steel rod with the other end of the magnet in the same way. When finished, test for magnetism in the steel rod by using it to pick up tacks, paper clips, or pins. Have students test other solid objects to see if they can be magnetized.

Testing the transparency of some solids to magnetism. Have students pick up different objects with a magnet, then place a sheet of paper between the magnet and the objects and try again. Test various thicknesses of the paper (add more sheets) to see what limit there is to the magnetic force. Students can test other materials to see which allow the magnetic force to pass through. **18 Ec**

Generalization III. Solids can be classified on the basis of their physical characteristics.

Contributing Idea A. Some characteristics of solids can be seriated.

Seriating solid objects by sight. Have students order or sequence objects on the basis of shape (e.g., one-sided to many-sided), size (small to large), or color (light to dark). Collections of insects, rocks, leaves, and so forth can also be seriated. Have students give rationales for the arrangements they make. **19 Aa Da**

Seriating solid objects by touch. Students can use a Feely Box (see *111.05*) to order samples of papers by their textures (smooth to coarse), nails by their sizes (short to long or small to large), bags of sand by their weights (light to heavy), and so forth. When finished, give them another object, and have them place it in the series they have arranged. **20 Ab Da**

Contributing Idea B. Some characteristics of solids can be classified.

Classifying solid objects by sight. Give students an assortment of buttons. Tell them to sort them in any way they wish. Some may group them by color, size, shape, or other characteristic; some may use combinations of characteristics. Let the students give a rationale for their arrangement and share the different ways of sorting. Stress that none of the ways is incorrect—all are correct as long as they are sorted on the basis of some rationale. When finished, take the groupings apart, and have them regrouped on another basis. This activity can be repeated using mixtures of breakfast foods, parquetry blocks, or natural objects such as rocks, animals, or leaves. **21 Aa Dc**

Classifying solid objects by touch. Prepare a Feely Box (see *111.05*). Let students sort objects that might be grouped in more than one way. For example, if the certain objects (stars, squares, circles, etc.) were placed in the box, a student might group them by shape (all circles in one pile, squares in another, etc.). When finished, let the students give their rationales, then take the groupings apart and ask the students to group them again but in a different way. This time they might be grouped by size. **22 Ab Dc**

Classifying solid objects on the basis of tests. Indicators can be used to test the properties of some solid objects, and the objects can be grouped by the way they react (e.g., dissolve vs. don't dissolve; acid, neutral, or alkaline). Other tests pertaining to the strength of solids can be tried as follows. **23 Bb Dc Ec**

a. Students can roll sheets of typing paper into a tube, holding it together with a piece of tape. Stand the tube on a bathroom scale and push straight down on it with a hardbacked book to see how hard they must push before the paper crumples *(a)*. Students can experiment by roll-

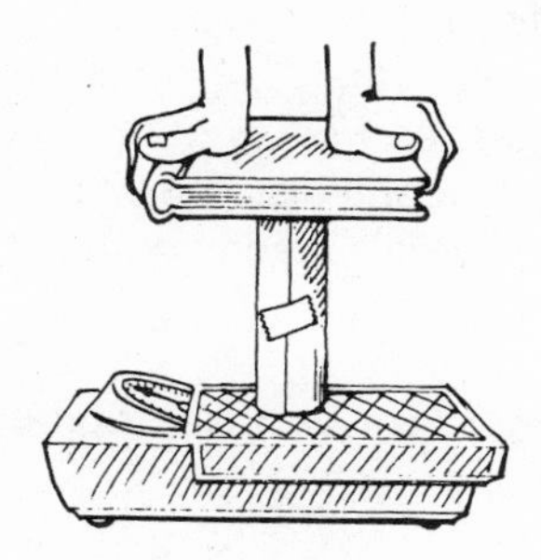

FIGURE 111.23a

ing skinny and fat tubes, square ones, triangular ones, and rectangular ones. They can try several tubes at once to see if quantity makes a difference. Let them keep records of their judgments. They can test and classify objects on the basis of findings (e.g., weak vs. strong). You might tell them that the measurement of a solid's ability to withstand crushing is called *compression strength*. Finally, give them two challenges: (1) ask them to find and discuss pictures of man-made structures that make use of columns for support; (2) ask them to test if it takes twice as much compression to collapse two tubes half the size of one tube.

b. Students can cut ¾ inch by 8 inch (2 cm by 20 cm) strips of different papers. Make loops of the strips and hold them in place with glue or a piece of tape. Attach a strong rubber band to the top of a ruler with a tack. From the band, hang a paperclip. Note where the lower end of the paperclip is on the ruler, then attach one of the loops and slowly pull it downward until it breaks *(b)*. When it breaks, have students record how far the

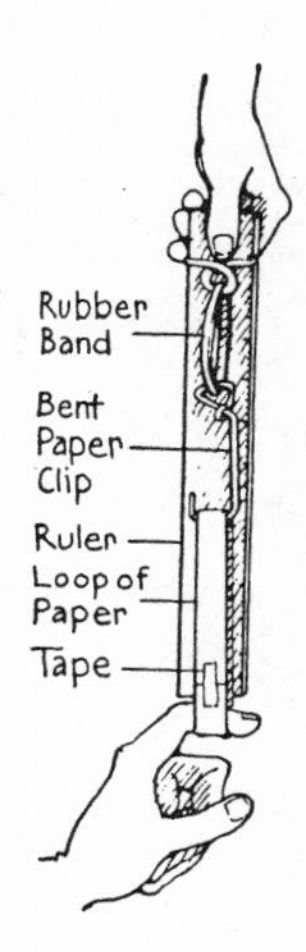

FIGURE 111.23b

paperclip moved. Repeat the test for other loops of paper to see which is the strongest and which is least strong. Students can sequence or classify papers on the basis of strength or compare brands advertised on television to see if strength claims are correct. You might tell them that the measurement of a solid's ability to withstand breaking is called *tensile strength*. To extend this activity, cut ¾ inch by 8 inch (2 cm by 20 cm) strips of different shapes but from the same kind

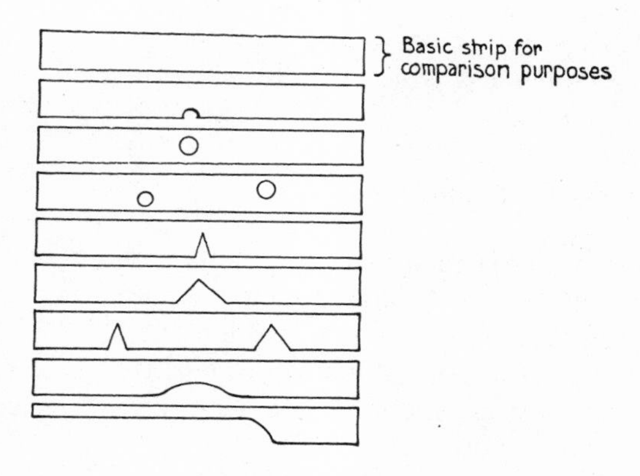

FIGURE 111.23c

of paper *(c)*. Students can guess where each strip will break when pulled, then check their guesses by pulling the strips apart by pulling on the ends. Finally, ask them to find pictures of manufactured structures that make use of strands for strength (e.g., cable supports on bridges).

112 INTERACTIONS

Generalization. Certain dissolvable and colloidal solid materials can pollute water and air.

Contributing Idea A. Some solids dissolve in water, some do not.

01 Aa

Observing some solids that dissolve. After students examine the characteristics of solid pieces of salt and sugar with a hand lens, let them place 1 teaspoonful (one 5 ml spoonful) of each in different jars of water. (The materials will dissolve.) Next, set the jars in the sunlight or near a heat source to allow the water to evaporate. When the water has evaporated, the students can examine the solid materials again with a hand lens. They will realize that water can contain unseen, dissolved solid materials. Encourage them to test other solid materials to see which dissolve and which do not.

Contributing Idea B. Solid impurities can pollute water.

02 Aa

Detecting solids in water. Students can collect running water from a stream or gutter after a rainfall. Set the water in a widemouth shallow pan, and allow the water to evaporate. Examine the residue with a hand lens to see what materials the water was carrying. Identify any manufactured materials found.

03 Af Ca

Comparing the effects of some solids upon water. Obtain several jars with lids. Place a small amount of various solid materials into the jars (e.g., salt, sugar, coffee, soap, pieces of food). Label each jar for its contents, then cover them. After several days, open the jars, and let students smell, feel, and look at the contents. They can discuss how some materials readily pollute the water.

Contributing Idea C. Solid impurities can pollute air.

04 Bc Ca

Detecting solids in air. Students can make devices for collecting airborne particles in several ways.

a. Coat baby food jars with Vaseline, set them inside large cans, and suspend or place them in various locations. Particles will adhere to the Vaseline, and the cans will protect the jars from ground dirt.

b. Bend a heavy wire, such as a coat hanger, into the shape shown *(a)*. In the

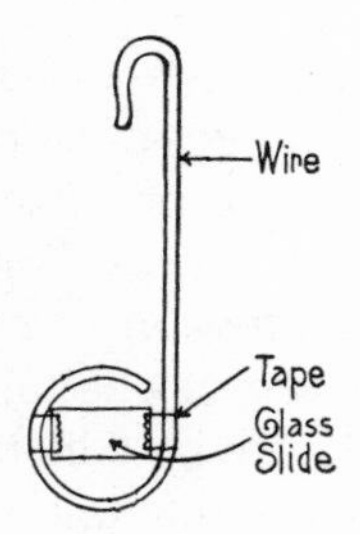

FIGURE 112.04a

lower loop of the wire, fasten a strip of tape to write on. To the sticky side of the tape, fasten a glass slide that has been thinly coated with Vaseline. This device can be preweighed, then weighed after particles have collected on it for several weeks. The area of the exposed surface

TABLE 112.04. Detecting Air Pollution

Location	*Time Exposed*	*Description of Particles*	*Amount of Pollution*

and the accumulated weight can be computed to indicate the amount deposited per square mile (kilometer).

c. Cut a 2 inch by 6 inch (5 cm by 15 cm) strip of heavy tag-board and fold it lengthwise. Punch five holes through one side of the tag with a standard hole punch. Place a long strip of transparent tape, gummed side up, on a flat surface and set the tag on the tape so that the sticky surface shows through each hole *(b).* Make

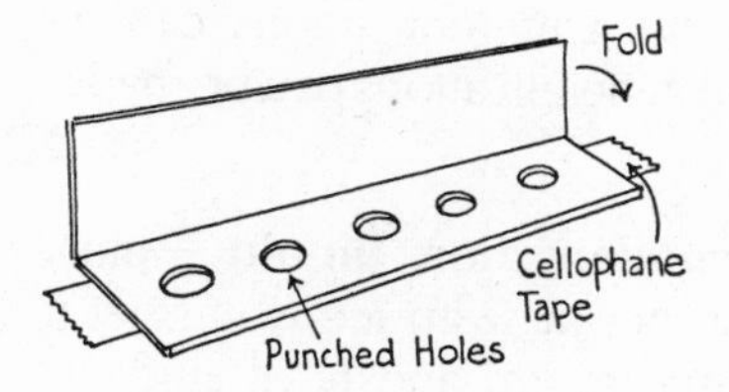

FIGURE 112.04b

the strip of tape longer than the tag so that the top half can be sealed over the lower half to keep particles from adhering to the tape before it is ready for use. The device can also be placed in an envelope to protect it further. Students can expose one or more of the devices to the air in different locations. The exposure time and location of sampling can be recorded for each device along with other conditions such as the relative wind velocity or number of days since the last rainfall. Samples can be taken in a variety of places (indoors, on windowsills, or outdoors, in trees, on porches, etc.) and in different positions (high, low, horizontal, vertical). A table such as the one shown (Table 112.04) can be helpful in organizing observations. From it, students can determine the hours of the day or the days of the week when most pollution occurs, the area in which most pollution is collected, and the amount produced by people or natural means. Students can also examine the collected particles with hand lenses or, under the low power of a microscope, identify the particles. Let them discuss how pollution might be reduced.

Observing particles being added to the air. Have students collect pictures of factories, furnaces, fireplaces, etc., that spew smoke into the air. Have them identify such contributors in their hometown, discuss the problems that are created by such pollution, and plan how the problems might be solved.

05 Aa

120 liquids

121 CHARACTERISTICS: LIQUID STATE

Generalization I. Liquid is one of three states of matter.

Contributing Idea. Liquids evaporate into gases when heated; gases condense into liquids when cooled.

Observing evaporation (liquid → gas). Have some students keep a daily check on the water level of a room aquarium or other container filled with water. Others can mix paints with water, paint a picture, and check its wetness several hours later. You might explain that the term used to describe what happened is *evaporation.*

01
Aa
Bb

Graphing evaporation rates. Students can measure how fast a liquid evaporates from a material by hanging a wet piece of the material from an arm balance. Balance the arm and observe at regular intervals. As the water evaporates, the material end of the balance arm will rise. At the regular intervals students can rebalance the arm and keep track of how many units it takes to balance the arm each time. When the material is dry, the data can be graphed. To extend this experience, test and graph other materials or make tests to speed up the evaporation (e.g., fanning, heating). Different graphs can be compared and generalizations can be made.

02
Bf

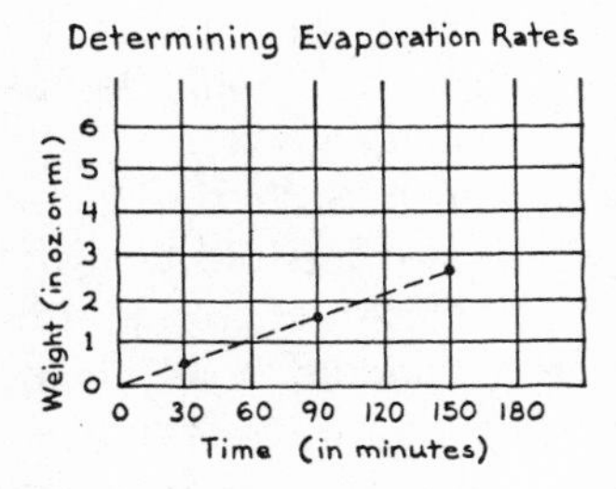

GRAPH 121.02

Observing condensation (liquid ← gas). Fill a quart or liter jar with ice water, let it stand out in the room, and have students observe the moisture that collects on the outside of the jar. Add some food coloring or ink to the water in the jar, and they will realize that because the moisture on the outside is clear, it probably did not come through the glass. Students can also exhale against the glass and watch the moisture form. Tell them what they observed is called *condensation.* You might make analogies to common experiences such as the formation of moisture on mirrors and windows.

03
Aa
Bb

Generalization II. Liquids have identifiable characteristics.

Contributing Idea A. Our senses can be used to identify liquids.

Observing some physical characteristics of some liquids. Students can compare and describe the appearance of separate jars of water obtained from different sources (e.g., pond, lake, sea, and/or stream water; distilled water; tap water). Note the colors and materials in the waters. Similarly, small quantities of various liquids (e.g., orange juice, tomato juice, apple juice, etc.) can be placed into clear glasses and compared using other senses, such as smell and touch. Or several similarly appearing liquids of different densities and viscosities (e.g., white syrup, mineral oil, kerosene) can be put in identical jars for students to observe characteristics other than color.

04
Af

Feeling some characteristics of liquids. The sense of touch can be used to explore the properties of liquids by having students feel various liquids (e.g., alcohol, water, shampoo, molasses) with their fingers. Students can describe what they feel to develop useful vocabulary words (e.g., wet, sticky, slippery).

05
Ab
Ba

Comparing liquids by taste. Have students taste various liquids (e.g., sugar water, salt water, lemon juice, quinine water). **(Caution: Students should be made aware of the danger in tasting unknown liquids.)** Group the liquids by similarities in taste or classify them by the headings: sweet, sour, salty, bitter. Taste other liquids and add them to the classifications.

06
Ae
Ca

Contributing Idea B: Indicators can be used to identify liquids.

Using litmus paper to identify liquids. Obtain some blue and red litmus paper. Have students first use the blue litmus paper to test different liquids such as water, orange juice, lemonade, vinegar, and liquid bleach. Let them also test as many other liquids as possible, making a list of the liquids that changed the color of the litmus paper and a second list for the liquids that did not change the color. Retest the same liquids using the red litmus paper and make new listings. When they compare the lists, students will discover substances that turn the blue litmus paper red do not affect the red litmus paper and vice versa. They will also discover the neutrality of water.

07
Fb

Contributing Idea C: Liquids generally take the shape of their container.

Describing the shape of liquids. Pour some water into a tall container and ask students to observe and describe the shape of the water. Pour the water from the container into a low, flat container; ask for another description, then pour the water into a third, dissimilar container. From their descriptions, students can generalize that water has no shape of its own.

08
Ba

Contributing Idea D: Some liquids have the ability to dissolve certain other materials; some materials dissolve in liquids and some do not.

Identifying materials that dissolve in liquids. Fill each of four beakers with ½ cup (100 ml) of water. Let students put 1 level teaspoonful (one 5 ml spoonful) of different materials (e.g., sugar, salt, iron filings, dirt) into different beakers. *(Note: A level spoonful can be obtained by scraping off any excess with the edge of a ruler.)* Separate spoons should be used to stir each beaker. Stir thoroughly and observe the contents. (Some materials will "disappear" or dissolve; some will remain visible.) Test other

09
Aa
Bc

materials to see which dissolve and which do not. Students can compile lists under such headings as: dissolves, partially dissolves, and does not dissolve. Let them test other liquids.

Contributing Idea E: Liquids occupy space and have weight.

Determining that liquids occupy space and have weight. Obtain two empty half-pint (half-liter) milk cartons. Ask students to fill the cartons with a liquid such as water. Hang the cartons from an arm balance and adjust until it balances. Let one student hold an empty container under one carton and, with a small nail, puncture the bottom of the carton. As the water runs out, students will observe a change in the balance and realize that the water has weight. **10 Cf**

Contributing Idea F: Nearly all liquids expand when heated and contract when cooled.

Comparing the expansions of different liquids. Fit three or four bottles with one-hole rubber stoppers. Place 12 inch (30 cm) lengths of glass tubing in each stopper. Fill each bottle completely with a different liquid (e.g., water, alcohol, liquid detergent, kerosene). Place the stoppers into the bottles so that some of each liquid rises in each tube. With a medicine dropper, add some liquid to the tubes making the liquids equal in height. Mark the heights with a piece of thread, string, or rubber band. Immerse the bottles in a large pan of hot water. Students can observe the rise of the liquids inside the tubes and make some judgments about the differences in expansion. Next, place the same bottles in a pan of ice cubes, and observe them again. (The liquids will contract and their height in the tubes will be below the markers.) Students can compare the contractions. **11 Cd**

Contributing Idea G: Liquids have surface tension.

Comparing the shapes of droplets of different liquids. Use various liquids such as water, honey, cooking oil, and alcohol, and have students compare the appearances of single drops of each on waxed paper. (From the side, drops approximate a hemisphere.) Students can make profile charts of what they see. Two drops of the same liquid can be pushed together with a pin or toothpick. Results can be compared with the action between two drops of water. Students will notice that there is a tendency for each liquid to pull in its open surface so that the surface is as small as possible. Tell them that this characteristic is called *surface tension.* **12 Bb Be Ca**

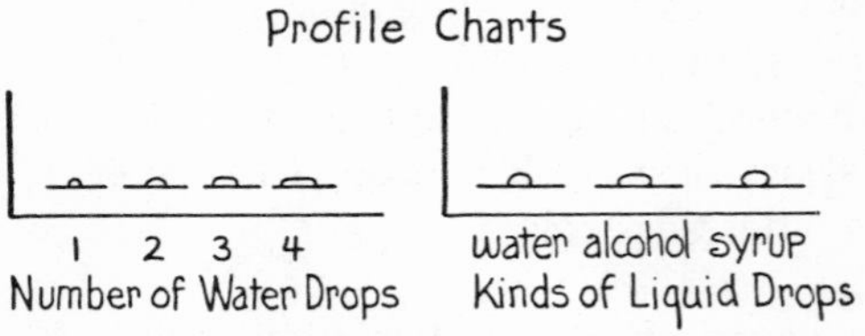

Profile Charts

1 2 3 4
Number of Water Drops

CHART 121.12a

water alcohol syrup
Kinds of Liquid Drops

CHART 121.12b

Picturing the shapes of droplets on different surfaces. Have students experiment with different nonabsorbent surfaces (e.g., aluminum foil, glass, paraffin, waxed paper) to see what effect each has on the shapes of drops. Profile charts as suggested in the previous activity can be made and used for descriptive comparisons. **13 Be Ec**

Observing surface tension in soap films. Make various copper wire forms such as those shown. Have students dip them into **14 Aa**

a thick soap solution. (Place 4 level teaspoons (20 ml) of soap flakes or powder into 4 cups (1 liter) of hot water. Let stand for several days before using.) As the wire forms are withdrawn from the solution, the students can observe the films that stretch between the wires. Repeated dipping can cause more than one film to form. Students can observe the shapes and behaviors of the films, then state explanations for any regularities they see. *(Note: If a slider is attached to the square form (e), the slider can be pulled gently, and the film will be stretched. When the slider is released, it will be pulled back by the strength of the surface tension.)* Students can be encouraged to invent other wire forms.

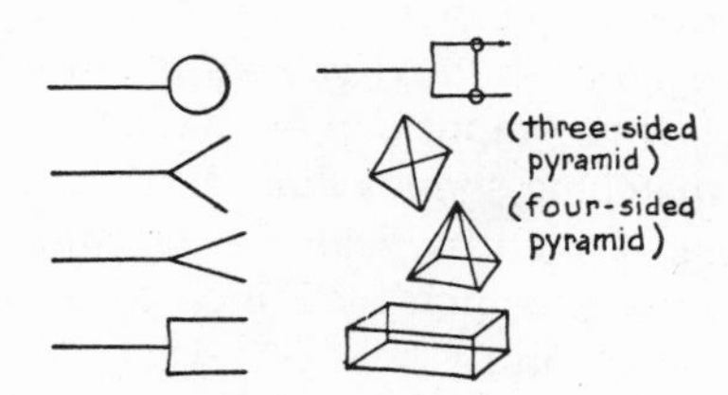

FIGURE 121.14

Observing that the surface of a liquid acts like a skin and can support weight. 15 Aa Ca The surface tension of water can be used to float some objects that would normally sink. Each of the following activities can be done using different liquids, although water is used for descriptive purposes. If various liquids are tried, comparisons can be made.

a. A student with a steady hand can, from a very low height, drop a paperclip, razor blade (double edge), or needle upon the surface of some water in a widemouth glass or similar container. The object will seem to float on the surface. (Actually it will be supported by the surface tension.) Students can observe the skinlike surface bending under the weight of the object.

b. A student can float a piece of tissue paper upon some water, then carefully place an object to be floated upon the paper without sinking it. As the paper absorbs the water, it becomes heavy and sinks, leaving the object at rest on the surface.

c. Objects can be gently lowered onto the surface of some water with the aid of a fork or bent paperclip.

Observing that surface tension allows liquids to build up without overflowing. 16 Aa Each of the following activities can be tried with different liquids.

a. Give each student a small paper cup, a medicine dropper, and a glass of water. Ask each student to put as much water into the cup as he or she can without having it overflow. Students will find that the water will rise considerably higher than the rim of the cup. Let them describe and discuss what they see. (The surface seems to have a skinlike appearance.) A sheet of dark paper held behind each cup will improve the observations.

b. Students can fill paper cups with water until the water is level with the edge. Using medicine droppers, they can find out how many drops of water it takes to make the level cups of water overflow. As drops are added, have them observe the rise of the water above the edge.

Making and using a tensiometer. 17 Bf Ga Gc Students can realize, in a physical sense, the concept of surface tension by preparing a paper device called a *tensiometer.* The tensiometer is made by cutting and folding a piece of wrapping paper as shown *(a).* Thoroughly moisten the base of the device (keeping the stem dry), and bring its lower edge close to the surface of a liquid such as water. Students will see the base being drawn to the water. The student holding the device will sense a definite "pull." Other students can make tensiometers and test the pull of various liquids in the following ways.

a. Attach tensiometers to one end of an equal-arm balance. The base of each tensiometer should rest horizontally (flat) on the surface of the liquid in a container. Students can count how many washers or other units of measurement it takes to pull the tensiometer from the surface *(b).*

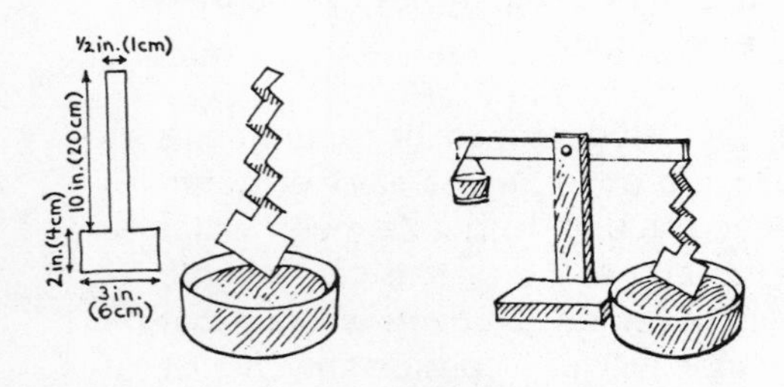

FIGURE 121.17a **FIGURE 121.17b**

FIGURE 121.17c

(Note: With repeated trials the number of washers may vary—the base of the tensiometer might have touched the edge of the container; air bubbles might have been trapped under the flat base; etc. Such variables can be explored and discussed with the students.) Other tests can be carried out using tensiometers made from other materials such as tagboard, cardboard, stiff plastic, or glass. The latter can be attached to the balance arm with thread glued or taped to each base *(c).*

b. Students can design tests to answer many of their own questions concerning tensiometers and surface tension. Variations on the previous activities are almost limitless. For example, if a student wonders whether all liquids have the same "pull" (surface tension), different liquids can be tested and compared by washer weight to see which has the greater tension. If a student wonders whether the shape of the tensiometer makes any difference, differently shaped bases can be prepared, tried, and compared. The liquids can be seriated and the results can be graphed as shown. *(Note: Differently shaped bases should have identical surface areas so that valid comparisons can be made.)* Other graphs can be made to compare matching shapes that increase in area (e.g., 2 inch by 2 inch or 2 cm by 2 cm square; 3 inch by 3 inch or 3 cm by 3 cm square).

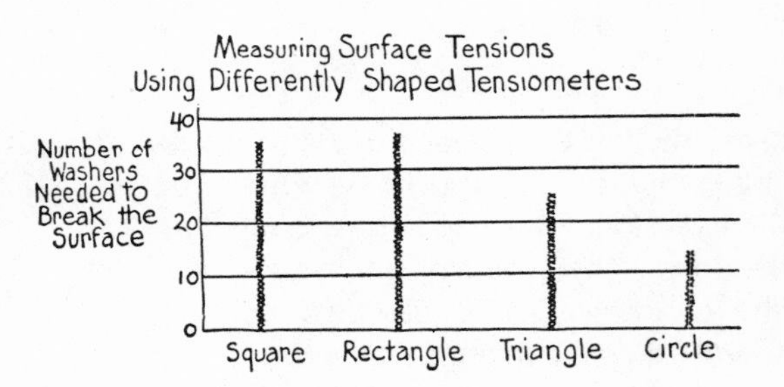

HISTOGRAM 121.17

Testing the surface tensions of water and soapy water. Several of the following activities can be done using liquids other than water. **18 Ea Eb Ec**

a. Give students large (8 inches or 20 cm across or more) plates (not paper). Be sure the plates are clean and soap-free. Fill them with water, and give each student a strand of lightweight thread about 4 inches (10 cm) long. Tie the thread into a loop and set the loop on the surface of the water near the center of the plate. Touch the water in the center of the loop with a bar of soap or place a drop of soapy water in the center of the loop. Discuss observations. (The loop will enlarge to form a nearly perfect circle because the surface tension within the loop is reduced.)

b. The previous activity can lead to a large number of experiments that can answer questions the students raise. For example, a student might wonder if the thread loop was pushed or pulled out in a circle. To find out, the activity can be repeated and the soapy water can be added outside the loop. The observed result can be compared to the original procedure by placing soapy water in the plate and adding plain water to the loop.

c. To make a comparison of the surface tensions of water and soapy water clearer, sprinkle some fine powder onto the surface of a layer of water in a large, open-faced plate. Various powders work well: facial powder, talcum powder, pepper, cinnamon, nutmeg, powdered tempera paint. When the water is calm, have the students use a medicine dropper to place a drop of soapy water onto the surface. Observe what takes place on the surface. (The surface tension of the water will break up and draw together away from the drop.)

Observing that the surface tension of a liquid can be reduced. Fill a large-mouth container with cold water. Allow the water to become calm, then sprinkle the surface with a fine powder to make viewing easier. Place a small piece of soap into the center of the container. Let the students observe what happens. **19 Aa**

Reducing surface tension to propel small boats. Students can observe and make use of the reduction of surface tension by making surface tension boats out of cardboard. Cut the tiny boats (about 1 inch by 2 inches or 2.5 cm by 5 cm) from cardboard or stiff paper. Along the stern of each boat, spread a thin line of airplane glue *(a).* Place the boats in a container of water. (The depth is not important.) Students will observe that the glue reduces the surface tension behind the boats. (Less force is exerted at the rear of the boat than in front.) If a fine powder is sprinkled on the water in front of the boats, the action will be more apparent. For a variation, a drop of glue can be placed on **20 Ga**

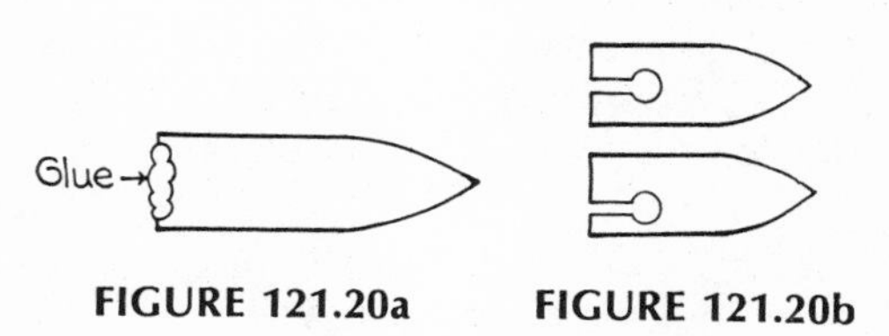

FIGURE 121.20a **FIGURE 121.20b**

the end of a wooden or plastic toothpick. If several toothpicks are placed on a large surface of water at the same time, some interesting interactions can be observed (Some toothpicks will seem to repel others.) If a slot is cut in the stern of each boat, the boats can be "refueled." Into a chamber in the slot, place some soap, detergent, rubbing alcohol, or small piece of gum camphor. Slots can also be cut in various ways to cause different boat movements *(b).*

Generalization III: Liquids can be classified on the basis of their characteristics.

Contributing Idea A: Some characteristics of liquids can be seriated.

Seriating liquids. Have students use medicine droppers to place drops of different liquids upon waxed paper. Let them seriate the liquids on the basis of the physical appearance of the drops. Liquids can also be seriated on the basis of other characteristics such as color, oiliness, stickiness, viscosity, solubility, density. **21 Da**

Seriating liquids on the basis of surface tension. Fill several identical containers with various liquids (e.g., vinegar, milk, alcohol). Make sure each container is filled until the liquid is level with the edge. Students can use medicine droppers to add drops of each liquid to the containers, compare the "piling up" of the liquids above the rims, and order the liquids by the number of drops needed past the level point to make each overflow. Tensiometers can be used to seriate the liquids more accurately. To make a tensiometer, see *121.17.* **22 Da**

Contributing Idea B: Some characteristics of liquids can be classified.

Grouping liquids by sight, smell, and taste. Present students with six sealed transparent vials, two vials each of three different clear liquids such as white vinegar, water, and sugar water. Ask students to match the vials—first without opening them, then by using only their sense of smell, and finally using their sense of taste. **(Caution: Students should be told about the danger of tasting unknown liquids.)** **23 Af Db**

Classifying liquids. Although liquids can be classified in numerous ways (color, smell, taste, viscosity, density, solubility, etc.), most are classified as acids (e.g., vinegar, citrus fruits, tomato juice), bases or alkaline liquids (e.g., ammonia, lye, washing soda), or neutral liquids (e.g., water). Blue litmus paper tests for acids (it turns various shades of red). Red litmus paper tests for bases (it turns various shades of blue). To give students an opportunity to utilize such indicators as an aid in identifying and classifying liquids, have them bring various liquids from home to test. After testing, group the liquids according to the results of the tests. When varying shades of color change are considered, the liquids can be ordered by degrees of acidity or alkalinity. **24 Dc**

122 INTERACTIONS

Generalization I: The volume of a liquid is a measure of the space it occupies.

Contributing Idea A: The space a liquid occupies can be measured.

Measuring liquid volumes. The volume of a liquid can be measured in a number of ways by using any small container to fill a larger one. Each filling from the small container represents a unit of measure. **01 Cg**

a. Arrange containers of various sizes and shapes for students to see. Fill a drinking glass to the brim with water or another liquid (the liquid should be colored so that it can be easily observed). Ask students to estimate how high the liquid would be if it were poured into one of the other containers. Empty the contents into one container, refill the glass, and repeat for several containers. Discuss the various levels observed. *(Note: Liquids take on the shape of the containers that hold* *them. This fact often misleads children, because a tall, narrow, container seems to contain more liquid than a shallow, wider one.)* Ask how the different levels might be measured. (The amounts have already been measured; the same tumbler was used in each case.)

b. Fill a large container to see how many tumblers, or other units of measurement, it will hold.

c. Partially fill a large bucket with water or other liquid. Ask students to determine how much water is in the bucket. They can use tumblers or other containers to find out the volume of the water (the water can be withdrawn in equal units). Possible measuring containers range from thimbles to buckets. Students can choose the best size unit to use. (A thimble might take all day to empty a large container, while a bucket might be too large to give useful information.)

Making instruments for measuring liquid volumes. To measure the volumes of liquids, each student can make his or her own measuring device by obtaining a tall bottle. Each bottle can be turned into a graduated cylinder by scratching a scale on its side with a file. Each mark on the scale can represent any small measuring unit of liquid (e.g., a thimble, a jigger glass, a top from an aerosol can). **02 Gc**

Using standard liquid volume measuring instruments. Obtain several common measuring containers (e.g., baby bottles, measuring cups, various milk cartons). Fill each measuring device with water or some other liquid. Let students pour the liquid back and forth among different containers to find the relationships of amounts each holds. **03 Ga**

Contributing Idea B. Liquid volumes have weight.

Feeling the weight of water. Have students heft several differently sized, empty milk cartons. Fill them with water and heft again. Discuss the differences felt. Similarly, have them lift a dry wash rag, then soak it in water and lift again. Students will realize that water has weight. They can look at the weight printed on bottle labels of different liquids such as syrup, cola, bleach, and oil. **04 Ab**

Weighing liquids. To help students realize that any weighing is a comparison of a known amount (mass) whose weight has been decided against an unknown amount (mass), ask them to determine the actual weight of an amount of water (mass) in a carton that is filled to the brim. They can use arbitrary units of measurement (e.g., washers, nails, pennies, marbles) or standard units. Compare results with the weights of other liquids in identical cartons. **05 Cf**

Comparing the weight of water to the weights of other liquids. Compare the relative weights of various liquids to water by obtaining six or more plastic pill bottles with tight-fitting caps, filling an aquarium one-fourth full of water, then filling one plastic bottle with water and capping it so that no air remains in the bottle. Set the bottle in the aquarium and observe the level at which it floats. (The water level in the bottle may be slightly below the surface of the aquarium water level due to the weight of the plastic bottle.) Now fill the other bottles with different liquids (e.g., vinegar, alcohol, soda, milk) and float them in the aquarium. Students can compare the level at which each floats with the level of the floating bottle of water. Order the liquids from the lightest to heaviest on the basis of observations. **06 Cf Da**

Weigh each container to see if the observations agree with the actual weights.

Generalization II: The specific gravity of a liquid is a comparison of its density to the density of water.

Contributing Idea A. Each liquid has a different density.

Determining the density of liquids. The weight of a volume of one liquid differs from the weight of an equal volume of another liquid. The density of water (equal to 1) is used as a base or standard for measuring the densities of other liquids. (Alcohol is less dense or lighter than water; molasses is denser or heavier than water.) Several activities provide students with comparisons. After one or more of the experiences, explain that the relationship between a material's weight and its size is called *density.*

07
Bb
Cf
Da

a. Students can measure and compare the weights of a pint or half-liter of water with pints or half-liters of alcohol, syrup, milk, oil, etc. Order the liquids from lightest to heaviest.

b. Place equal volumes of various liquids into identical plastic pill containers. Measurements can be made with a commercial or homemade arm balance scale. Order the containers from lightest to heaviest.

c. The density of liquids can be measured fairly accurately by placing each container of liquid on one side of an arm balance. The weight of the liquid and its container can be balanced by adding units of weight to the other side of the balance arm. To find the weight of each liquid, the students must weigh the containers *before* adding the liquids, then subtract this weight from the total. You might explain that the ratio between the density of a material and the density of water is called *specific gravity.*

Contributing Idea B. The hydrometer is an instrument used to measure specific gravity.

Making a hydrometer. Simple hydrometers can be made in several ways.

08
Gc

a. Put a thumbtack in the eraser end of a pencil. Float the pencil vertically in a beaker of tap water. Mark the water level on the pencil. Remove the pencil and dissolve 3 teaspoonsful (three 5 ml spoonsful) of salt in the water, and float the pencil again. Compare the water level with the mark made on the pencil. (The pencil should float higher in the salt water because it has a greater density.) Test other liquids with the pencil hydrometer.

FIGURE 122.08a

b. Seal one end of a waterproof drinking straw with melted wax and place some sand into the straw until it floats in a vertical position. A drop of melted wax or glue placed in the straw above the sand will keep it in place. Float the straw in ordinary tap water and mark the water level on the straw. Other liquids can now be tested and compared with the water mark. *(Note: The watermark line represents the specific gravity of water—the* *specific gravity of any other liquid tells how its density compares with the density of water.)* If the specific gravity of water is considered to be 1 (unity), the submerged part of the straw to the mark can be considered as a whole unit (1), and this unit can be scaled into tenths. Another unit can be extended above the water line and scaled into tenths. Other liquids can be tested and compared very accurately with this scaled hydrometer.

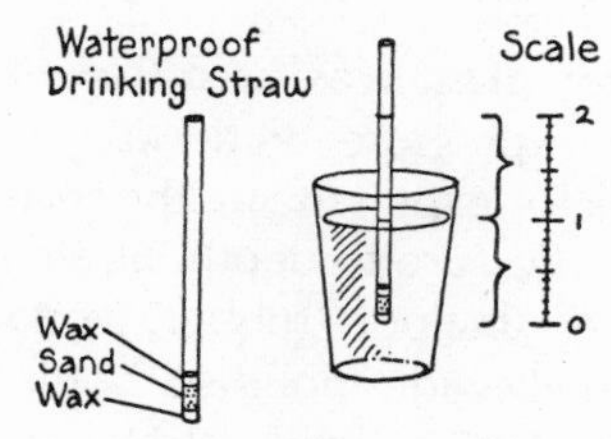

FIGURE 122.08b

Comparing liquids by their specific gravity. Place a glass tube (¼ inch or 1 cm diameter or more) into a glass container of water. Pour a liquid that is lighter than water into the tube until it forces the water to the lower end. Students can compare the height of the liquid in the tube to the height of the water outside the tube. *(Note: The difference in height is a measure of the specific gravity of the water, as shown by the water level.)* If a liquid heavier than water is used, fill the container with the liquid and pour water down the tube until the liquid is forced to the bottom. The tube can be marked and scaled similarly to the hydrometer in the previous activity. Measurements taken by the two devices can be compared. **09 Cg**

Generalization III. Liquids exert pressure.

Contributing Idea A. The pressure of a liquid increases with depth.

Observing that the pressure of water increases with depth. Punch three holes with a small nail, equidistant and vertical to each other, in the side of a one-quart (1 liter) or one-gallon milk carton. Cover the holes with tape and fill the carton with water. Hold the carton over a sink or large container and quickly pull the tape away. Have students observe the flow of water from the holes. (The flow will be improved if small pieces of glass or metal tubing about ⅛ inch (or.3 cm) in diameter, ½ inch (1.2 cm) long are tightly fitted into the holes.) Repeat the activity several times. Let students draw inferences about what they see. (The pressure of the water at the base of the container is greater than at the top.) If the carton is held high enough above the sink, so that the streams of water have a chance to fall through their full trajectory, the streams will appear as they do in Fig. 122.10. **10 Ca**

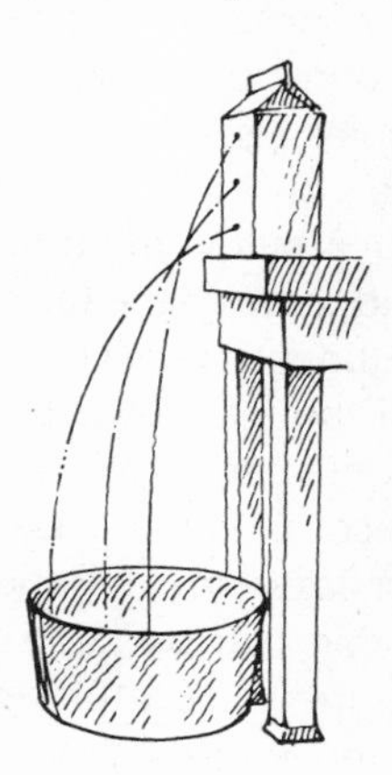

FIGURE 122.10

Measuring the pressure created by the weight of water. Remove the top only of one milk carton. Remove both the top and bottom from two other milk cartons. Tape two cartons together, one on top of the other that has only the top removed. Punch three holes in the side of the bottom carton, **11 Bc Cd Fa**

TABLE 122.11. Measuring Water Pressure

Number of Cartons	*Length of Jet 1*	*Length of Jet 2*	*Length of Jet 3*
3			
2			
1			

equidistant and vertical to each other. Place tape over the three holes, fill the lower carton with water, hold over a sink, then quickly remove the tape. Use rulers to compare how far each of the three water jets travel. Now retape the holes, fill both cartons and repeat the activity. Next, tape the third carton on top of the other two and repeat. If the lower cartons bulge from the weight, put tape around them. Now tape each of the three holes separately. Place a yardstick against the bottom carton so that it extends away from it below the holes. Fill the three cartons and remove the tape from the bottom hole. Students can check and record the length of the jet. Cover the bottom hole, fill the containers again, remove the tape from the second hole, measure the jet, and record the measurement. Repeat for the third hole. A table can be kept of the data. Next, remove the top carton and repeat the activity. Remove the second carton and do the same. From their observations and the data, students can draw inferences about the height of a column of water and its relationship to the pressure the water exerts.

Contributing Idea B. The manometer is an instrument used to measure the pressure liquids exert.

12 Ca

Comparing water pressures at different depths. A simple pressure bottle can be made by students for experiencing the relationship of water pressure to depth. Stretch a piece of rubber (balloon rubber works well) across a small, wide-mouthed bottle. Make sure the rubber is tightly stretched. Fasten it securely with a rubber band. The bottle can now be submerged to different depths in a tank of water. Students can observe differences in the rubber membrane at different depths. (The deeper the bottle, the greater the pressure.) They can also turn the pressure bottle in various directions at the same depth and observe what happens. (The pressure at any particular depth is equal in all directions.)

13 Gc

Making a manometer. A *manometer* is a U-tube pressure gauge that can measure the pressure that liquids exert. To make a manometer, connect two transparent soda straws (or straight glass tubing) with a short piece of rubber tubing. Attach the arrangement to a stand as shown. Fill the tubing with colored water until it rests at the same level in each straw. To one straw, fasten about 1 yard or 1 meter of rubber tubing. On the end of the tubing, attach a small funnel, thistle tube, or pipette with a piece of rubber stretched tightly across its face (balloon rubber can be used). Have students press on the rubber and observe the liquid in the U-tube. (It should rise when pressure is applied.) Place the funnel end of the rubber tubing just below the surface of some water in a large (deep) container. Students can observe the change in the liquid level in the manometer. Move the funnel deeper, and observe again. Next, put the funnel to the bottom of the container. Students will realize that the pressure increases with depth.

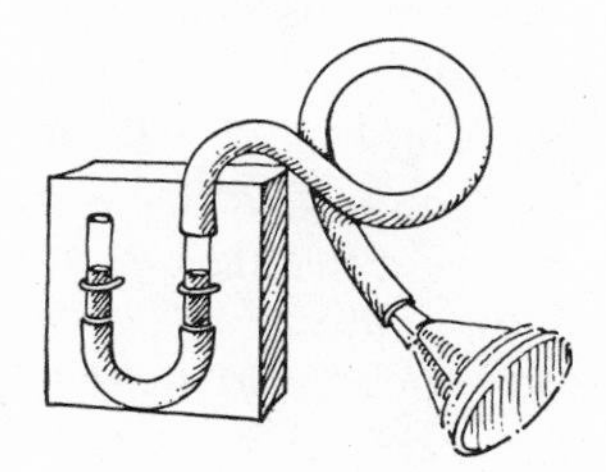

FIGURE 122.13

Using a manometer. To make accurate measurements with a manometer, a scale can be added to the device described in the previous activity. First, mark the level of the colored water in the U-tube on the tubing itself or on a card set behind the open-ended tube. Submerge the funnel below the surface of some water and measure, in inches or metric units, how deeply the funnel is submerged. For every unit in depth, mark the change in level on the U-tube. A record can be kept as shown in Table 122.14. After the scale is completed, other liquids can be tested and compared with the pressure of water. The data should produce a straight-line graph from which students can predict the pressure at other depths by extrapolation. (The information will also indicate that the pressure of water is directly proportional to its depth.) By extending the length of the rubber tubing and replacing the open-ended tube with several feet of glass tubing, the

14
Bc
Ga

TABLE 122.14. Measuring Water Pressure

Depth of Funnel	*Rise of Water in Open Tube*
0 inches	__ inches
1 inch	__ inches
2 inches	__ inches
3 inches	__ inches
↓	↓

manometer can be used to test depths in a swimming pool. From graphed data, students can predict the pressure at a particular depth, then use the revised manometer in a swimming pool to check their predictions.

Contributing Idea C. The pressure exerted by a liquid depends upon the density of the liquid.

Comparing the pressures of different liquids. Obtain two one-gallon size containers into which the funnel end of a manometer will fit. Fill one container with water and one with alcohol or some other liquid less dense than water. Measure and compare the pressures at the bottom of each liquid. Repeat with a liquid that is heavier than water. **(Caution: Alcohols are inflammable and should be kept away from open flames.)**

15
Cg

Contributing Idea D. The pressure of a liquid is the same in all directions at any given depth.

Comparing pressures of liquids in all directions. Fill an aquarium with water. Use a manometer to measure the pressure of the water in all directions. A depth mark on the side of the aquarium will enable students to hold the manometer in different positions at

16
Cg

the same depth *(a)*. Since an immersed hand will change the depth of the water greatly, tape or tack the manometer funnel to a small, thin board that is attached to a ruler or yardstick with a bolt and small screw or wing nut *(b)*. The funnel position can then be adjusted easily and immersed.

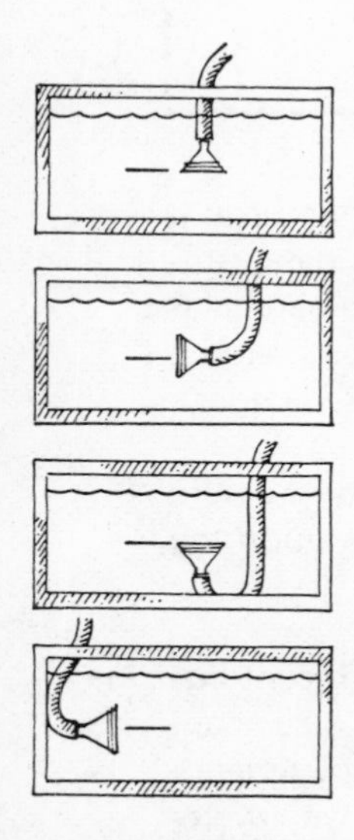

FIGURE 122.16a

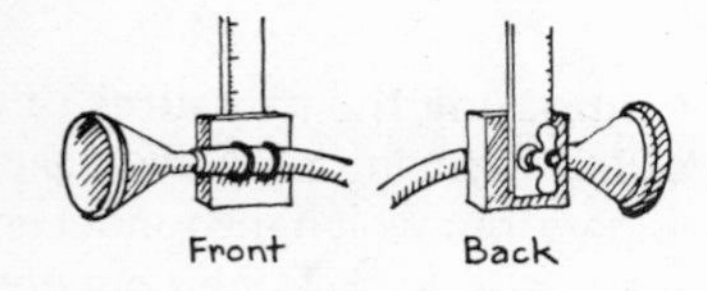

FIGURE 122.16b

Generalization IV. Buoyancy is the upward force exerted upon an immersed or floating object by a liquid.

Contributing Idea A. Buoyancy is an upward force.

Feeling and describing buoyancy. Objects float because water pushes up against them. Several activities can let students feel the push or buoyancy of a liquid.

17
Ab
Ba
Bb

a. Students can hold a block of wood under water, then describe what they feel. You might tell them what they feel is called *buoyancy*.

b. Float an empty can in a pan of water. Have students put one hand in the can and push downward without sinking the can. Ask them to describe what they feel.

c. Obtain a large cork. Have students push it to the bottom of a bucket or other suitable container filled with water. Have them describe how it feels and the amount of force (push) needed to hold it under the water. Put the cork in a bottle and repeat the activity to see if any difference is made in the effort needed to hold the cork under water.

d. Blow up a balloon and have a student push it to the bottom of a bucket of water. Compare the force needed to hold other objects under water.

e. Obtain an empty coffee can (or small plastic pill bottle) with a tight-fitting lid. Have students push the covered container, top down, to the bottom of a bucket of water. Have them describe how the can feels, then quickly release it and observe what happens. Repeat this action at different depths and compare differences.

Contributing Idea B. Buoyancy makes objects seem lighter.

Feeling how buoyancy affects an object. Students can feel the lightness of an object due to the buoyancy of water by filling a coffee can with water and covering it. Cut a string or cord about 1 yard or 1 meter long. Double the string for strength and attach it to the can so that it can be held by the loop. Have several students lift the can of water (it should feel heavy), then lower the can into a bucket of water and describe how it feels (it will feel much lighter).

18
Ab

Weighing objects in water. Several activities help students realize that objects in water seem to weigh less due to buoyancy.

19
Cf

a. Obtain a small empty jar (or plastic pill bottle) with a tight-fitting lid. Fill the jar with water and put the lid on it. Suspend the jar from a spring scale or rubber band. Lower the jar into a container of water and record any change in the spring scale or in the stretch of the rubber band. Students will see that the buoyancy of the water pushes up on the jar and lessens the weight.

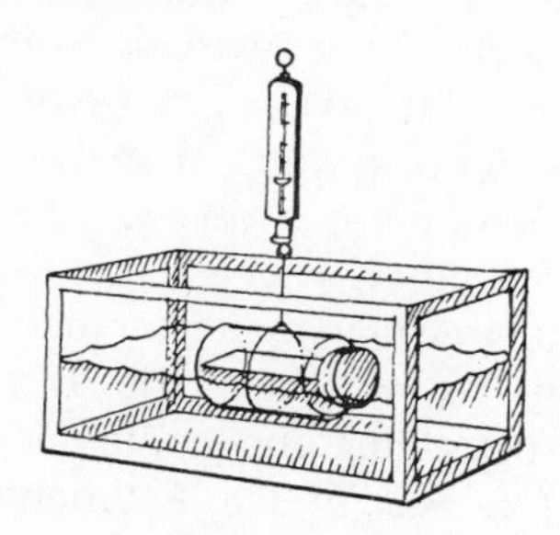

FIGURE 122.19

b. Weigh a large stone with a spring scale. Record the weight and let students suspend the stone from the scale into a container of water. Record the "new" weight. (The stone will seem to weigh less in the water due to the buoyancy.)

c. Test the materials in the previous activities again, using other liquids such as alcohol, salt solutions, or sugar solutions. Compare the buoyant effect each has upon objects.

Contributing Idea C. Buoyancy of a liquid is related to the density of the liquid.

Increasing the density of water. Place an egg in a glass of tap water. Increase the density of the water by stirring in 3 to 5 teaspoonsful (three to five ml spoonsful) of salt. Let students observe what happens to the egg. (The density of the water is increased in relation to the density of the egg, and the water becomes more buoyant.) A similar way to increase the density of water would be to empty two tea bags and fill them with salt. Fill a beaker with warm water. Drop four or five unshelled peanuts into the beaker. (Peanuts have different densities, thus some will sink while others float.) Hang the two tea bags on opposite sides of the container and have students observe what happens. A sheet of black paper can be placed behind the container to make observations clearer. (Students should see the salt going into solution near the bottom of the beaker. A layer of the denser solution can be seen rising in the beaker. As it rises the peanuts will rise with it, supported by the buoyant force.) **20 Ca**

Contributing Idea D. Buoyancy is related to the weight and volume of objects.

Comparing and sorting floating objects by weight and volume. Collect objects such as paperclips, bottle caps, spoons, cans, candles, corks, marbles. Sort them by weight (light, medium, and heavy categories). Sorting can be done with a weighing device or by letting students heft each piece to make a judgment about its placement. Drop the objects, one at a time, into an aquarium filled with water. Let students describe what they see. (Some objects will float; others will sink.) Keep a record of what each object does, then compare with the weight groupings made earlier. Ask if weight has any relationship to whether objects sink or float. Students can make other inferences by sorting the same objects by size (small, medium, and large categories). **21 Cf Cg Db**

Contributing Idea E. An object will float if it displaces a volume of water whose weight is the same as its own; an object will sink if it weighs more than the volume of water it displaces.

Comparing the weight of displaced water to the weight of an object that floats. 22 Cf
Weigh a large dishpan and record its weight, then place a one quart (one liter) can in the dishpan. Fill the can to the very top with water. Wipe the outer surface of the can and the dishpan dry. Weigh a large block of wood or other object that floats. Place it in the can. Students can observe what happens. (The water will be displaced and overflow into the dishpan.) Now weigh the dishpan with the water in it. Calculate the weight of the water by subtracting the weight of the dishpan and compare it to the weight of the object. (The two weights will be the same. An object floats when it weighs less than an amount of water that it displaces.) Repeat this activity with several other objects that float.

Making and using an overflow container. 23 Ga Gc
An overflow container can be made from a large plastic bucket or tin can fitted with a rubber tube or a troughed spout made from a 3 inch (8 cm) piece of metal. Seal the spout securely to the container with wax or silicon rubber. Fill the overflow container until the water barely runs out the spout. (It should then be level with the spout.) Weigh and record the weight of this filled bucket. Place a second smaller container underneath the spout to catch overflowing water. Now obtain a piece of wood or other object that will float in water. Weigh the piece of wood on a spring scale and record the weight. Let the block of wood float in the water while suspended from a spring scale. (If the object floats, the weight is 0.) Find the weight of the overflow water and compare the loss of weight on the spring scale with the weight of the overflow water. (They will be the same; the buoyant force is equal to the weight of the water an object displaces.) Students can repeat the activity using other objects that float (e.g., different woods such as cork, maple, mahogany, or ebony cut to the same size and shape) or liquids other than water.

Comparing the weight of displaced water to the weight of an object that sinks. 24 Cf
Hang an object, such as a heavy rock, from a spring scale. Record the weight of the object, then immerse it in the water of an overflow container. Students can observe and record the object's "new" weight on the spring scale. Let them weigh the water in the container and find the weight of the overflow water by subtracting the container's weight from the total weight. Have students make some judgments about the apparent loss of weight of the object in relation to the weight of the water in the overflow container and about the original weight of the object in relation to the weight of the water displaced. (Objects sink when they weigh more than the water they displace.) In the same way, students can test other objects that sink. *(Note: Any material, no matter how heavy, can be made to float if it can be shaped to displace enough water.)*

Generalization V. Moving liquids have identifiable characteristics.

Contributing Idea A. Some characteristics of liquids can be observed as the liquids fall.

Listening to moving water. Whenever possible, have students listen to and describe the sounds made by various kinds of moving water, such as ocean waves, waterfalls, running brooks and streams, light and 25 Ac heavy rainfalls, and dripping and flowing faucets.

Observing and measuring columns of falling liquids. Obtain several clean, empty, plastic glue bottles. With a hot needle, make one small hole in the base of each bottle. Hold one finger over the hole and fill the bottle with water. Place the cap on the bottle and turn it upside down over a sink or large container. Remove your finger from the hole and have students observe the column of water as it streams from the opening in the bottle cap. Students will first see a column of water that tends to break up and bead some distance from the bottle opening. They will hear the difference in the continuous or broken portion of the column if the bottle is raised and lowered above the container. Repeat the activity using other liquids, and have the students compare what they see and hear. Next, draw a line near the top of a chalkboard so that a bottle can

26
Af
Cd
Da

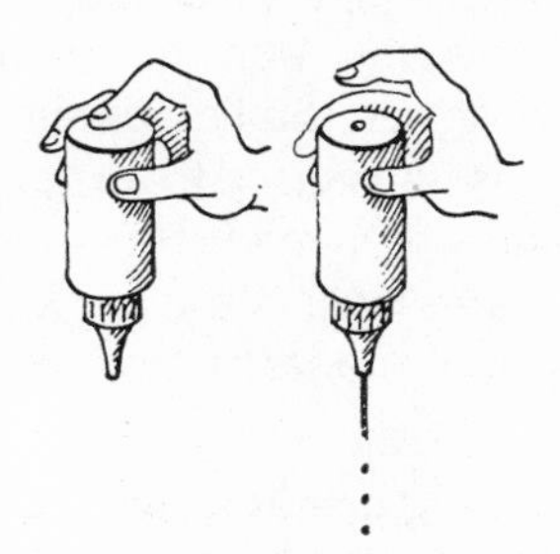

FIGURE 122.26

be held at the line. Fill one bottle with water, hold it at the line, release the hole, and let students mark on the chalkboard the place where the column of water breaks up into beads. Label the mark on the board, then test other liquids such as alcohol, molasses, honey, syrup, and soapy water. Compare the marked columns. The data can be ordered from the shortest column to the longest. Students can compare these observations with observations made during surface tension activities (see *121.12–121.20*).

> *Contributing Idea B.* Viscosity is the characteristic of a liquid that makes it resist flowing.

Observing and describing viscosity. Obtain several small plastic pill bottles with tight-fitting caps. Fill one bottle with oil, one with water, and one with a thick shampoo. Cap the bottles, leaving a small air bubble in each. Invert the bottles simultaneously. Students can observe how long it takes the bubbles to rise in each bottle. Ask them to explain what this tells them about each liquid. (They may describe some liquids as being "thicker" than others.) You might introduce the term *viscous,* and have the students test other liquids. The tested liquids can be ordered on the basis of their viscosities.

27
Aa
Ba
Bb
Da

Comparing viscosities. Fill several small identical jars half-full with liquids that differ greatly in viscosity. Cap each jar and roll the jars down a slightly inclined plane. Students can observe what happens. Ask them to explain what their observations tell them about each liquid. (The more viscous the liquid, the more slowly it rolls.) It is possible to time the rolling of each jar if start and finish positions are set up and if care is taken that each jar starts its roll with equal force. This can be done by slightly raising a board about 3 feet (1 m) long. A starting line can be drawn near the top and a finish line near the base. (A good distance between the lines would be 2 feet or 60 cm.) A small stick can be held across the starting line with the jar resting against it. At a signal from a timer, using a stopwatch or a watch with a second hand, the stick should be raised and the jar will

28
Ch
Da

begin to roll. As the circumference of the jar first touches the finish line, the timing stops. The liquids can be compared and ordered by the recorded times. The times will vary in relationship to the viscosity of each liquid. *(Note: It is best to take an average of several trials.)*

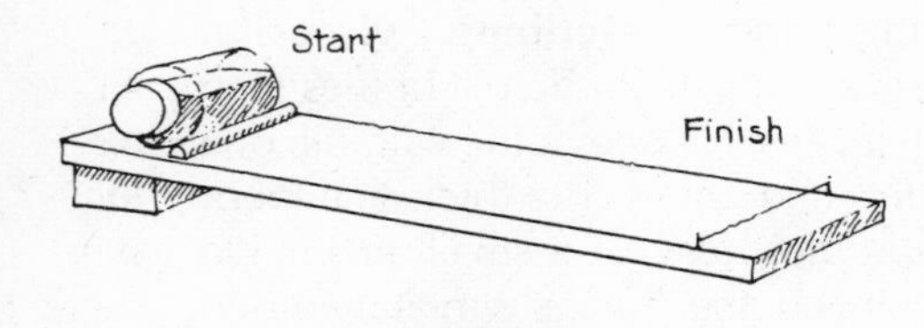

FIGURE 122.28

Generalization VI. Moving liquids have force and can do work.

Contributing Idea A. Liquids flow from higher to lower levels.

Observing that water flows from higher levels to lower levels. Prepare the apparatus that is shown. Remove the top of a one quart (one liter) milk carton. Make a hole near the bottom and insert a one yard (one meter) length of rubber tubing to fit snugly. Place the free end of the tubing in an empty aquarium. Use a clothespin to pinch the tubing closed near the carton. Fill the carton with water. Remove the clothespin. Students can test the carton at various levels to see which level is best for water to flow from the carton to the aquarium. (The flow will diminish as the height of the water column decreases.) Since water in most city water systems flows under the influence of gravity, an analogy might be made between the parts of this apparatus to watertanks or dams (carton), valves (clothespin), and pipelines (tubing).

29 Aa Fc

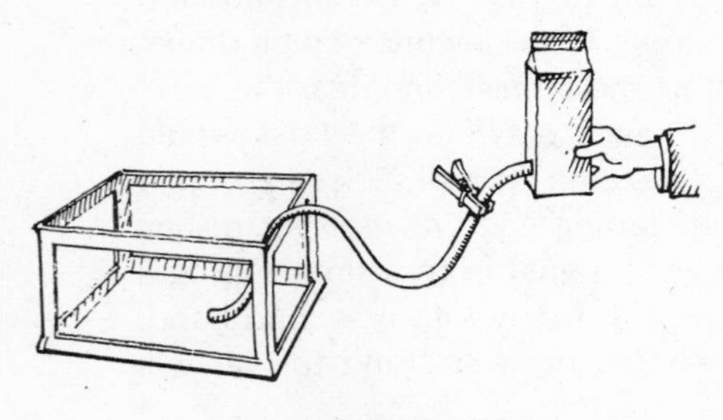

FIGURE 122.29

Observing that a liquid will cease to flow when it reaches its own level. Place two-hole stoppers in two bottles. Fill one bottle with a colored liquid and attach the two bottles with rubber and glass tubing. Hold the bottle of colored liquid above the empty bottle, place a finger over the hole without the tubing, and invert the bottle. As soon as the liquid flows into the empty bottle, set the full bottle beside it. (Students will observe that the liquid continues to flow until it reaches the same level in each bottle.) This activity can be repeated with bottles of different sizes. As a variation, a simple siphon can be started by sucking upon the tubing in a bottle of liquid. When the liquid enters the mouth, the tube is transferred to an open pan or jar *below* the water level. The water will then flow until the level is the same in each container or until the jar is emptied.

30 Aa

Measuring and making predictions about the flow of water from a steady source. A marked vessel (in one-half ounces or metric units) can be used to collect a volume of water at a drinking fountain. Measurements can be made of the collected amounts in ten-second intervals. After five or six measurements, the data can be graphed and predictions made of the time needed to fill the vessel or other container. Drinking fountains in different locations can be compared.

31 Bf Cg Ch

Contributing Idea B. The force of moving liquids can be used to do work.

Feeling and using the force of moving water. Let students feel the force of water flowing from a faucet, a drinking fountain, or a carton filled with water. This force can be put to work to move an object by preparing a simple waterwheel as illustrated. Place the cardboard pieces together and tie them; then mount the nails through the cardboard pieces as an axle. Have students work in pairs, one holding the waterwheel at the axle while the other pours water against the blades to turn the wheel. 32 Ab Ga

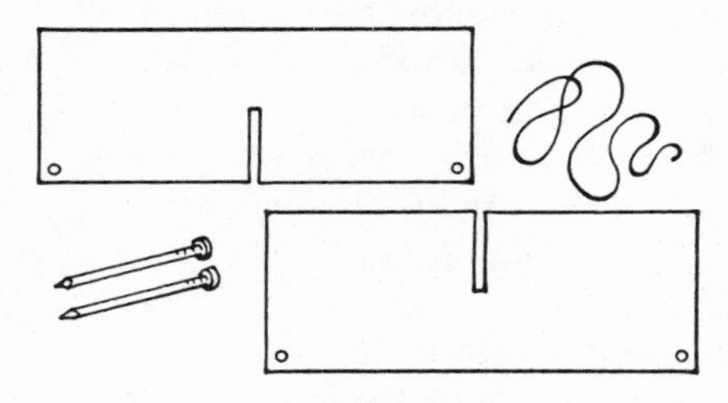

FIGURE 122.32a

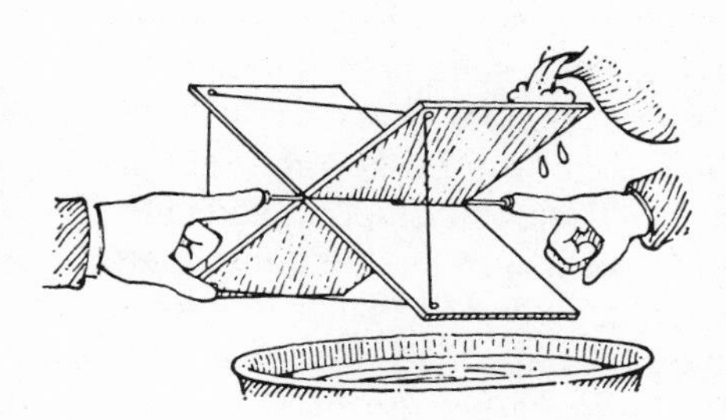

FIGURE 122.32b

Using a waterwheel to do work. Wind the end of a string around one of the nails in the center of a waterwheel. Pass the other end of the string through two eye-screws on a board at the edge of a table. Tie a weight (e.g., paperclip or fish weight) to the string so that it rests on the floor. By pouring water on the wheel, the weight will be lifted from the floor. As a variation, the wheel can be used to pull a plastic toy car or wagon across the table. 33 Ga

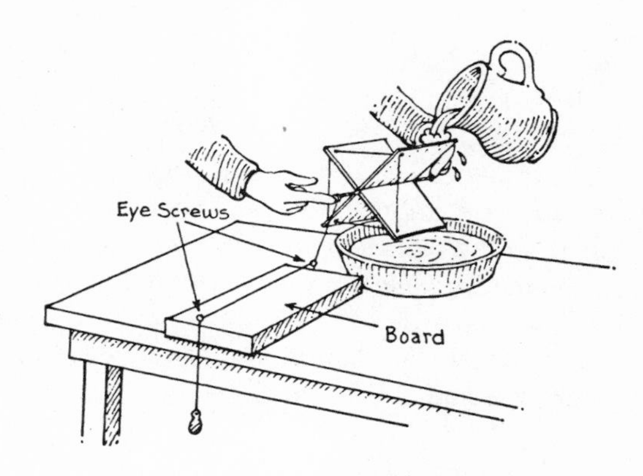

FIGURE 122.33

Generalization VII. Capillarity is the action by which the surface of a liquid, in contact with a solid, is elevated or depressed.

Contributing Idea A. Capillary action is primarily due to narrow spaces in a material and the attraction between the liquid and the material.

Observing capillary action. Using such material as a sponge, blotter paper, or cloths, have students observe how they soak up water. The water can be colored to improve observations. To observe the fibers and narrow tubes in materials that absorb liquids, obtain hand lenses. Students can examine closely the torn edges of blotter paper, cloths, and other absorbent materials, and describe what they see. (They should see tiny, closely packed fibers.) If a microscope is available, use it to observe colored liquid rising through the fibers. You might tell them that what they see is called *capillary action.* 34 Aa Bb

Comparing influences on capillarity by the amount of spacing in a material. Obtain two clear pieces of glass or plastic (lantern slide size or larger). Spread a film of 35 Cd water over one side of one piece and place the other piece on top. Secure one end with a rubber band. Separate the pieces at the other end with pieces of tag board so that the separation is not more than 1/32 inch (.2 cm) thick. Set the pieces of glass in a pan or saucer of water. Have students observe what happens over a period of time. (The water will rise by capillary action, but it will rise higher on one side of the apparatus than

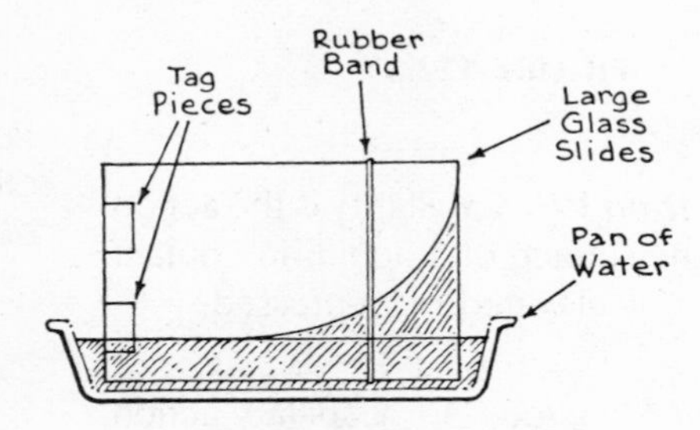

FIGURE 122.35

the other due to the change in closeness of the pieces of glass.) Students can try this activity using other liquids and making comparisons of the heights to which they rise. A grease pencil can be used to trace the waterline on one sheet of glass. This line can be used for making comparisons with other liquids.

Generalization VIII. A solution is a homogeneous liquid containing more than one material.

Contributing Idea A. Some materials dissolve in liquids; some do not.

Observing materials dissolving in water. Have students bring several tea bags to school. Cut a hole in the top of each bag and remove the tea. Pour dissolvable materials into each tea bag (for example, salt, sugar, photographic hypo). Place each bag in a separate beaker of warm water. Let students observe and describe what they see. *(Note: A sheet of black paper placed behind the beakers will aid the observations.)* Test other materials in a similar way. Students will find that some dissolve and some do not.

36
Aa

Determining how much of a material will go into solution in a given amount of water. Measure one-half cup (120 ml) cold tap water into each of four beakers. To each beaker, add 1 level teaspoonful (one 5 ml spoonful) of one of the following: salt, sugar, photographic hypo, or dirt. Stir vigorously, label the beakers, and observe the results. Students will see that some crystals are more soluble than others—they dissolve completely—and that none of the dirt seems to go into solution. (Some small particles may remain in suspension for a long time, and the water will appear cloudy.) Continue to add spoonsful of salt, sugar, and hypo to the appropriate beakers, stirring after each entry, and observing the results until the salt no longer goes into solution. (Some salt will remain at the base of the beaker after about 7 spoonsful.) At this point, students can infer that there is more salt in the container than can go into solution and the salt is less soluble than the other materials. You might introduce the term *saturation* at this time. Students can continue adding spoonsful of the other materials until they find the saturation point for each. A table of data can be kept.

37
Bb
Cc

Contributing Idea B. Some materials can dissolve in one liquid but not in another.

Identifying solvents. Have students try to dissolve small but equal amounts of various materials such as salt, sugar, grease, petroleum jelly, cocoa, flour, and soda. Have them observe and compare the rates at

38
Fa

which the materials dissolve in water, then let them try dissolving the same materials in other liquids such as alcohol, milk, kerosene, oil, and vinegar. They can infer that water might be the most universal solvent and that some materials will dissolve in some liquids, but not in others (for example, grease will not dissolve in water, but it will in alcohol).

Contributing Idea C. Hard water contains dissolved materials.

Making hard water. Hard water contains minerals dissolved from rocks. It can be obtained from a stream or prepared artificially in the classroom as in the following procedures: 39 Gc

a. Drop lime tablets, available from drugstores, into water or mix some lime from a hardware store with water. Let it stand overnight. Siphon off the clear lime water. When carbon dioxide is bubbled through lime water, temporary hard water will be produced.

b. Stir some plaster of paris, Epsom salts, or flakes of gypsum into water and let it stand overnight. Permanent hard water will be obtained when the solution is filtered.

Making hard water soft. Removing dissolved minerals from hard water will produce soft water, as soft water contains few or no dissolved minerals. To make soft water from temporary hard water, simply boil it. To make soft water from permanent hard water, add sodium carbonate (washing soda) or borax to the water. 40 Gc

Testing the washing power of soaps in hard water and soft water. Prepare two identical dirty cloths. Have students keep track of how much time it takes to wash one cloth in soft water with soap until it is clean. Wash the other cloth in hard water with the same soap for the same amount of time; then hang both to dry. Compare the differences between them. Samples of different detergents can be brought from home and their effects on hard and soft water can be compared. 41 Ec

Contributing Idea D. Some minerals in solution affect the freezing and boiling temperatures of liquids.

Comparing the times at which different liquids freeze. Place a jar or beaker of each of the following liquids in the freezing compartment of a refrigerator: water, alcohol, kerosene, syrup, and antifreeze. Wrap each container in newspaper or plastic bags, in case the containers break when they freeze. Have students observe the containers every half-hour for a 4-hour period. List the liquids in the order by which they froze. *(Note: Some materials may not freeze.)* 42 Ch Da

Comparing the times at which different solutions freeze. Give students four or five unsealed jars or beakers containing water in which different amounts of salt have been dissolved (e.g., 0 teaspoonful, 1 teaspoonful, 2 teaspoonsful, or use 5 ml spoonsful). Have them place the jars in plastic bags and set the bags in a freezer. Compare the times it takes for the different solutions to freeze by checking them at half-hour intervals. *(Note: Some may never freeze.)* Remove the jars after four hours and record the results. The jars can be left in the schoolroom to see which frozen solution melts first. Students will see whether or not a salt solution inhibits the melting of water. This activity can be repeated using different solutions such as sugar or photographic hypo. 43 Ch

Contributing Idea E. Temperature affects the solubility of most materials.

44 Ca

Determining the effect of heat upon solubility. Fill a beaker with half a cup (120 ml) of water. Add salt, stirring in 1 level teaspoonful (one 5 ml spoonful) at a time until you cannot get any more salt to dissolve. (At about 7 spoonsful, some salt will remain at the bottom of the beaker.) Place the beaker over a heat source and slowly heat it. Let students observe what happens to the salt at the bottom of the beaker. (It will dissolve.) Allow the beaker to cool and observe again. (Some salt will reappear.) Repeat this activity with sugar. Students will realize that when a liquid is heated, more material will dissolve in it.

Contributing Idea F. Crystals can be made from solutions.

45 Aa

Observing the formation of crystals. It is possible to remove some or all of the water from a solution and leave the dissolved substance. To do this, have students pour a small amount of salt solution into a saucer and place the saucer on a sunny window ledge. Put a beaker of the solution on a heating unit to evaporate the water. Check the saucer periodically until all the water is gone. Students will see that salt crystals remain in both instances. Compare the appearance of the crystals in both containers. (There will be a difference due to the speed at which the water evaporated.) Repeat the activity using sugar or photographic hypo.

46 Gc

Making crystals. Students can prepare a saturated solution of sugar—about 26 level teaspoonsful (twenty-six 5 ml spoonsful) in one-half cup (120 ml) of water. Suspend a nail, nut, or other such object midway in the solution. Set the container on a sunny window ledge. As the water evaporates, observe the formation of crystals on the object. Dissolved substances tend to crystalize on objects within a solution. Another container can be prepared with a sugar solution sprinkled with a very small amount of fine dust or sand. Let students observe where the crystals form.

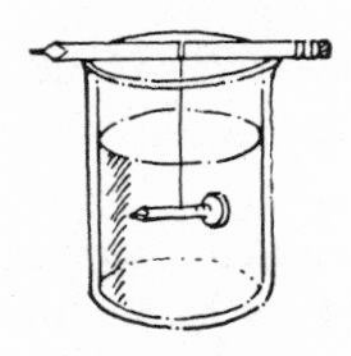

FIGURE 122.46

47 Bb Gc

Growing large crystals. It is possible to grow large single crystals in a solution. To do this, fill a beaker with two cups (475 ml) of water. Saturate the water with salt (about 7 level spoonsful.) Heat the solution until it boils, then add more salt until the water is again saturated. Allow the solution to cool. Students can observe salt crystals formed at the base. To remove the crystals, pour the solution through a filter or a piece of cloth folded over several times. Collect the liquid in several small jars. The cloth will catch the crystals, and the liquid in each jar will be a saturated solution. Now have students collect several of the largest crystals, called *seed crystals,* and drop one in each of the jars. If the lids are kept off the jars, the crystals will grow larger in the next few days due to the evaporation of some water. Still larger crystals can be made by filtering the solutions again to collect the large crystals. Tie a thread to each of them to provide a base on which other crystals can grow. Heat a beaker of the saturated salt solution and add several more spoonsful of salt. As the beaker is allowed to cool, suspend the crystals in the solution from their threads. Observe them daily. They will grow in size equal to the "added" amounts of salt. If foreign particles have entered the solution, clusters of crystals will form. The crystals can be made larger by repeating this process several times—reheat the saturated solution, add more salt, allow it to cool with the crystals in suspension. Some students might try placing the beaker in a refrigerator to see what happens. (As the temperature drops, more salt will come out of solution and form crystals.) Students can hold a contest to see who can grow the largest crystal in a certain amount of time. Other crystals can be made by the same procedures using different materials. The resulting crystals of sugar, copper sulfate, Epsom salts, alum, photographic hypo, or Rochelle salt can be compared by their characteristic shapes.

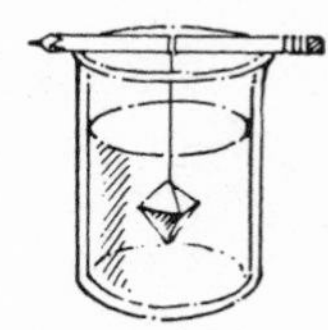

FIGURE 122.47

Generalization IX. Certain materials can pollute liquids.

Contributing Idea A. Impurities can pollute liquids.

Determining how water can become polluted. Obtain a dozen baby food jars with lids. Place small quantities of many different materials into different jars (e.g., salt, sugar, sand, ink, oil, beans, coffee, wood, soap, candy, chalk, carrots). Have students half-fill each jar with water and label the contents. Let them keep records of the sealed jars for one week. At the end of the week, open the jars to observe, smell, and feel the contents. **(Caution: Do not let students taste the water.)** Pour all the contents into a one quart (one liter) jar. Let students look at and smell the mixture. An analogy can be made to polluted rivers and lakes and to the problems facing living organisms from the spreading contamination of water. Students can speculate whether or not they would like to drink the liquids. **48 Ad Ga**

Contributing Idea B. Impurities can be removed from liquids.

Making a device to remove some impurities from water. Put a wad of cotton in the base hole of a flower pot and cover the cotton with several inches (10 cm) of clean sand. Pour some muddy water into the pot and collect it in a container. Observe the liquid as it emerges, and compare it with some remaining muddy water. **49 Gc**

Making a filter. An efficient filter can be made by sequencing the following materials in a funnel, tall tube, flower pot, or inverted bottle that has had the bottom removed. Put one-half inch (1 cm) of cotton in the base (or a layer of filter paper) to retain the sand. **50 Gc**

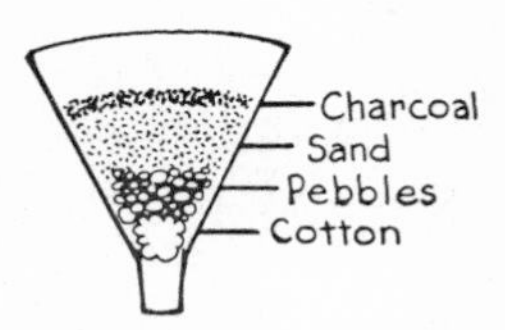

FIGURE 122.50

Add 2 inches (5 cm) of pebbles, then one inch (3 cm) of coarse, clean sand. *(Note: Charcoal can aid the filtering process—pieces of charcoal can be added in a layer beneath the pebbles or it can be made into a paste by grinding up wood charcoal and mixing it with a little water and spreading it over the top layer of sand.)* Pour muddy water through the filter. Collect and observe the filtered water. **(Caution: The collected water might not be pure enough to drink.)**

Using filter paper. Filter paper can be used to filter some suspensions, but not solutions. Commercial filter paper can be obtained or ordinary paper toweling can be cut into circles with 6 inch (15 cm) diameters. Fold a circle in half, then in half again. Place the quarter-circle into a funnel. Spread one leaf of the paper against one side of the funnel and three leaves against the other side. The paper will tend to pop out until it is wet. Slowly pour a liquid to be filtered into the funnel and collect it below. If different liquids are tested, students will see that in a solution, the substance that dissolves cannot be separated by filter paper. *(Note: Use a different filter paper for each test.)*

51
Ga

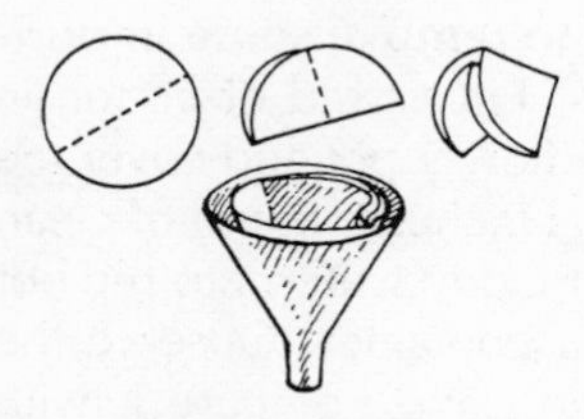

FIGURE 122.51

Separating materials from water by evaporation. Place a salt or sugar solution in a tea kettle. Heat the kettle to a boil and collect the steam from its spout on a flat pan. If the pan is tilted, the water vapor will collect and run off the pan. It can be caught in another container. Students can taste the collected liquid to see if it still contains any salt or sugar. Now have students taste the water remaining in the kettle.

52
Ga

130 gases

131 CHARACTERISTICS

Generalization I. Gas is one of the three states of matter.

> *Contributing Idea.* Gases condense into liquids when cooled; liquids evaporate into gases when heated.

Observing condensation. Have students watch water condense on the outside of a jar of water that you seal after adding ice to it. Ask where the water comes from. If students think that the droplets come from the ice water through the glass, prepare an identical jar, but fill it with warm water. They will see that droplets do not form on this jar. They will realize that there is water in the air in the form of an invisible gas and that when the gas cools, as it does against the glass, it takes on the form of a liquid. You can explain that the term *condensation* describes what they saw. Some students might tell about dew that they have seen and recall the time of day and conditions when it was observed (usually on cool mornings). Others can describe the droplets that form on windows, mirrors in bathrooms, and eyeglasses. Challenge them to recall the conditions that might have caused the droplets. The surface upon which the droplets formed was probably cool. Students will begin to realize that water in a gaseous state will condense as a liquid against cooler surfaces. **01 Aa Bb**

Observing evaporation. Students can take turns dipping a finger or hand in water and waving it until it dries. Let them describe what they feel, and ask them where the water went. Explain that when liquids, like water, seem to disappear, they have changed into an invisible gas. You might tell them that the term *evaporation* describes what they observed. Next, set a saucer of water in direct sunlight and have students check it periodically as the water evaporates. **02 Aa Ab Bb**

Observing changes of state: water (gas ⟷ liquid). Place a saucer of water inside a large glass jar or aquarium. Put some Vaseline around the rim of the container, and cover it with a pane of glass. Be sure it is tightly sealed. Students can observe the saucer periodically to see what happens to the water. (It will evaporate.) Ask where the water went. (It can be seen as droplets on the glass sides of the container.) Ask if anyone saw the water move from the saucer to the glass. Students will begin to understand that when water disappears (evaporates), it becomes an invisible gas, and when it reappears (condenses), it usually becomes a liquid again. Students can repeat this activity using other liquids such as oil or acetone, or solids such as cubes of ice. **03 Aa**

Observing changes of state: mothballs (gas ⟷ solid). Place some mothball (naphtha) crystals in a saucer. Have students smell the pieces to realize that they are changing into a gas that cannot be seen, but can be detected by smell. Now place some naphtha crystals into a test tube. Use a test tube holder to hold the tube over a candle flame. Keep rotating the test tube in the flame to keep it from becoming overheated in one spot. **(Caution: The material is flammable.)** Students will see fernlike crystals form and realize that the solid crystals changed into a gas (sublimed), then changed back into a solid (frosted). Dry ice can be used to show sublimation (solid → gas).

04
Af

Generalization II. Gases have identifiable characteristics.

Contributing Idea A. Our senses can be used to detect and identify gases.

Inferring the presence of a gas: air. Obtain two identical sheets of typing paper and ask students if one is heavier than the other (the weights are identical). Crumple one and leave the other flat, then ask again if one is heavier than the other (the weights are still the same). Now stand on a chair and drop both papers at the same time. Challenge students to explain why one takes longer to fall when both are the same weight (one has more surface area and must push more air aside as it falls). Since air offers resistance to objects moving through it, students can have fun playing with a whirlybird. Cut the whirlybird from mimeograph or similar paper and fold as shown. By tossing the heavy end into the air, students will see that air affects the whirlybird's fall.

05
Fc

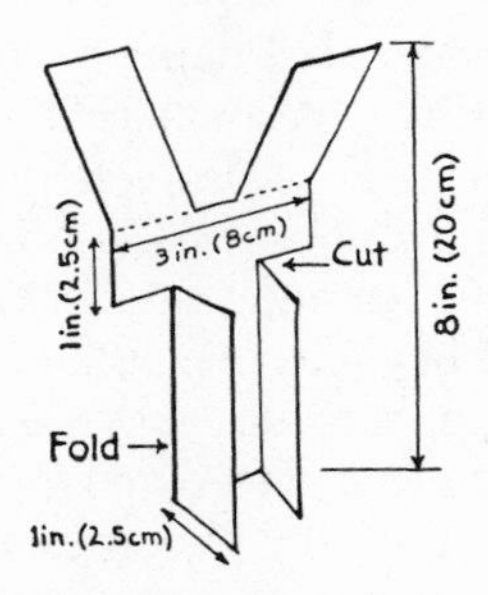

FIGURE 131.05

Observing some physical characteristics of a gas: air. Let students catch some air in a plastic sandwich bag by holding the mouth of the bag open and swishing it through the air. Quickly close the mouth and twist it tightly to trap the air. Fasten the bag with a rubber band. Students can: (1) squeeze the trapped air and describe how it feels; (2) sit on the bag or on several bags at a time to see if the bags can support weight; (3) look through the bags to see if they can see each other; (4) open their bags a little and smell the air to see if any odor can be detected that is different from the plastic bag; (5) take the bag outdoors, open it, empty out the air, fill it with outdoor air, bring it back inside, and explain how they can be sure it is filled with outside air and not inside air; (6) collect samples of air from other places around the school (e.g., in the basement, closet, hallway) and do all the things with the bag that they did with the classroom air.

06
Af

Seeing some characteristics of a gas: air. Students can "see" air in several ways:

07
Aa

a. They can blow through a straw into a large jar or aquarium of water and watch the bubbles coming from the end of the straw. If they fill a glass with water and invert it in an aquarium, they can blow bubbles and catch them in the glass.

b. Students can fit a rubber tube over the end of a glass funnel, hold it upside down on the bottom of an aquarium, and blow air into the tube to slowly force the water out of the funnel. Other students will be able to "see" the air pushing the water out.

c. Students can quickly submerge an empty soda pop bottle, top up, into a large jar or aquarium of water. As the air bubbles rise from the bottle, the students will realize that the bottle was not "empty," but that it was filled with a gas that we call air.

Feeling some characteristics of a gas: air. Have students swing their hands through the air quickly and describe what they feel. Next, have them hold stiff sheets of cardboard and swing their hands again. They will sense that the cardboard pushes against something. Now have them fill a balloon with air and hold the neck closed with one hand. With the other hand, squeeze and knead the balloon and describe how the contents tend to push back, but the balloon moves if pushed hard enough. When the neck of the balloon is released, they can feel the air rush out.

08
Ab

Smelling some characteristics of gases. Show some liquid perfume in a sealed bottle. Uncap the bottle when the air in the room is still and have students raise their hands when they smell the aroma. They will realize that a gas comes from the bottle. (The gas cannot be seen, but it can be detected by the sense of smell.) Students can note the order in which hands are raised and see that the aroma is gradually diffused throughout the room. Have them note other smells that indicate the presence of a gas (e.g., fumes near a gas station pump; aromas of flowers).

09
Ad

Contributing Idea B. Indicators can be used to identify gases.

Producing carbon dioxide. Carbon dioxide is an odorless, colorless gas. It does not burn or explode. It is not poisonous. It is heavier than air, used to carbonate soft drinks, and is important to the life of plants. Students can produce carbon dioxide easily by several methods.

10
Bb
Gc

a. Pour about two inches (6 cm) of vinegar into a soda pop bottle, flask, or other container. Add two teaspoonsful (two 5 ml spoonsful) of baking soda. Students will see that bubbles are produced. Baking soda is the common name for a chemical compound called sodium bicarbonate. It consists of the elements carbon, hydrogen, oxygen, and sodium. Vinegar consists of a mild acid called acetic acid. When the two are mixed, a chemical action occurs which sets free the gas called *carbon dioxide.*

b. Cover the bottom of a soda pop bottle or other container with a layer of eggshell pieces or broken bits of limestone or marble. Fill the container about three-fourths full of vinegar. Students will see bubbles (carbon dioxide) being produced.

c. Pick up a few small pieces of dry ice with gloves or tongs and place them in water. **(Caution: Dry ice is extremely cold—handle it with care.)** Observe the bubbles (carbon dioxide) coming from it. *(Note: The fog that is produced is condensed moisture in the air and not carbon dioxide.)*

d. Mix yeast with sugar and water. The gas produced is carbon dioxide.

e. Drop an Alka-Seltzer tablet in a glass of water to produce bubbles of carbon dioxide.

Collecting carbon dioxide. Carbon dioxide can be collected by several methods, the selection of which depends upon what is to be done with the gas. **11 Ge**

a. As soon as some carbon dioxide is produced in a soda pop bottle or flask by any of the methods described in *131.10,* have students immediately pull the neck of a balloon over the mouth. The balloon will inflate. When filled, tie off the neck. Students now have a quantity of carbon dioxide gas in the balloon.

b. Produce carbon dioxide in a flask and collect it by placing a one-hole stopper tightly into the flask. Extend from the stopper, a short glass tube to which is attached about 18 inches (50 cm) of rubber tubing (long enough so that it can be placed under test tubes or bottles as shown). Students will see the carbon dioxide as it bubbles through the water.

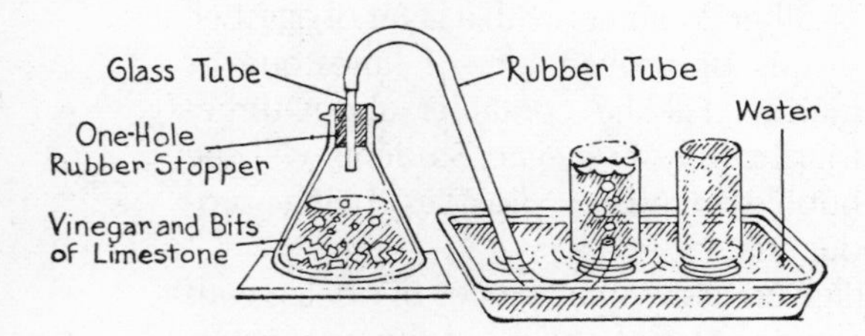

FIGURE 131.11

c. Pieces of dry ice (or vinegar added to baking soda, eggshells, or limestone) can be sealed in plastic sandwich bags. The bags will inflate and can be sealed with a rubber band.

d. An Alka-Seltzer tablet can be placed at the bottom of a balloon or plastic sandwich bag. Tie it tightly with rubber bands or string, then fill the remainder with water and seal it. The gas can be produced by loosening the bands and allowing the water to reach the tablet.

Using an indicator to test for the presence of carbon dioxide. Several tests may be performed to detect the presence of carbon dioxide. **12 Fb Ga**

a. Collect some carbon dioxide in a test tube, place a thumb over the tube, and hold the tube upright. Carefully pour a small amount of clear limewater into the tube and observe what happens. (The limewater will turn milky indicating the presence of carbon dioxide.) To show that carbon dioxide is present in air, place a few spoonsful of clear limewater in an open container for several days and note that white deposits of limestone form around the edge of the container at the surface of the limewater. (This is the result of carbon dioxide in the air combining with the limewater.) To have students see that their own breath contains carbon dioxide, have them exhale through a straw into a glass of clear limewater. Students will find that it takes a great deal of blowing to turn the liquid milky. (Exhaled breath contains very little carbon dioxide—about 3 to 4 percent.) Students can also invert a jar over a burning candle or germinating seeds, then test the candle and seed "airs" to see if they contain carbon dioxide.

b. Collect carbon dioxide by letting dry ice sit in a wide-mouth container for several minutes. A glowing splint or lighted wooden match placed into the container will go out because carbon dioxide does not support combustion and can keep air from a fire.

Producing hydrogen. Hydrogen is an odorless, colorless gas. It is the lightest of gases and can be used to float balloons. Hydrogen, however, explodes easily and is dangerous. The gas should be prepared with careful guidance. Begin by cutting a ½ inch (1½ cm) strip of zinc from a flashlight cell battery with heavy shears. Cut the strip into several smaller pieces and drop them into a test tube. Cover them with 1 inch (3 cm) of vinegar and put a one-hole stopper in the test tube. Insert a glass tube through the stopper so that it is about 1 inch (3 cm) above the liquid. Place the test tube in a holder and gently heat it over a flame, keeping the tube slowly moving back and forth to distribute the heat. Be sure that none of the vinegar enters the tube. **(Caution: Keep the tube pointed away from the body and other students.)** When the liquid begins to boil, stand the test tube in an empty can or holder to cool. Do not remove the stopper. The tube now holds a small quantity of a gas called *hydrogen*.

13
Bb
Gc

Testing for the presence of hydrogen. The presence of a small amount of hydrogen collected in a test tube or other container can be detected by its explosive characteristic. Hold the container of hydrogen upside down (hydrogen is lighter than air), light a wooden splint, turn the container so that it is slanted away from the body, and remove the stopper. Quickly put the burning splint into the contents. The pop that is heard indicates the presence of hydrogen. Students may wish to research the hazards of using hydrogen in balloon flights.

14
Bd
Fb
Ga

Producing oxygen. Oxygen is an odorless, colorless gas. It makes up about 20 percent of the air. It is slightly heavier than air and is a necessary component in supporting life. Whenever anything burns, it unites with oxygen. Oxygen can be produced by students in several ways.

15
Bb
Gc

a. Place one tablespoon (three 5 ml spoonsful) of dry yeast in the bottom of a heat-resistant bottle or flask. When 1 tablespoon (three 5 ml spoonsful) of hydrogen peroxide (6 percent) is added, a bubbling of *oxygen* gas will result. The gas can be collected in quantity by the collecting procedure described in *131.-11b*. If bubbles cease to be generated, simply add more peroxide. *(Note: The peroxide should be fresh. Since it is light sensitive, it may decompose within a few weeks and become primarily water.)*

b. Put a heaping spoonful of pieces of activated charcoal into a container such as a heat-resistant baby bottle. Cover the charcoal with 6 percent hydrogen peroxide and cover the bottle with a sheet of cardboard to trap the oxygen that is produced.

Testing for the presence of oxygen. Several tests and activities can be performed to detect the presence of oxygen.

16
Fb
Ga

a. Hold a container of oxygen right side up and keep it sealed until ready to test. Blow out a flaming splint of wood so that only a tiny spark remains. Remove the cover on the container and insert the glowing splint. (It will glow more brightly or burst into flame, thus indicating the presence of oxygen.) For a variation, twist a small piece of steel wool, hold it with a pair of tongs in a candle flame to bring it to a glow, then drop the wool into a bottle containing oxygen. *(Note: To keep the bottle from cracking during the burning, add a small amount of sand after the oxygen has been collected, but* *cover the bottle quickly to keep the gas from escaping.)*

b. Iron can be used as an indicator for the presence of oxygen because it chemically interacts with the gas and produces rust (a form of burning). To prepare the testing conditions, obtain a box of steel (iron) wool from a grocery store. Remove one roll, find its loose ends, and gently unroll it. **(Caution: Use gloves to protect hands from steel wool slivers.)** Clip the knotted ends away with scissors and divide the remainder into fourths. Cut each fourth into fourths and roll each piece into a loose ball about ¾ inch (2 cm) in diameter. Soak the balls in white vinegar overnight to remove the coating that protects them from rusting, then rinse them with tap water. Now wedge a ball of wet steel wool into the bottom of a test tube or other gas-collecting container. Wedge it just tightly enough to hold it in place when the tube is inverted. Collect oxygen in the container as described in *131.15a* and leave the container inverted in a pan of water for forty-eight hours. If the steel wool rusts, oxygen was present in the container. An estimate of the amount of oxygen can be made by measuring how much water moved into the container from the pan and subtracting this volume from the total volume of the container. If three or four identical collecting containers can be obtained, students can do experiments to answer most questions that naturally arise from this test. For example, if students think that the distance between the steel wool and the water made a difference in the rise of the water, let them put a steel wool ball in the base of one container, three-fourths of the way into another, one-half of the way into a third, and one-fourth of the way into a fourth. Leave the containers inverted for forty-eight hours, then observe the water levels. If students want to know if two steel wool balls would cause the water to rise higher than one, let them place two balls in two containers and one ball in each of two other containers. Observe after forty-eight hours.

Contributing Idea C. Gases have no shape of their own.

Observing the shape of a gas. Fill a balloon with air and seal the neck of it with a rubber band. Squeeze and make different shapes with the balloon. Ask students what shape the air inside has. They will realize that air can be compressed, and its shape can be changed. (Every time the balloon undergoes a change in shape, the air in it does, too.) If students repeat this activity with a balloon filled with carbon dioxide, they will realize that all gases are shapeless, but take on the shape of their container. **17 Aa**

Contributing Idea D. Gases occupy space and have weight.

Observing that gases occupy space. Turn a glass full of air upside down and push it down into a pan or wash basin which contains 3 or 4 inches (8–10 cm) of water. Students will see that the air in the glass takes up space and holds the water from entering the glass. Now let them slowly tip the glass and observe the air bubbles as they leave it. When placed down again, they will see that the water has now entered the glass. Several variations of this activity can be performed. **18 Aa**

a. Push a piece of dry, crumpled, paper toweling into a dry glass. Be sure that the toweling cannot fall out. Invert the glass and push it to the bottom of the water in the pan. The toweling will remain dry because the air kept the water from entering.

b. Float a small cork in the pan and invert the glass over it. Note the level of the floating cork and let students explain why it changed. If a bottomless glass or jar can be obtained, it can be set over another floating cork next to the first glass, and the levels of the two corks can be compared side by side. Challenge students to give explanations for the differences they see.

c. Hold two inverted glasses, one filled with water, one with air, underwater in an aquarium. Pour the air back and forth between the glasses by tipping one to the side and trapping the rising air bubbles with the other.

Weighing a gas: air. Using a balance scale, have students weigh a balloon before they blow it up. Record the weight. Blow up the balloon and weigh it again. Compare the two weights and have students explain what the measurements tell them about air. Similarly, the weights of inflated and deflated playground equipment (e.g., footballs, basketballs, volleyballs) can be compared. **19 Cf**

Weighing a gas: carbon dioxide. Use a commercial balance scale with glasses balanced on each side or balance a yardstick with two small empty paper bags of identical size as shown. Fill a jar with carbon dioxide and slowly pour it into one of the glasses or bags. Students will see the balance tip as the container receives the carbon dioxide, and realize that this gas is heavier than air. **20 Cf**

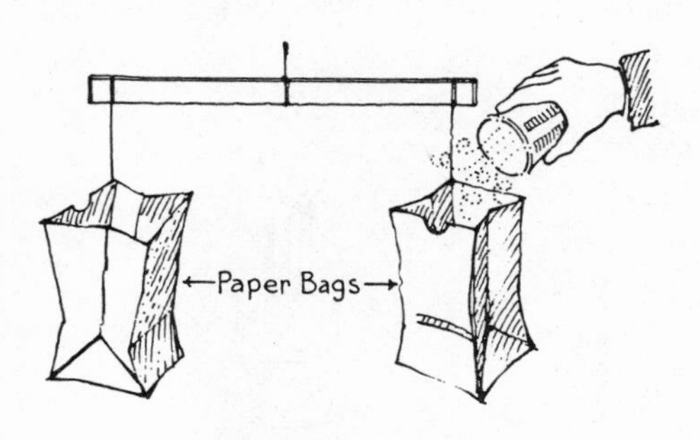

FIGURE 131.20

Using a heavy gas to put out a fire. Use tongs or wear gloves to place several pieces of dry ice in a dry beaker. **(Caution: Do not handle dry ice with bare hands.)** Stand a short candle in another beaker and light it. Be sure the candle is below the top edge of the beaker. After a few minutes, hold the beaker containing the dry ice over the candle and tilt it as if to pour something out *(a)*. Students will see that the candle flame is extinguished. You might tell them that the dry ice is frozen carbon dioxide. As it changes to a gas, it remains in the beaker because it is heavier than air. When poured on a burning candle, it pushes the air out of the way and smothers the flame. Similarly, a jar containing vinegar and soda solution can be tipped so that the gas produced (carbon dioxide) will flow down a creased strip **21 Ga**

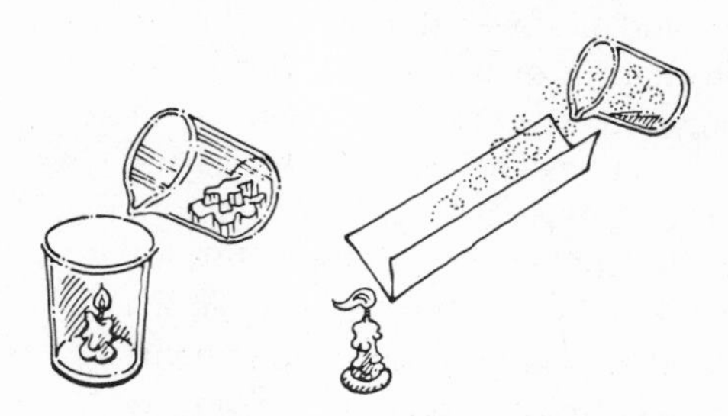

FIGURE 131.21a **FIGURE 131.21b**

FIGURE 131.21c

of paper aimed directly at the candle flame *(b)*. An alternative method would be to light a candle that is set in a wide mouth jar that contains an inch (3 cm) of water *(c)*. Add several teaspoonsful (5 ml spoonsful) of baking soda to the water, then pour vinegar into the solution. As the bubbles rise and the amount of carbon dioxide collects, the candle will go out. Students will realize that the carbon dioxide not only puts out fires, but it is heavier than the surrounding air. Many fire extinguishers use carbon dioxide to put out fires. If possible, ask the school custodian to demonstrate one to the class.

Contributing Idea E. Gases expand when heated and contract when cooled.

Observing that gases expand and contract when heated or cooled. Cut a piece of rubber from a balloon and stretch it across the mouth of a flask or bottle. Secure it with a rubber band and place the container near a warm radiator or in a pan of warm water. Students can observe changes in the rubber membrane and discover that when the air inside is heated, it expands and forces the balloon outward. Now cool the container in a bowl of ice cubes and observe what happens. Similarly, several bottles of different sizes can be cooled in a refrigerator for about an hour, then capped with balloons and returned to the classroom. Judgments concerning expansion in relation to volumes can be made. A reverse procedure can be carried out by heating the bottles over a radiator, capping them with the balloons, and allowing them to come to room temperature. **22 Aa**

Measuring the expansion and contraction of a gas when heated and cooled. Blow up a balloon, measure its circumference with a tape measure, and place it over a radiator (or by some other heat source) for about ten minutes, then measure its circumference again. Remove it from the heat source for about ten minutes (or place the balloon in a bowl of ice cubes or refrigerator) and measure its circumference. By comparing the three measurements, students will be able to judge that the heat caused the air inside to expand and when the heat was removed, the air contracted. **23 Cd**

132 INTERACTIONS

Generalization I. The volume of a gas is a measure of the space it occupies.

Contributing Idea A. The space a gas occupies can be measured.

Observing that gases occupy space. Carefully push a funnel through a one-hole stopper and place the stopper tightly into a bottle. Slowly pour some water into the funnel and observe what happens. Repeat using a funnel in a two-hole stopper *(a)*. Challenge students to explain the reason for the differences they note. (In the first test, air occupied the space in the bottle and the water could not enter; in the second test, the air was able to escape through the second hole as the water entered.) For a variation of this activity, insert a U-tube or piece of plastic **01 Aa**

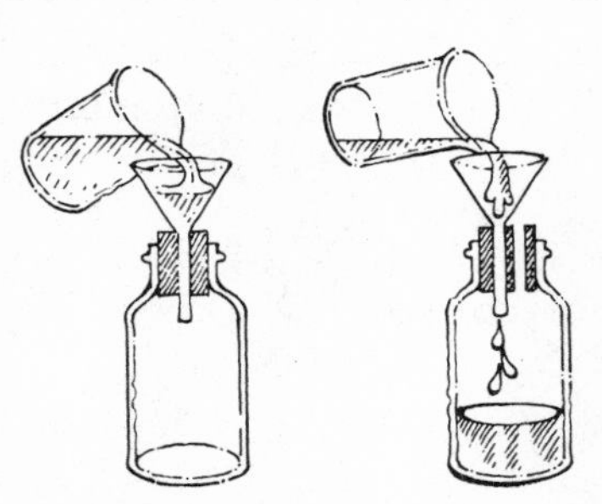

FIGURE 132.01a

tubing in the second hole of the two-hole stopper. Fit a balloon over the end of the tube and tie it securely. Slowly pour water into the funnel until the bottle is filled *(b).* Ask for explanations of the balloon's size increase as the water is poured. Remove the balloon, empty the bottle, and let a student hold a finger over the end of the tube. Fill the funnel with water again. Move the finger to open and close the tube and observe what happens. Ask students to tell about experiences they have had trying to pour or drink liquid from a can with only one hole punched. They will begin to realize that two materials cannot occupy the same space at the same time.

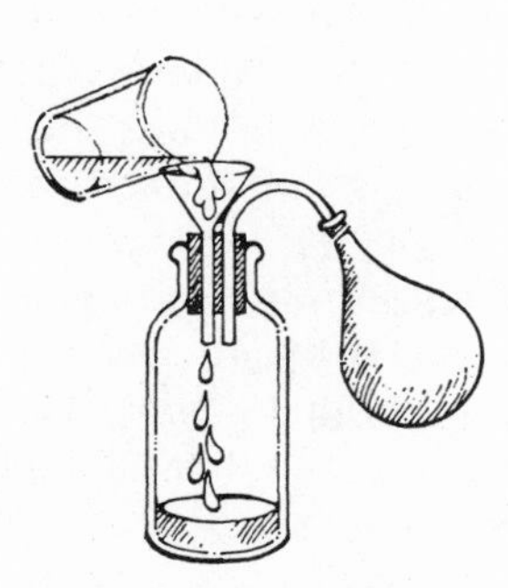

FIGURE 132.01b

02 Bf Ca Cg

Measuring and graphing air volumes in relation to burning times. Bring to school several jars of different sizes. Divide the class into groups, each with a birthday cake candle and a lump of modeling clay. Have one student be responsible for lighting the candle properly. *(Note: Have a can of water available for the matches after the candle has been lit.)* Let another student be a timer, using a watch with a second hand. Have each student, in turn, set the jar over the lighted candle when the timer says "go." Let each record how long it takes to burn, then measure the volume of space in the jar by filling it with water from a measuring cup. Each test should be repeated several times to check on measurements. *(Note: It is best to swing the jar through the air a few times before each test to fill it with fresh air.)* The collected information can be placed on a graph like the one shown. Students should place their measurements for their own jars on the grid. From the final graph, they will be able to predict how long a candle will burn for any size container.

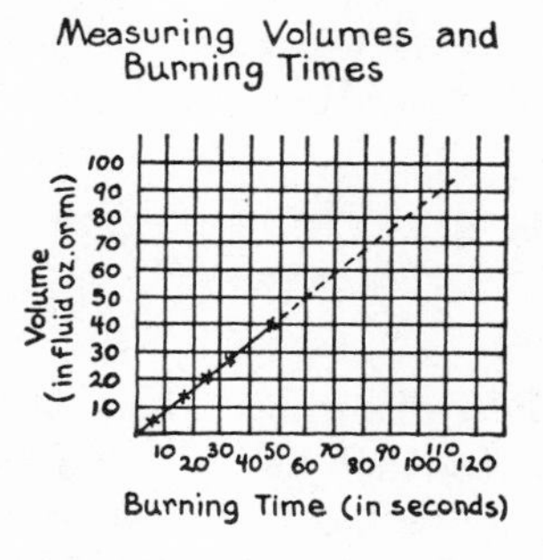

GRAPH 132.02

Contributing Idea B. Gases have weight.

03 Cf

Comparing gases by weight. Collect carbon dioxide in a balloon, then blow up a second balloon with air to the same size. Toss the balloons into the air and let students observe what happens. If helium can be obtained from a chemical supply house, fill one balloon with it, another to the same size with carbon dioxide, and a third with air. Release all three at the same time. Students will be able to judge that different gases have different weights.

04 Cf Cg

Comparing gases by weight and volume. Inflate a balloon and set it over a radiator or other heat source. After ten minutes, measure its circumference. Inflate a second balloon to the same size, but keep it at room temperature. Place the two balloons side by side and ask which contains more air and which is heavier. Let students check their guesses by placing the balloons on each side of a balance scale. Next, let the warm balloon cool and have the students compare the sizes. Students will realize which contained the most air and why one was heavier, even though both were once identical in size. (Identical gases of the same volume may differ in weight due to temperature differences.)

Generalization II. Gases exert pressure.

Contributing Idea A. The pressure of a gas increases with depth.

Feeling the weight of air. Let each student fill a plastic sandwich bag with air, either by blowing up the bag or by pulling it through the air quickly with the mouth held open. They may help each other tightly tie the air-filled bag upside down over the mouth of a jar with a string or rubber band. Have them press down on the bag, lean on it, rest a book on it, and describe what happens. Now ask them to remove the bag, empty it of air, and tie it again—but this time with the bag inside the jar. When the bag is secure, have them hold the jar and try to pull the bag out. They will find that they cannot do this. (The plastic bag has air resting in it—this air weighs much more than the students can lift.)

05
Ab
Cf

Measuring the amount of air weight (air pressure) upon a surface. Place a yardstick, a slat from a wooden crate, or other very thin piece of wood about 1 yard (1 meter) long and 2 inches (6 cm) wide on a table, allowing about 1 foot (30 cm) to extend over the edge. Have a student strike the end of the stick with his or her hand just hard enough to knock the stick off the table. **(Caution: Do this activity away from other students and be sure no one is in front of the stick.)** Now replace the stick in the same position and lay an open section of newspaper over the part that is on the table. Let the student strike the end of the stick as before. Students can discuss what caused the stick to break. Next, fold the newspaper into a small bundle, approximately 6 inches by 4 inches (16 cm by 12 cm) and tape the edges so that the bundle lies flat. Let the student repeat the test using the folded paper. Challenge students to explain what might account for the results this time. As an alternative or supplementary experience, mark off one side of the open newspaper into 1 inch (3 cm) squares before placing it on the stick. When the paper is folded, many less squares or surface areas are exposed to the air. By reminding students that air has weight (at sea level, the weight of the air on 1 square inch or 3 square cm of area is about 15 pounds or 7 kg—all the air directly above 1 square inch or 3 square cm of surface), they can compare the weights on the squared surfaces. You might tell them that this weight is called *air pressure.* Next, mark off a piece of notebook paper into 1 inch (3 cm) squares, then have students draw an outline of their hands on the paper and calculate how many pounds of air are pressing down on their hands. Have students estimate how many 1-inch squares could fit on the outside of their bodies and how many pounds (kilograms) of air they think their bodies support. If students wonder why they are not crushed by all this pressure, you might explain that our bodies, like most objects, have air in them and this air pushes outward with the same force. When the forces are equal, we cannot feel the pressure. When they are unequal, we feel the pressure. Students can

06
Bb
Cf

relate the sensation of their ears "popping" to this experience. (The sensation occurs when either the outside or the inside pressure in their ears is different.)

Contributing Idea B. The altimeter is an instrument used to measure the pressure air exerts at different altitudes.

Making an altimeter. Draw a line lengthwise across the middle of a white card and mark off ¼ inch (1 cm) spaces along the line. Insert the short arm of an L-shaped tube through a one-hole stopper and place a drop of colored water in the center of the long arm. Hold a finger over the end of the long arm while students insert the stopper into the mouth of a gallon plastic jug. Be sure the stopper is air-tight (some clay or Vaseline will seal the stopper). Now tape the card to the long arm of the tube and tie a string around the neck of the bottle leaving a loop

07
Bb
Gc

FIGURE 132.07

so that it can be carried without fingers touching the jug. If your school has several floors, students can carry the altimeter to the highest floor and notice any change in the water marker. (Variations in air pressure upon the jug will cause the water to move in the tube.) You might tell them a device that indicates the weight or pressure of the air at different altitudes is called an *altimeter.* If possible, let them carry the altimeter in an elevator to observe what happens.

Contributing Idea C. The pressure of a gas is the same in all directions at the same altitude.

Testing to see if air pressure is the same in all directions. Cover the top of a funnel with a piece of rubber from a balloon. Be sure the rubber is stretched tightly and held in place with a rubber band. Students can take turns placing a finger over the small open end of the funnel and turning it in different directions to see if the rubber membrane remains level. Ask what would happen if the outside pressure was greater than the inside pressure. Now suck some air from the funnel and quickly cover the end with a finger. Ask why the membrane is depressed. (The air pressure inside the funnel is less than the air pressure outside.) Again turn the funnel in different directions to see if there is a change in the rubber membrane *(a).* Students will realize that if the outside

08
Ec
Fa

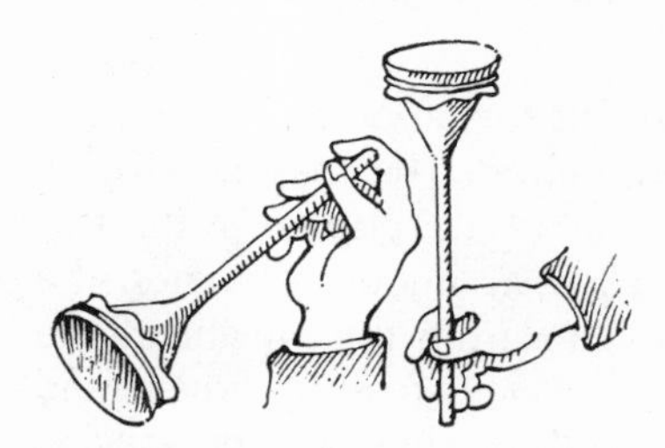

FIGURE 132.08a

pressure were less at any position, the membrane would be less depressed or would bulge. If possible, test this device at different altitudes to see if air pressure is the same in all directions. Similarly, students can fill a glass to the brim with water and press a plastic dish cover over it *(b).* A thin piece of cardboard can be used, but a dish cover is impervious to water. Let one student carefully turn the glass upside down over a sink or large pan and slowly remove his or her hand. (The cover will remain in place due to the outside air pressure pressing upward against it.) Now hold the glass sideways and turn it in various directions. Ask students what this tells them about air pressure and what would happen if the pressure inside the glass were greater than the pressure outside. As an extension to this experience, you might ask if less water in the glass would make any difference in the test; then let students find out by retesting with the glass only partially filled with different amounts of water. They will discover that no matter how

FIGURE 132.08b

much or how little the amount of water, the cover will not fall off. The reason involves a factor that might be explained to the students: With a partial amount of water, the cover sags slightly from the weight of the water inside, the weight of the cover, and the pressure of the air inside. The sagging slightly enlarges the space the air inside occupies, the almost imperceptible enlargement being enough to reduce the pressure inside the glass (pressure decreases as volume increases), and the greater outside air pressure being enough to stabilize the situation and hold the cover in place.

Contributing Idea D. The pressure of a gas decreases as the amount of space available for the gas increases; the pressure of a gas increases as the amount of space available for the gas decreases.

09 Aa Fa

Observing that the expansion and contraction of a gas increases and decreases its pressure. Invert a bottle or jar over a burning candle that is set in a tray of water. The candle is used to heat the air inside the container. As the air is heated, bubbles will be driven from the mouth of the container. (The air inside expands and the increased pressure forces some of the contents out.) When the candle is extinguished and the air inside cools, water will enter the container. Students will realize that this is caused by a reduction in pressure (The outside air pressure was greater than that inside the jar, thus its push on the surface of the water forced some of it into the container.)

10 Gc

Using the pressure of a gas to extinguish fires. Students can make simple fire extinguishers that work from the pressure created by the production of carbon dioxide.

a. Clean an ink bottle that has a well inside. Put a small hole in the cap and insert a glass tube or plastic straw (the smaller the opening, the better). Let the tube extend no more than ¼ inch (.5 cm) into the bottle and seal the fitting with clay, candle wax, or several coats of sealing wax. Now fill the well inside with sodium bicarbonate (baking soda) and the remaining portion about one-quarter full of vinegar. Screw the cap on tightly *(a)*. A small fire or burning candle can be extinguished by tilting the bottle upside down and aiming the tube or straw at it.

b. Empty a tea bag and fill it with sodium bicarbonate. Suspend it in a bottle that is half-filled with vinegar water (50 percent vinegar and 50 percent water). Insert a one-hole stopper into the bottle. Extend from the stopper a piece of glass or plastic tubing with a short piece of rubber tubing attached *(b)*. The rubber tubing can be used as a hose to direct the spray when the bottle is inverted. To be effective, all fittings must be tightly sealed. Nozzles made from medicine droppers can be used on the end of the rubber tubing to increase the range of the spray, and different amounts of baking soda and vinegar can be tried to see which produce the strongest stream of water

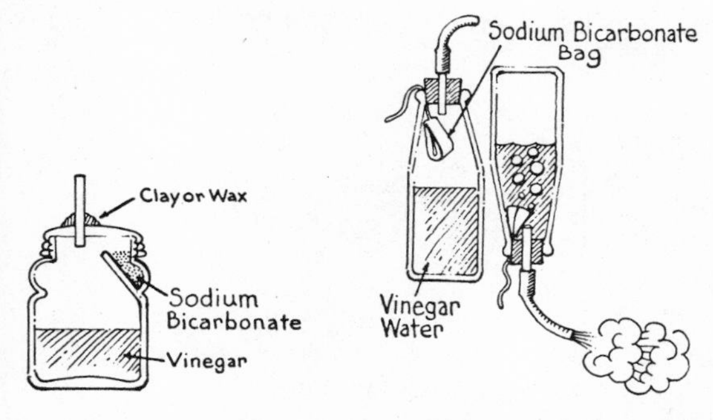

FIGURE 132.10a FIGURE 132.10b

Observing the effects of increasing air pressure. Blow up a balloon until it is slightly larger than the mouth of a large empty bottle. Tie the neck of the balloon with a rubber band, then wet the balloon and the mouth of the bottle to make them slippery. Have students take turns trying to force the balloon into the bottle without letting the air out of the bottle. Now deflate the balloon and place it inside the bottle, a cup, or a glass with the neck up. Blow up the balloon as much as you can and lift it by its neck. Students will see that the sides of the bottle compress the balloon which, in turn, increases the air pressure inside. The pressure is great enough to hold the object firmly so that it can be lifted. 11 Aa

Observing the effects of reducing air pressure. Blow up a balloon part way, tie it with a rubber band, and squeeze it into a transparent plastic jug. Prepare a stopper with a glass tube about 10 inches (25 cm) long and place it in the mouth of the jug. Attach a piece of rubber tubing to the glass tube and tighten the stopper in the jug. Have a student suck some air from the jug using a clothespin to squeeze the rubber tube together whenever the student stops to exhale. While the student continues sucking the air, have the others try to explain what causes the balloon inside to enlarge. 12 Aa

Contributing Idea E. Changes in air pressure can be measured.

Observing that air can support a column of water. Have students fill an olive jar or other tall thin jar with water and place a piece of cardboard over the open end. While keeping the cardboard in place, invert the jar in a bowl of water. When the open end of the jar is below the level of the water, remove the cardboard. Ask students to explain why the level of the water remains higher in the jar than in the bowl. (The weight of the air pressing upon the surface of the water in the bowl is sufficient to support a column of water in the jar.) 13 Aa

Making and using a water barometer. Place a bottle with a long slender neck (e.g., wine bottle or hair tonic bottle) inverted in a quart or liter jar that is half-full of water. Let a student warm the bottle with both hands until several large bubbles of air escape. Let the bottle sit undisturbed for several minutes. Students can be challenged to explain why the water rises in the bottle as it cools. Place the jar and bottle in a spot where the temperature remains constant. (A thermometer might be placed nearby to make sure the temperature is constant.) Number a large index card lengthwise with lines ¼ inch (1 cm) apart. Tape the card, 14 Bb Ga Gc lined side against the glass, on the side of the jar. Students can periodically read and record the level of the water inside the bottle and realize that the water level is an indicator of the pressure of outside air (providing the pressure inside the bottle does not vary as a result of temperature changes). You might tell them that the device they have made is called a *water barometer.*

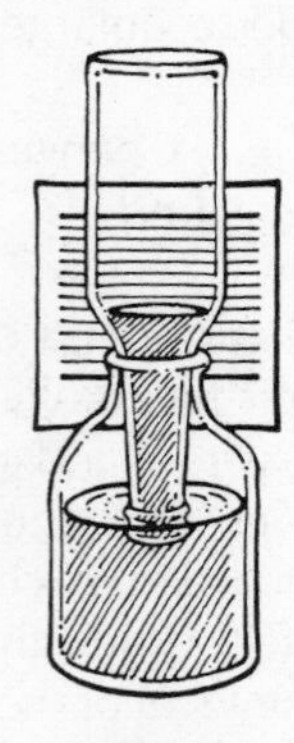

FIGURE 132.14

Making and using an air barometer. Obtain a large, clean, dry bottle and place it in warm water for a few minutes. Be careful not to let any water enter the bottle. Over the top of the bottle, stretch a piece of rubber from a balloon and secure it tightly with rubber bands. Glue a soda straw to the rubber top so that the straw extends horizontally from the center. Hold the straw firmly against the glue until it dries, then flatten the free end by pinching it. Set a white card behind the straw and place a mark on it at the level where the straw touches it. Mark this position "0," then at ⅛-inch (.5 cm) intervals mark "1," "2," and "3" above and below the "0." Ask students to figure out in which direction the straw will move when air pressure increases or decreases. Since the bottle contains a closed volume of air under a certain pressure, any changes in the outside pressure will cause the rubber top to move in or out slightly, causing a rising or dipping of the straw that can be noted on the scale. Now set the bottle away from direct sunlight, radiators, or other heat sources so that heat will not influence the operation of the instrument. You might tell them that this device is called an *air barometer.* Keep records of changes in air pressure. If a water barometer is also constructed, comparisons of readings between the two can be made.

15
Bb
Ga
Gc

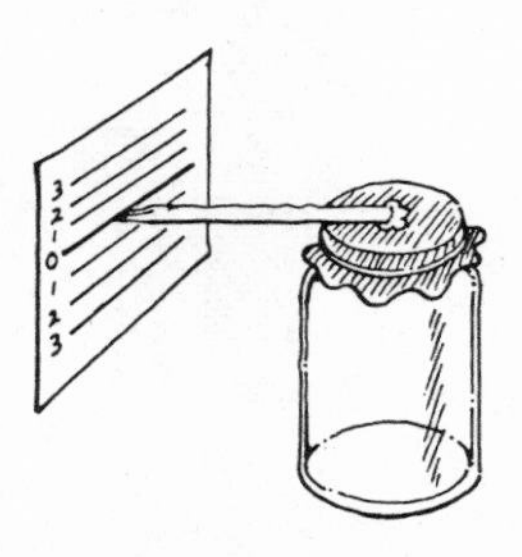

FIGURE 132.15

Generalization III. Heat causes gases to move.

Contributing Idea A. The movement of gases can be detected.

Detecting the diffusion of gases. Put a spoonful of inexpensive cologne in a jar. Inflate a balloon, secure it with a rubber band, and push it inside the top of the jar just enough to seal it. After twenty-four hours, remove the balloon and take off the rubber band. Let students describe the smell of the escaping air. They will realize that the cologne, as a gas, moved without the aid of wind currents through the air in the jar and through the balloon. You might explain that this type of movement is called *diffusion.* Next, close all the classroom windows and when everyone is still, open a bottle of perfume or other material of strong scent. Have

16
Aa
Bb

students raise their hands when they detect the smell so they can see how the gas diffuses throughout the room.

Constructing devices to detect air movements (currents). Have students observe the movement of leaves on trees, dust on the ground, or flags on poles. Ask them to notice other indicators that show when air is in motion. You might tell them that moving air is called *wind.* Students can construct various kinds of air movement indicators.

17 Bb Ga Gc

a. On a white card, make three concentric circles with diameters of 2½ inches, 2¼ inches, and ½ inch (6 cm, 5 cm, and 1 cm). Set the ends of a compass so that they are 1¼ inches (3 cm) apart. Place the point of the compass anywhere on the outer circle, and make six consecutive marks around the outer circle. Using a ruler, connect opposite marks with a straight line *(a).* Carefully cut out wind foils along the dark lines and fold along the dotted lines as illustrated. The detector can be mounted on the eraser end of a pencil with a pin *(b).* Be sure it spins freely. The device can be used to detect wind currents in different parts of the school grounds or schoolroom.

b. Stick a pin through the middle of a plastic straw and attach it to a cork. Turn the straw on the pin a few times so that it swings freely. Next, pinch together one end of the straw and fasten a paperclip to it. On the other end, glue or tape a light index card in a horizontal position. Level and balance the straw by moving the paperclip. Use a protractor to make a scale like the one illustrated. Attach the scale to a large side of a small breakfast cereal box. Glue the cork to the center of the scale. When the box is held in a vertical position, students will see that the detector arm moves up and down and indicates the relative force of the rising or falling air. When held in a horizontal position, it measures air that moves horizontally. Hold the detector in various locations (e.g., over lighted bulbs, below open refrigerator doors, over radiators, below open windows). Measurements can be recorded and compared.

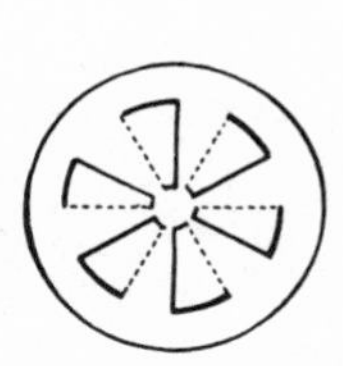

FIGURE 132.17a

FIGURE 132.17b

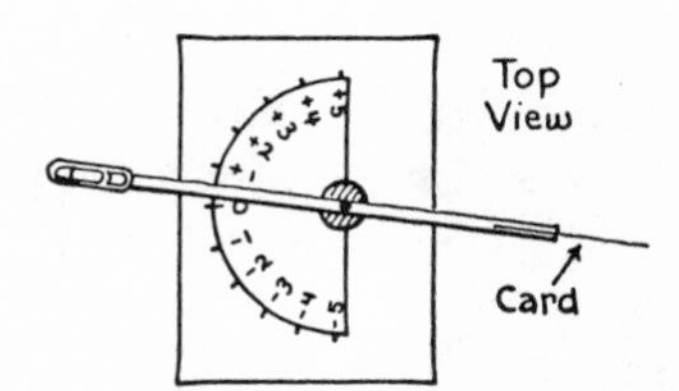

FIGURE 132.17c

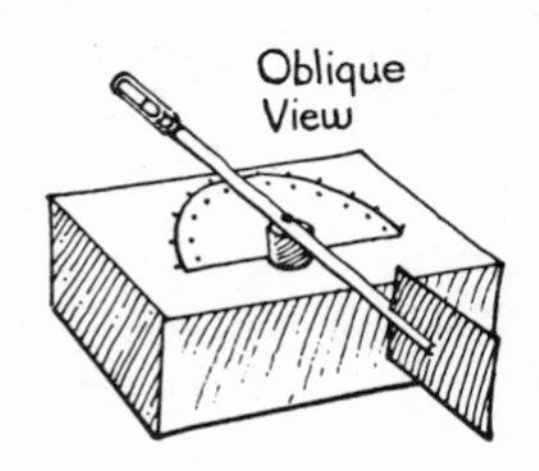

FIGURE 132.17d

Contributing Idea B. The exchange of heated and less heated gases creates convection currents in gases.

Observing that warm air rises and cold air falls. Blow some soap bubbles or bring a burning candle flame near a hot radiator.

18 Aa

Students can observe the movement as the bubbles or flame are placed in different positions. They will realize that the air above the radiator seems to travel upward while the air near the floor moves toward the base of the radiator. Similarly, the bubbles or flame can be brought near an open window to see which way the air moves. If observations are coupled with temperature measurements, the students will find that hot and cold airs tend to trade places. *(Note: The detector described in 132.17 can be used instead of the soap bubbles or candle flame.)*

Generalization IV. Moving air has force and can do work.

Contributing Idea A. Moving air has force.

Observing that moving air has force. The force of moving air can be observed in many ways. **19 Af**

a. Have students cut a hole in one end of a shoebox or similar cardboard box. Squeeze the sides of the box and feel the air coming from the hole. Squeeze the box to blow bits of paper or feathers with it.

b. Float a cork, a piece of wood, or a matchbox in a pan of water. With a sheet of cardboard or a fan, fan the object and move it across the water.

c. Fill a balloon, then let it go. Students will realize that when the gas moves out of the neck, the balloon is forced in the opposite direction.

d. Smooth a paper bag and put it near the edge of a table with the opened end over the side. Place a book on the bag and blow the bag up. Students will see that the air has enough force to lift the book.

Contributing Idea B. The force of moving air can be used to do work.

Finding examples of the force of air being put to use. For each of the experiences in *132.19,* let students find pictures in magazines that show how people have made use of the force of air to do work. For example, air compressors and vacuum cleaners would exemplify the first experience, sailing vessels the second, rockets the third, and air pumps the fourth. Students can also make various pinwheels and study how windmills can do work much like waterwheels. **20 Ga**

140 geology

141 CHARACTERISTICS

Generalization I. Rocks have identifiable characteristics.

> *Contributing Idea A.* Igneous rocks are formed from hardened magma.

Identifying the characteristics of magma. Obtain a stick of red sealing wax. Let students feel it, then set it in a hot place (near a heater or sunlit windowsill). When warmed, let them examine it again, and describe what changes took place. Now hold one end of the sealing wax in a candle flame, and let the wax drop onto a sheet of cardboard. Point out that the characteristics of the sealing wax are similar to those of thick, hot, molten, pliable rock within the earth. You might tell the students that the hot plastic rock within the earth is called *magma,* but when it comes to the surface it is called *lava.* Explain that when magma cools, it hardens (like the sealing wax) and forms masses of hard rock.

01
Aa
Bb
Fc

Making models to show how magma rises to the earth's surface. Obtain a partly used tube of toothpaste, screw the cap on tightly, and flatten the tube so that the toothpaste is evenly distributed inside. Provide the analogy that the metal side of the tube represents the hard crust of the earth, and the toothpaste represents the molten rock or magma beneath. Demonstrate how magma flows under pressure by pressing on the lower half of the tube. Students will note how the upper end enlarges under the pressure. After you explain that magma sometimes finds breaks or weak places in the crust, puncture the end near the cap with a pin. Students will realize that the toothpaste flowing up through the hole due to pressure is similar to magma flowing from the earth and forming volcanoes. This idea can be further explored by making a volcanic cone from a sheet of paper and placing it over the open end of the toothpaste tube. Cut the point from the cone, and squeeze the tube. The toothpaste will flow from the cone like magma from a volcano. You might tell students that when magma flows on the surface of the earth, it is called *lava.*

02
Fc

Identifying the characteristics of igneous rocks. Bring to class several examples of each of the following igneous rocks: granite, basalt, obsidian, pumice. Explain that each was formed from magma and that because of this, some people call them "fire rocks." You might tell students that scientists call such rocks *igneous rocks.* Students can describe the characteristics of each—their

03
Aa
Bb

shapes, colors, and textures. Some students can break pieces of each kind with a hammer for closer examination with a hand lens and to see if each breaks in a regular way. **(Caution: Place each rock in a bag or wrap it in a cloth to keep pieces from flying.)** Some of the following information on each type might be helpful:

a. *Granite.* Granite forms under the ground when magma cools and hardens in spaces between layers of rock. Students can note that this type of rock is coarse and contains large crystals. It varies in shading from white to gray, and is sometimes pinkish. If possible, show examples of the use of granite in making statues, monuments, and buildings.

b. *Basalt.* Basalt forms when magma pours out of the earth slowly and cools and hardens slowly. Students may note that this rock contains very small crystals. It is usually a dark, greenish-gray color. Some pieces have holes made by escaping gases.

c. *Obsidian.* Obsidian forms when magma pours out of the earth slowly like tar, but cools quickly. The students may note that this rock has no visible crystals. It has a black, glassy appearance. When fractured with a hammer, note what pattern the chips make. Hold pieces up to the light to see if light shines through. If possible, show examples of Indian arrowheads made from obsidian.

d. *Pumice.* Pumice forms when magma spills out of the earth so quickly that it is foamy. You might shake and squirt the foam from a soda pop bottle to illustrate how magma sometimes spills out. When cooled quickly, it forms pumice. Students can note that the grayish rock is very light, porous, and can be floated in a container of water.

04 Ca Fc

Comparing crystalline differences in igneous rocks. Obtain several examples of granite, basalt, and obsidian rocks. Let students compare samples of each with a hand lens. They will see that granite contains large crystals, basalt very tiny crystals, and obsidian no observable crystals. Ask why such differences might occur, then prepare solutions of powdered alum in two test tubes (2 spoonsful of alum in water—shake to dissolve). Sugar, salt, Epsom salt, or borax can be used in place of the alum. Boil the solutions over a candle flame and add another spoonful to each. Explain that the solutions represent magma in the earth. Now hang small nails from strings and suspend them in the test tubes. Set one test tube in a holder and allow it to cool slowly. Place the second test tube upright in a bowl of ice cubes to cool it quickly. In twenty-four hours the students will see differences in the formation of crystals—large crystals in the container that cooled slowly, small crystals in the one that cooled quickly. Students can make the analogy to rocks—slowly cooling rocks contain large crystals, quickly cooling rocks contain small crystals. Let them order their rock samples by the speed at which they probably cooled.

Contributing Idea B. Sedimentary rocks are formed through the depositing of sediments that are pressed and cemented together.

05 Fc

Making a model of sedimentary deposition. Half fill a quart-size glass jar with water and add several handsful of sand, gravel, and soil. (An alternative mixture might be made of powdered clay, sand, iron filings, and small brads.) You might point out that the mixture represents materials that a river might deposit in an ocean, sea, or lake. Shake the jar, then set it on a table where it will not be disturbed. Observe the jar periodically throughout the day until the water clears. Students will note the layered appearance of the settled materials and can determine which materials are on top, in the middle, or on the bottom. The action is analogous to that of natural materials being added to old layers year after year. Many mineral deposits are thought to have been formed in this way.

Feeling and seeing the effect of the buildup of subsequent layers of sediment. The compaction of layers of sediment can be shown by having several students put one hand, palm down, flat on a table top and placing a book on it. The hand and the book are analogous to layers of sediment. Now place a second and a third book on top of the first. Students can describe how the hand feels, and realize the analogy to the pressure placed on lower layers of earth as subsequent layers are placed on top. Similarly, students can experience compaction by pressure by having several take a handful of wet soil and squeeze it as tightly as they can. They will notice that the material seems to hold together, is harder, and is more compacted after the pressing. You might tell them that when small pieces of material are pressed and cemented together, they form a kind of rock called *sedimentary rock.* To make models of fossils in sedimentary rocks, see 151.15.

06
Af
Bb

Identifying the characteristics of sedimentary rocks. Bring to class several examples of each of the following sedimentary rocks: conglomerate, sandstone, shale, and limestone. Explain that each was formed underwater as a result of layers of deposited materials. Differences in kind depend upon the materials deposited. Tell students that particles that settle are called *sediments,* and that when they harden, they are called *sedimentary rocks.* Have them describe the characteristics of each—their shapes, colors, and textures—and note whether or not each breaks in a regular way. Examine the broken pieces with hand lenses. To identify the samples, some of the following information might be helpful:

07
Aa
Bb

a. *Conglomerate* forms underwater, made up of stones and pebbles cemented together. Students can see these pieces in the rock and can probably break some away from the specimen. Look closely at a sidewalk (an artificial conglomerate) to see how similar it is in composition.

b. *Sandstone* forms underwater, made up of tiny grains of sand cemented together. Students can clearly see the sand grains with a hand lens. By rubbing two pieces of sandstone together over a sheet of black paper, they can collect and study some of the grains. Sandstone may be of various shades of gray, yellow, or red. If it is brown or orange, it contains some iron.

c. *Shale* forms underwater, made up of clay or mud that is cemented together. By rubbing two pieces of shale together over black paper, the students can collect and study some of the tiny fragments with a hand lens. If a few drops of water are placed on the shale, it will give off an odor of mud or clay. Shale varies greatly in color and may be shades of gray, green, yellow, red, or black.

d. *Limestone* forms underwater, made up of shells and skeletons of animals that lived long ago. Often the shells are worn, making them difficult to recognize, but a vinegar test can be used to identify limestone. To do the test, set a piece of limestone in a dish and put two other rocks which are not limestone in two other dishes. Pour some vinegar on each. Students will see bubbles and hear a fizzing sound in the limestone dish, and realize that such a reaction takes place only with forms of limestone rock. To further demonstrate how vinegar can be used as an indicator of limestone, pour some on a piece of classroom chalk and note the reaction. Explain that chalk is a very fine-grained limestone. Natural samples of limestone may be white, gray, or red in appearance.

Contributing Idea C. Metamorphic rocks are formed when pressure and heat change igneous and sedimentary rocks.

Identifying the characteristics of metamorphic rocks. Bring to class several examples of each of the following metamorphic rocks: gneiss, schist, quartzite, slate, and marble. Explain that each was formed from igneous or sedimentary rocks that were subjected to great pressures and heated over long periods of time. Tell students that scientists call such rocks *metamorphic rocks.* Have them describe the characteristics of each—their shapes, colors, textures, and hardness—then cover samples with a cloth or bag and break each with a hammer. Students can note whether or not each breaks in a regular way. Examine fragments with a hand lens. If possible, compare the metamorphic rock form with an example of the igneous or sedimentary rock form from which it was made. The following information might be helpful:

08
Aa
Bb

a. *Gneiss* is formed by pressure and heat from igneous granite. The crystalline fragments within the two kinds of rock can be compared.

b. *Schist* is formed by pressure and heat from sedimentary conglomerate.

c. *Quartzite* is formed by pressure and heat from sedimentary sandstone. With a hand lens students can note how the sand grains have been reduced in size through the crushing action of great pressures.

d. *Slate* is formed by pressure and heat from sedimentary shale. Ask students to find out if wet slate smells like wet shale. (It does, but faintly.) They can examine the school blackboard to see if it is made of slate. (Most blackboards are made of slate, although some newer ones are being made of other materials.)

e. *Marble* is formed by pressure and heat from sedimentary limestone. Compare a piece of marble with a piece of limestone to note similarities and differences. Each can be subjected to a vinegar test. If the marble is first crushed with a hammer, it will react visibly in vinegar.

Making models of pressure and heat sources in the earth. Models of pressure and heat sources can be shown in several ways.

09
Fc

a. Have students open books on their desks. Tell them that the pages represent layers of sediment in the earth. Let them hold the center of the book with one hand while they push on the edges of the pages with the other. Explain that the bulging in the middle is similar to the building of layers of sediment in the earth when pressure is applied. You might explain that pockets formed beneath the layers sometimes fill with magma creating more pressure and heat on the layers.

b. Students can crumple a piece of paper, unfold it, and place it flat on a desk. When they push the ends of their papers toward each other, they will see that the pressure causes the surface to wrinkle up along certain crease lines. This suggests how mountains were formed along fault lines in the earth and that as they were formed, the extreme pressures may have changed the surrounding rocks.

c. Wrap alternate colored slices of white and brown bread together in waxed paper to make about six sedimentary layers. Place several very heavy books on the slices. Unwrap the slices after a day, and note what changes took place. Students can realize that in a similar way, sedimentary layers of rock are sometimes compacted by extreme pressures in the earth.

Contributing Idea D. Rocks can be identified by certain physical characteristics.

Identifying rocks by size, hardness, and weight. Bring a variety of rocks to class for students to observe, handle, and compare. Different activities emphasize certain physical characteristics of rocks.

10
Bb
Ca
Da

a. *Size.* Set out some rocks ranging in size from sand to large stones. Have students sort the rocks by size. You might tell them that small pieces are *sand,* the next size rocks are *pebbles,* the next are *stones,* and the very large are *boulders.*

b. *Hardness.* Show that some rocks scratch or crumble easily and some do not. Try scratching one rock on another to see which is the hardest. Students can sort the rocks by degrees of hardness.

c. *Weight.* Show that some rocks are heavy and sink in water while others are light (pumice) and will float. Explain that in spite of differences in size, hardness, and weight, all are called *rocks.* Ask in what ways all the rocks are alike or why they are all called rocks. Some guidance in deriving a definition might be provided by such questions as: "Do rocks rust, rot, or burn?" "Do they move on their own, breathe, or grow?" Ask how they can be sure about answers to the guiding questions (e.g., place a rock and an iron nail in jars of water for several days to see which rusts; place a rock and a piece of fruit on a strip of wax paper on a windowsill for a week to see which rots; place a rock and a piece of wood in a flame to see which burns). After such tests, ask again in what ways all the rocks are the same.

Identifying rocks by color and composition. Bring a variety of rocks to class and have students describe them by physical characteristics such as color and composition.

11
Ba
Ca
Db

a. *Color.* The colors of opaque rocks are best observed if the rock is scratched across a piece of unglazed tile. *(Note: The streak that is left may be different from the rock's color.)* Transparent and translucent rocks can be tested for their color or lack of color. Students can sort the rocks on the basis of colors.

b. *Composition.* Study the surface of the rocks with a hand lens, and note if they are made up of one or more kinds of material. If a rock is covered with a cloth or bag, it can be broken with a hammer, and the inside of the rock can be compared with the outside features. Rocks can be sorted by the size of the particles they contain, or whether or not the particles are crystalline. In breaking rocks you may find that some break in certain ways (along cleavages). Note the directions of breakage, if any (e.g., number of parallel sides or faces).

Contributing Idea E. Indicators can be used to identify some rocks.

Using indicators to identify some characteristics of rocks. Some rocks can be identified and sorted because they respond in certain ways to magnets or chemicals. Rocks that contain iron will respond to magnets. Rocks that have magnetic properties (e.g., magnetite, lodestone) will attract pieces of iron. Limestone and marble react to the vinegar test (see *141.07d*). **12 Db Fb**

Generalization II. Minerals have identifiable characteristics.

Contributing Idea A. Rocks are composed of minerals.

Observing minerals within rocks. Using a hammer, pulverize a piece of granite that is wrapped in a cloth or in a bag. Pour the fragments on a piece of paper and observe them with a hand lens. Have students sort some of the fragments. There are three main kinds of minerals: feldspar (may be many colors), mica (may be black or light), and quartz (may be many colors). Tell students that all rocks are made up of materials called *minerals.* Explain that all the rocks of the world are made up of various combinations of minerals and that minerals have a definite, unchanging composition while rocks may contain varying amounts of different minerals. Pulverize other rocks and sort out the different minerals that are found. (Some will be made up of only one mineral while others will be made up of many.) **13 Aa Bb Db**

Contributing Idea B. Minerals can be identified by certain physical characteristics.

Identifying the characteristics of some minerals. Divide the class into small groups and provide each group with several hand lenses, a small bag of sand, and some black paper or a paper plate. Have each group examine its sand and sort the pieces by colors. After sorting, students can use Table 141.14 to identify some of the minerals that commonly make up sand. Have different groups compare findings, then compare the composition of a sandstone rock with their sand samples. Students will realize that the various kinds of rocks are determined by the kinds of minerals within them. **14 Aa Db**

TABLE 141.14. Color Sorting Key—Minerals in Sand

Color	*Mineral*
white or colorless	usually quartz
pink	usually feldspar
black	hornblend
green	serpentine
black and flat	black mica
shiny and flat	light mica

Identifying minerals by a streak test. Rub a sample of pyrite or chalcopyrite across a piece of unglazed porcelain tile. Tell students that the greenish-black powder is from the mineral in the rock and that it is often different from the color of the rock. Rub other rocks against the tile and compare the color differences. Students will begin to understand that a color streak is useful in the identification of some minerals, but that **15 Aa Ca**

TABLE 141.15. Color Sorting Key—Minerals in Rocks

External Color	*Streak Test*	*Mineral*
blue-green	white	apatite
blue or white	white	calcite
brass yellow	greenish-black	chalcopyrite
green, purple, white	white	fluorite
lead gray	lead gray	galena
gray, red-brown	red-brown	hematite
brown	ochre yellow	limonite
gray or green	white	talc
black	black	magnetite
bright green	pale green	malachite
pale yellow	greenish-black	pyrite

white streaks are not. Table 141.15 lists the external coloring and the mineral streak made by some common minerals.

Generalization III. Soils have identifiable characteristics.

Contributing Idea A. Soils are made up of small pieces of rock mixed with plant and animal materials.

Observing the nonorganic and organic components of soil. Gather a handful of dry leaves and sticks. Have students note how easily the materials can be crumbled to resemble soil. Let them feel some rich topsoil and compare the feel with soil from a barren spot. Tell them that dead plant and animal matter decays and builds in thickness to become what is called *humus.* Mix some humus with the soil taken from a barren spot and note how the humus changes the color of the soil. If possible, take a nature walk, and have students notice where topsoil is built up and where it has been washed away. Ask in what kinds of places it seems to be deepest. Use rulers to see how deep the topsoil is. (It takes about 300 years to produce 1 inch (3 cm) of good topsoil.) Students can note that topsoil is darker than subsoil.

16 Af Bb

Sorting and identifying the nonorganic and organic components of soil. Place several handsful of soil in a large wide-mouth jar. Let students cover the soil with about 1 inch (3 cm) of water, cap the jar, and shake it thoroughly. As the materials settle, they will note that the heaviest settle first and the lightest last. (The layers will be in the order of the weight of the particles.) After the materials have settled, carefully pour off the water, disturbing the sediments as little as possible. Examine a small sample from each of the layers with a hand lens. Identify and list how many objects other than grains of sand, pebbles, and rocks can be found in a sample of soil (e.g., crumbled bits of plant material, insects).

17 Aa Db

Contributing Idea B. Rocks are broken down in the formation of soil.

Making models to show how rocks are broken down to make soil. The wearing away of rocks into smaller particles that make up soils takes place in a variety of ways. Depending upon the locality of the school, students can (1) observe the wind carrying and depositing dirt and dust; (2) observe a stream wearing away the earth, carrying sediments, and depositing them in other places; (3) observe how ocean shores are continually being worn away by water action; (4) observe how plants crack rocks. Some specific activities are suggested below.

18
Fc

a. *Wind and water.* Bring to class different kinds of rocks varying in degrees of hardness. Have students scratch one rock with another to see how rocks can be worn into soil. They can rub various other rocks together over a sheet of black paper and observe the soil-like grains as they fall. They can realize that wind and water carry abrasive materials that wear away softer materials in a similar way. You might have them find pictures of areas eroded by wind or water.

b. *Temperature changes.* Soak a rough piece of sandstone in water for about an hour, then pour the rock and water into a milk carton. Set the carton in a freezer. When frozen, tear away the carton, and let the ice melt naturally in a fine sieve or cotton cloth. Examine the remaining rock and particles. Students can make the analogy to rocks that get wet naturally, freeze, and thaw. Now have them heat small pieces of soft rock such as sandstone in a pan over a hot plate for several minutes, turning them to distribute the heat evenly. Drop the heated rocks into a bowl of cold water and observe changes. Explain that the changes in temperatures in each activity were to simulate natural (slower) changes between night and day or winter and summer. Help them realize that temperature changes break down rocks, contributing to the building of soil.

c. *Plants.* Soak some seeds in water overnight, then pack them tightly in a small plastic bottle with a screw cap. Cap the bottle and observe it daily. Students will see that as the seeds grow, the pressure exerted by them is great enough to break the bottle. Look for places in rocks, cement, or asphalt to find plants that are breaking down materials in a similar way.

Contributing Idea C. Plant and animal matter is broken down in the formation of soil.

Observing decay. Decay of organic materials can be observed in several ways.

19
Aa

a. Shove a small pointed stick into the ground in a shady spot near small plants or grasses. Impale an apple onto the stick so that it cannot be carried away by animals. Observe it day by day and keep a record of what happens (e.g., the first signs that something had been eating it, the first break in the skin, the first indication of decay odors, the appearance of insects or worms and where they came from, the appearance of molds or fungi). The record will indicate how many days it takes the apple to almost disappear and how many different agents worked on the cleanup job.

b. Place a ½ inch (1½ cm) square of sliced ham on the playground where it will not be disturbed. Mark the spot and watch from a short distance to note if the meat gets moved, how, and by what.

c. Have students explore the schoolyard to list all the discarded objects that they can see. A chart such as Chart 141.19 may

TABLE 141.19. Observing Trash

Substance Found	*Decays Easily*	*Decays Slowly*	*Does Not Decay*

be used. Determine what fraction of the trash will not return to raw materials for future life, what fraction will help to make up new soil, and what fraction will provide food for animals or plants. Discuss the problems caused by materials that do not decay.

Manipulating conditions that affect the rate of decay. Half-fill two glass jars with dry soil and leave a third jar empty. Add equal amounts of leaves and grass cuttings to each jar. Moisten the soil in one jar, cover the jars; then, after one week, students can compare the contents. Under which conditions did the materials seem to be changing into soil fastest? **20 Ch Ec**

Contributing Idea D. Some soils contain nutrients that contribute to plant growth.

Testing to see how soil influences plant growth. Several kinds of activities can be carried out to enable students to understand that some soils are very important for plant growth. **21 Ec**

a. Line two glasses with blotting paper. In one glass, hold the blotting paper in place with paper toweling. In the other, hold it in place with rich garden soil. Place radish or lima bean seeds between the blotting paper and the glass. Add equal amounts of water and set in a warm sunny location. Keep both moist during the growth, and keep daily records of what happens. Students will find that the plants in the glass without soil cannot sustain growth, and that the soil must be a contributing factor for continued growth.

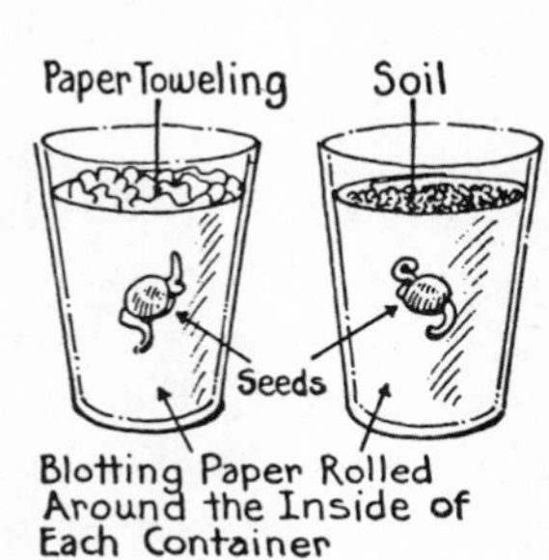

FIGURE 141.21

b. Fill one flowerpot with sand, another with clay (or subsoil), and a third with loam (or topsoil). Plant five lima beans in each pot and place the pots side by side where they will receive the same amount of sunshine. Add the same amount of water to each pot daily and keep a record of the growth for a month. Decide which kind of soil is best for growth. (The topsoil is usually best.)

TABLE 141.22. Hardness Sorting Key—Rocks

Hardness Scale (Soft to Hard)	*Hardness Test*	*Rock Example*
1	Scratches *easily* with a fingernail	talc
2	Scratches with a fingernail	gypsum
3	Scratches with a pin or copper penny	calcite
4	Scratches *easily* with a knife	fluorite
5	Scratches with a knife	apatite
6	Knife will not scratch rock and rock will not scratch glass	feldspar
7	Scratches glass easily	quartz
8	Scratches quartz easily	topaz
9	Scratches topaz easily	corundum
10	Scratches all other rocks	diamond

Generalization IV. Rocks can be organized by their physical characteristics.

Contributing Idea A. Rocks can be seriated.

Seriating rocks by hardness. Rocks vary in hardness and can be ordered from soft to hard. Table 141.22 is a commonly used hardness key. Students can set up their own hardness scale using the rocks they find simply by seeing if one will scratch another. (A rock that scratches another rock is harder than that rock.) They can then order their rocks by relative hardnesses. **22 Ab Da**

Seriating rocks by size. Have students bring to school rocks of different sizes. Line up all the collected rocks in a row from smallest to largest. Table 141.23 shows generally standard categories for the various sizes. **23 Aa Da**

A set of rocks representing the different sizes can be kept in the classroom for comparative purposes.

TABLE 141.23. Size Sorting Key—Rocks

Name	*Size (in inches)*
Boulder	more than 10 inches across
Cobble	2½ to 10 inches across
Pebble	⅛ to 2½ inches across
Granules	1/16 to ⅛ inches across
Sand	1/64 to 1/16 inches across
Silt	As fine as scouring powder
Clay	Particles can only be seen with a microscope

Contributing Idea B. Rocks can be classified.

Sorting and classifying rocks. Bring to class a variety of rocks. Let students group them in as many ways as they can (e.g., by size, shape, color, texture, or other properties). They can use screens with openings of diminishing size to sift rocks by size, then compare the kinds of rocks that comprise each size grouping. With knowledge gained **24 Dc**

from activities with igneous, sedimentary, and metamorphic rocks, students may be able to classify rocks on the basis by which they were formed. Young children can use egg cartons to separate specimens on the basis of simple observable characteristics (e.g., smooth or rough, flat or round, shiny or dull). Let the children decide where each new rock they bring to class should be placed.

142 INTERACTIONS

Generalization I. Wind physically changes the earth's surface by carrying and depositing soil, sand, and other debris.

Contributing Idea A. Wind is air moving over the earth's surface.

Observing wind moving across soil. Carry a vacuum cleaner outdoors and attach the hose to the end that blows air. Explain that the air from the vacuum represents wind that might blow across the earth. Point the nozzle at a spot of dry ground and turn the vacuum on. Students will see how the vacuum blows the loose particles of dirt, leaving a small depression. They can realize that wind erodes the land in much the same way—blowing dust and small particles from place to place. Have them look around the school grounds to find where wind blows up dirt and dust. See if they can also find places (e.g., in corners of buildings) where the wind deposits the dirt, dust, and other debris.

01 Aa Ga

Contributing Idea B. Wind carries and deposits materials.

Observing how wind carries and deposits materials. Make a mixture of materials ranging in size from clay to coarse sand. Set several shallow pans outdoors at different distances from an electric fan. Turn the fan on at a slow speed and slowly pour a small steady stream of the mixture into the wind from the fan. Students will see how wind blows the particles. Let them note what kinds of particles are collected in the different pans. (There will be a pattern with the heavier particles being nearer the fan.) Repeat the activity using faster fan speeds to see if the force of the wind makes any difference. Students will realize how this activity is similar to the way particles are actually moved across the earth's surface by the wind.

02 Ca Fc

Measuring the distances wind carries and deposits different materials. Mark a 20 inch by 30 inch (50 cm by 75 cm) sheet of paper or cardboard into 1 inch (2 cm) squares and tape it tightly to a table in front of an electric fan. Place a spoonful of rice on the edge of the paper directly in front of the fan. Hold a large sheet of cardboard between the fan and the rice. Turn the fan on. When it reaches a steady speed, remove the cardboard for 15 seconds, then turn the fan off. Students can measure how far the rice moved across the paper in that time and record their findings. Let them repeat the activity using puffed rice. Record measurements and compare differences between the two kinds of rice. Students might want to try an equal mixture of puffed rice and rice. Challenge students to find out if it is the size or the weight of the rice that makes a difference. Many variations of this activity can be tried. Students can test different kinds of materials (e.g., sand, sugar, salt, talcum powder, dry soil); they can try varying the wind power by using a larger or smaller fan, by increasing or decreasing its speed, or by changing its distance from the test material. After reaching some conclusions about size

03 Cd Ec

and weight and wind force, have them recall the greatest distance the rice (or other substance) moved in a previous test. At half the distance, place an eraser, then repeat the activity. Students will see that the rice collects against the object. Have them find examples around the school that show when something gets in the way of the wind, it lessens the wind's speed and the wind deposits some of the material it was carrying. You might explain that sand dunes are formed in this way.

04 Ca Ge

Collecting and comparing samples of wind-borne materials. On a windy day, set up a wide board in a field or open space on the school grounds so that the wind blows against it. Let students coat the board with some glue, paste, or strips of tape (sticky side out), and observe what collects against the board in a half hour. Note any differences in the kinds of particles found at the top of the board and those at the bottom. Comparisons can also be made by repeating the activity on a less windy day. You might let students describe the stinging sensation they have felt on their faces when walking into wind-whipped sand or dust.

Contributing Idea C. Wind-carried materials carve and erode land surfaces.

05 Ga

Finding examples of surfaces worn by wind-carried materials. Have students look at sand under a hand lens, and note that the edges of the grains (especially quartz) are sharp. Explain that these grains can be blown through the air by wind and that they can scratch and wear away rocks. Ask students to comment on how wind-blown sand and dirt feels against hands and faces. If any student has been in a sandstorm, let him or her relate the experience. Explain that the long-time effect of wind-carried particles gradually wears down the earth's surface. Have students find pictures from magazines of wind-worn desert rock formations. In some pictures they might note that the wearing seems to be greatest near the base. Let them hypothesize why. If possible, observe sandblasting to see how wind-driven sand is used to clean buildings or to take paint from surfaces.

06 Aa

Observing how soft rocks can be worn by hard rocks. Rub a piece of sandstone with a piece of limestone over a sheet of black paper. Students will see that tiny particles of rock are rubbed from the softer rock. Similarly, rub pieces of hardwood and softwood with sandpaper and collect the particles on separate sheets. Students will understand that these experiences are examples of how rocks can be eroded and that particles carried by the wind erode softer materials in a similar way.

Generalization II. Water physically changes the earth's surface by carrying and depositing soil, sand, and other debris.

Contributing Idea A. Some water is absorbed into the earth's surface and stored in porous soils and rocks.

07 Aa Bb Cf

Observing the porosity of soil and rocks. Absorption of water can be shown by dipping one end of a slice of bread or sugar cube in colored water. Explain that materials like these contain many empty spaces and are said to be *porous.* Now fill a glass three-quarters full of dry soil. Pack it in tightly, then quickly fill the rest of the glass with water. Observe the bubbles coming from the soil. Ask students to explain the derivation of the bubbles. (Soil is porous and filled with air pockets—as water fills the pockets, the air is driven out.) Next, show how individual rocks can absorb water by weighing a piece of granite and/or marble and a piece of sandstone and/or limestone, recording the weights, then placing the rocks in a jar of water. After one day, weigh each rock again. Students will find that some rocks are more porous than others, indicated by the increased weight of the absorbed water. For fun, let them try to wipe the rocks dry.

08 Cg

Measuring amounts of airspace in different soils. Put some soil in a glass jar and carefully, but quickly, pour in some water. Let students note the air bubbles which come from the soil as the water seeps through. Students can use this procedure to compare visually amounts of air in different soils. They will realize that the amount of space (porosity) influences the drainage and retention of water on land. More precise measurements of airspace can be made by several methods.

a. Fill a large glass jar with dry pebbles from a beach or creek bed. Let students estimate how much space there is between the pebbles, then see how near their estimates are by filling the jar to the brim with measured amounts of water. Other soils such as sand or gravel can also be measured.

b. Place 8 fluid ounces (237 ml) of water into a quart (liter) jar, then add dry sand, 1 fluid ounce (30 ml) at a time, to the water until the sand's surface is even with the water's surface. Count the fluid ounces (ml) of sand that were added and compare the two measurements (for example, there are 8 fluid ounces (237 ml) of space in 32 fluid ounces (948 ml) of sand—25 percent of the sand is airspace).

c. Put a two-hole stopper into a wide-mouth jar filled to the brim with a soil sample. Try samples of subsoils as well as topsoils. Tap the jar to settle the soil, then insert the stopper, leaving no airspace above the soil. In one hole of the stopper, place a thistle tube that extends to the bottom of the jar. In the other, place a U-tube that has some rubber tubing attached to it. Invert a second jar of water in a pan of water and place the end of the rubber tubing under it. Add water to the thistle tube until the jar of soil is filled to the stopper. Compare the amount of air collected in the second jar with the amount of water poured into the first. Keep a record of the amounts, and let students test other soils in the same way.

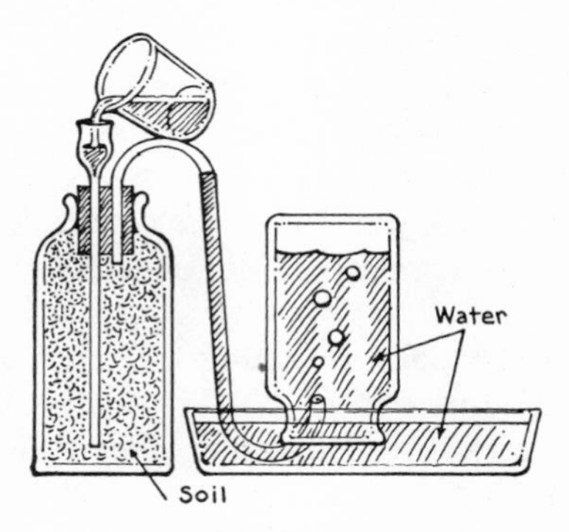

FIGURE 142.08

09 Cf

Measuring amounts of water in different soils. Place some soil in a tin can and cover it with a glass jar. Set the can in sunlight or in a pan on a hot plate and heat it. Have students note the moisture that collects inside the jar. Using this procedure, they can visually compare amounts of water in different soils. A more precise measurement of amounts can be made by weighing out a pound of garden soil on a scale, then placing it in sunlight for several days or in an oven for several hours to dry out. When dry, reweigh it to see how much weight was lost due to the evaporation of water. Similarly, students can take samples of different kinds of soil and make small wet mud balls (about the size of a pea). Let them weigh each mud ball sample on a scale, record the weight, and set it aside on a piece of wax or foil paper to dry. When dry, weigh the balls again and figure out how much water each held. Students can examine the characteristics of the different soils with a hand lens to judge why some hold more water than others.

Testing the compaction of different soils. 10 Ec
Fill several large juice cans with soil, each with a different type. Press a sharpened 1/4 inch (1 cm) diameter dowel into each can to see how far it can be inserted. Let students keep records of the results. Next, have them wet the soils and try this again. When the soils dry, try a third time. Students can make comparisons among the soils and among the three tests, and make judgments on which types of soil were the most or least compact.

Making a soil compaction gauge. 11 Gc
Have students sharpen one end of a 10 inch (25 cm) dowel in a pencil sharpener. With a ruler, measure 1 inch (2 cm) from the point, and make a ring around the dowel. Measuring from the unsharpened end, mark and number ten lines 1/2 inch (1 cm) apart. Fasten a wide rubber band to the top of a wooden spool with tacks or staples. Slip the dowel into the spool and pull down on the spool forcing the point of the dowel into the soil up to the ring line. The number at the top edge of the spool will indicate the relative compaction of the soil. Students can try the gauge in other places (e.g., lawns, gardens, baseball infields, paths, sloped hillsides), find where the soils are most and least compacted, and observe how the relative compaction compares with the number and kinds of plants growing nearby.

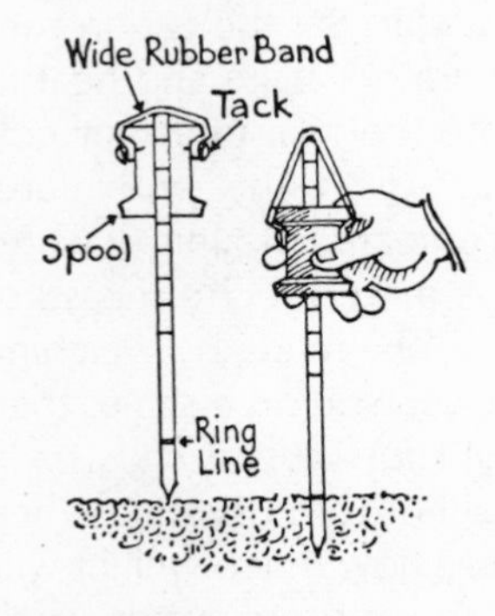

FIGURE 142.11

Comparing and measuring the absorption of water by different soils. 12 Cb Ch
Have students tie a piece of cloth over one end of each of several lamp chimneys. Fill each chimney three-fourths full with a different type of soil (e.g., sandy, loamy, fine gravel, clay). Place the lamp chimneys in a flat pan of water and observe the type of soil in which the water rises the highest or fastest. Have students make some judgments about the types of soil and their ability to absorb water. Rates of absorption can be determined by timing the absorption for 1 minute and measuring the height of the water in each lamp. The relationship of height to time will provide the absorption rate (e.g., 1 inch in 30 seconds). Clear plastic drinking straws can be used as substitutes for the lamp chimneys.

Contributing Idea B. Water accumulates and flows from high places to lower places over the earth's surface.

Making a model to show the cumulative flow of water. 13 Fc
Place a large plate in a sink; crumple up some waxed paper and put it on the plate. Place a small object under one edge of the plate so that it is slightly tilted. Tell students that the waxed paper represents a mountain range, then pour a little

"rain" in different places on the uphill end of the waxed paper. Let students observe where it goes. They will see that each trickle of water tends to collect and to work its way downhill to the lowest point. They will realize that, similarly, the water on mountains collects to form streams, and streams collect to form rivers that collect and eventually empty into oceans and seas.

Measuring the speed of flowing water. 14 Ch
The speed at which water flows can be determined by several methods.

a. Students can float an object in moving water (e.g., a stream or gutter flow) and measure the distance it travels in 1 minute or in some specified number of seconds. For example, a leaf or piece of paper placed in moving water might travel 88 feet (16.5 m) in 1 minute. This information can be calculated to determine the speed in miles (k) per hour (88 feet X 60 minutes = 5,280 feet per hour or 1 mile per hour; or 17 m X 60 minutes = 1020 m per hour, about 1 k per hour).

b. Mark a 100 foot (50 m) distance along a straight-flowing portion of a stream. Float an object between the marks and record the time. If the object travels 100 feet (50 m) in 25 seconds, then dividing by 25 gives 4 or 2 (the number of feet or meters traveled in 1 second). The speed per hour can be calculated: 3,600 seconds (number of seconds in one hour) X 4 feet (2 m) = 14,400 feet (7,200 m) (about 2.7 miles or 7.2 k per hour).

Making a clinometer. A *clinometer* is an instrument used to measure the slope of a hillside. One can be made by obtaining a plastic protractor or drawing a protractor on an index card and mounting it on a wooden ruler or stick. An indicator line made from a thread and weight (such as a paperclip) can be suspended by a thumbtack from the center of the protractor's diameter. When the instrument is tipped upward, the number of degrees from the horizontal can be read on the protractor. A soda straw attached to the top of the ruler aids sighting of some object at the top of a hill. The angle read on the protractor indicates the slope of that hill. 15 Gc

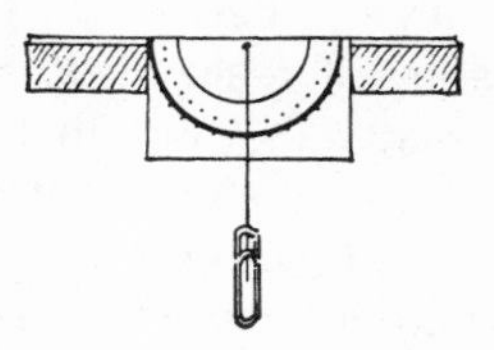

FIGURE 142.15

Determining the steepness of a slope. 16 Cd Ch
Have one student stand at the bottom of a hill and sight through the straw on a clinometer an object or another student standing at the top of the hill. Have another student read the angle indicated on the clinometer. If students compare the slope of a hill to the speed of water flowing down it and the amount of materials carried by the water, they will discover a relationship between the steepness of the slope, the speed of flow, and the amount of erosion taking place.

Contributing Idea C. Water carries and deposits materials.

Observing how rain carries away soil. 17 Aa
Fill several flower pots or cans with loose soil until the soil is just level with the edges. Place some small stones or bottle caps on the surface of the soils and set the containers outside or in a sink. Water them with a watering can to represent rain, gradually increasing the flow. When finished, let

students notice how the unprotected soil is splashed away, leaving columns of soil under the stones. After a rain, have students look for the same effect in an unplanted area.

Collecting and comparing samples of water-borne materials in different locations. 18 Ca Ge After a heavy rainfall, have students take samples of running muddy water in glass jars. Label the jars with the location of the water. Let the water stand for several hours until the contents have settled. From their observations, students will realize that moving water carries materials, and that the materials settle out of the water at different rates. Compare the kinds of sediments collected from the different locations.

Testing to see how water carries and deposits materials. 19 Ec Prepare a long U-shaped cardboard trough filled with coarse sand. Hold the cardboard in a slanting position over a pan and pour water, a little at a time, on the sand at the upper end. Have students note that as the water flows slowly, it carries small grains of sand with it into the pan while the larger and heavier pieces stay behind. Now pour the water with a greater force. Students will see that both the sand and larger pieces are washed downstream and will begin to realize that moving water carries objects. Students can experiment with the trough by raising and lowering it to see what effect the slope has upon the carrying and depositing of materials. They can experiment by placing a large object, such as a rock, in the flow of the water to see what effect it has on the deposition of sediments.

Testing to see how water-borne materials are deposited to build up land forms. 20 Bb Ec Cut a V-shaped notch in one end of an aluminum foil pan. Weigh some dry soil and fill the pan with it. Set the pan at one end of a larger pan and raise the uncut end by resting it on a small block of wood. Each day for a week, let students sprinkle a half cup of water over the soil at the top of the pan. At the end of the week, have them examine the soil that was washed into the larger pan. They will see that the smaller, lighter materials fan out the farthest from the opening. Explain that this happens because the flowing water slows down as it fans out, thus depositing the heavier materials first and the lighter ones last. Tell students that the materials deposited from rivers as they slow down are called *sediments.* If possible, show pictures of delta regions (e.g., the Mississippi River) to help students realize that such deposits take place at the mouths of rivers. You might challenge them to hypothesize what happens to sediments that are blocked by dams. To check hypotheses, build a dam in a gulley or gutter of running water or block it by curving a large sheet of cardboard. As the water rises behind the dam, observe what happens to the materials it carries. Students can make the analogy to actual dams that span rivers. Discuss the problems of sediment buildup and possible solutions.

Contributing Idea D. Water and water-borne materials carve and erode land surfaces.

Using a model to test the erosive power of raindrop splashes. 21 Ec Fc Have students fasten a sheet of white paper to a piece of cardboard. Set it on the floor, and from a foot above the paper, drop colored water (to represent raindrops) from a medicine dropper. Note the size and shape of six splashes. Carefully remove the paper and replace it with a fresh sheet. Tilt the board slightly and repeat. Students can compare these splashes with the others. Continue the activity by gradually increasing the slope of the surface, using a new sheet of paper each time. When dry, the papers will be a visual record for display. Other tests can be made by varying the height of the falling drops and by varying the height and the slope at the same time.

Measuring differences in soil erosion caused by raindrops. Make several splash boards out of shingles or similar board pointed at one end. A covering over the top will keep sheets of graph paper attached to the splash boards from being hit directly by raindrops. When the weather forecaster predicts rain, have students drive the pointed ends into the soil in several different locations (e.g., under trees, on lawns, on bare paths). Be sure the graph paper is level with the earth surface. After a rain, measure the height to which the soil splashed in the different locations. Students can judge which surface seems to protect the soil best from being carried away by raindrops.

22
Cd

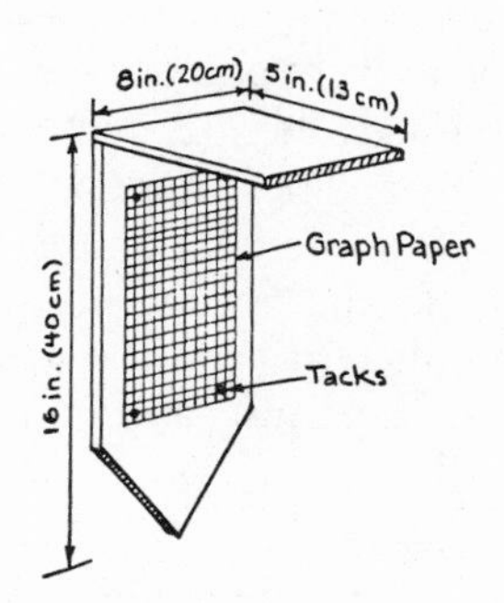

FIGURE 142.22

Making a model to show wave action along a shoreline. Obtain a large, 10 inch by 15 inch (25 cm by 30 cm) baking pan, about 5 inches (12 cm) deep. Mix some soil with water to make mud. Fill about one-third of the pan with mud at one end to a height of 4 inches (8 cm). When it drys, place a 2 inch (4 cm) layer of sand in the remaining portion of the pan and add enough water to submerge the sand. Place a wooden board into the pan at the sandy end and move it back and forth to make small waves. Explain that the model represents the ocean battering a shoreline. Students can observe the waves as they splash against the land mass and note how a beach gradually forms, how the particles are tumbled by the waves, how the land mass is eroded, and how the eroded portions are distributed.

23
Fc

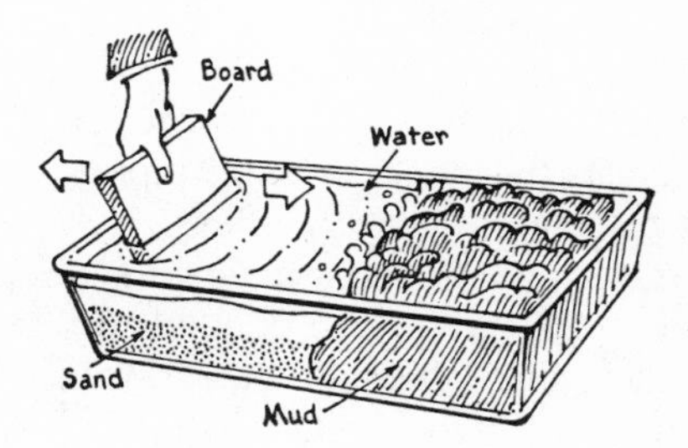

FIGURE 142.23

Contributing Idea E. Glacier-carried materials carve and erode land surfaces.

Making a model to show that pressure melts glaciers. Have students arrange two groups of ice cubes (four in each group) and place a square of cardboard on top of each group. Place a brick or a pile of heavy books on top of one group and observe which group melts faster. Students can decide whether or not pressure makes a difference. Explain that, similarly, the bottoms of glaciers melt due to the weight of the upper layers of ice. Although the movement of glaciers is very complicated, some theories suggest that the continuous process of melting,

24
Fc

refreezing, and melting again contributes to the movement.

Making a model to show glacial erosion. 25 Fc
Have students rub an ice cube across the painted surface of a board, noting that the ice is not hard enough to abrade the surface. Now let them press the cube into a dish of sand. When the ice melts a bit, place the dish and the cube into a freezer. After it refreezes, remove the cube and rub it across the board again. (It will have picked up pieces of sand and will scratch the board's surface.) Explain that glaciers similarly melt, pick up sediments, and refreeze. Ask what might happen when soil, sand, rocks, and boulders become frozen into the edges of a moving glacier. Ask what happens when a rock rubs against other rocks and soil. If possible, show pictures of glacial valleys. Students can note how sliding glaciers have ground away rocks and soils to deepen and widen the valleys through which they travel. Notice that the valleys are always U-shaped.

Generalization III. Temperature changes and plant growth physically change the earth's surface by breaking larger materials into smaller ones.

Contributing Idea A. The alternation of hot and cold temperatures breaks rocks.

Testing a model to show how a change from a hot to cold temperature can affect materials. 26 Ec Fc
Heat several glass marbles on a plate in an oven or on a pan over a hot plate. Move them about to heat them as evenly as possible. When very hot, remove them with tongs and plunge them into a container of very cold water. Students can observe the results and hypothesize what will happen when you repeat this activity using small rocks. (Some rocks, like those containing large amounts of quartz, will break into various sized pieces.) Students will realize that repeated heating and cooling (day-night; summer-winter) slowly causes rocks to break into smaller pieces.

Contributing Idea B. Water breaks rocks when it freezes.

Testing a model to show how water freezes and expands to break rocks. 27 Ec Fc
Fill two small jars to the brim with water. Cap one tightly, but leave the other uncapped. Students can pretend the jars are rocks and that the water has filled cracks in them. Place the jars in plastic bags and seal the bags with rubber bands. Set them in a freezer overnight. (Be sure the open jar does not spill.) Have students examine the results the next day. (One jar may break while the water in the other may expand above the rim without breaking the jar.) Now have students soak several pieces of porous rock (e.g., sandstone, limestone, pumice) in water for an hour. Place the pieces in a plastic bag and set the bag in a freezer overnight. Challenge the students to hypothesize what will happen to the rocks, then examine the rocks the next day. By analogy, students can realize that water from rain and melting snow flows into cracks and pockets in rocks during the daytime, then at night the liquid sometimes freezes, expands, and cracks pieces from the rocks.

Contributing Idea C. Growing plants break rocks.

Observing the breaking of rocks by plants. 28 Aa
Cut one side from a milk carton, and half-fill it with moist soil. Plant some soaked corn or lima beans just under the surface of the soil, and pour about ¼ inch

(1 cm) plaster of paris on top of the soil. Set the carton aside and examine it daily. Students will soon note that the germinating seeds push up under the plaster with enough force to break it. Similarly, students can plant the seeds and place a sheet of heavy glass over them. They will soon see that the germinating seeds will push the glass upward. (The glass allows the students to observe the process directly.) Students will learn that germinating seeds can either cause rocks to move out of the way or crack them into smaller pieces. Explore the school grounds to find places where plants have broken through asphalt or cement.

Generalization IV. Water chemically changes some materials on the earth's surface.

Contributing Idea D. Water dissolves and deposits some materials.

Observing that some materials are dissolved and deposited by water. The following activities demonstrate the formation of mineral deposits on the surface and interior of the earth:

29
Aa
Fc

a. Salt deposits can be found in various locations on the earth's surface. To show how they may be formed, have students put a small amount of salt in a jar of water and shake it. Ask what happens to the salt. (It seems to disappear as it dissolves.) Ask students to name other materials that dissolve in water. Now let them continue adding spoons of salt until some remains in the bottom of the jar even after a thorough shaking. Pour the water into a shallow pan and set the pan in a sunny place to let the water evaporate naturally. Discuss the results. Ask where on the earth's surface similar salt deposits have been found (e.g., Great Salt Lake, Utah). Explain that salts in the earth are continually being dissolved by water. Ask where the water usually takes the salts (lakes, seas, and oceans). Let students describe the ocean's taste and relate their descriptions to this activity.

b. Stalactites and stalagmites build up in caves from the dissolving and depositing of minerals by water. To illustrate the formation of stalactites, which hang down from the ceiling of some caves, and stalagmites, which stand upright on some cave floors (usually beneath stalactites), have students dissolve as much Epsom salt in a container of water as they can. Fill two smaller containers with the solution, set them on paper toweling, and suspend a thick string between them, one end in each container. Let the string remain for several days. Students will see that water soaks the entire string and drips off at the low point between the containers. They will also see that deposits form where the water drips—both from the string and on the toweling. They can infer that the deposits are carried in solution to the drip point and that through evaporation of the water, the mineral is left behind. They will realize that this formation is similar to the slow formation of stalactites and stalagmites in some caves.

151 CHARACTERISTICS

Generalization I. Ocean water has identifiable characteristics.

Contributing Idea A. Ocean water is a solution of many dissolved gases and solids.

Observing that ocean water contains dissolved materials. Obtain a quart (liter) of ocean water, or simulate ocean water by dissolving 1 ounce (28.35 grams) of table salt in a quart (liter) of water. Pour the water into a clean, shallow pan. Allow it to evaporate by placing the pan in the sunlight or by gently heating it over a hot plate. When the water is gone, let students inspect the residue. If simulated ocean water is used, students can weigh the residue to see if the original amount of salt remains. *(Note: By weight, ocean water is about 3.5 percent salt.)* **01 Aa Cf**

Making a model to show how rainwater dissolves materials in the earth. Wash some aquarium sand several times, then set the sand out to dry. When dry, mix a teaspoonful (5 ml) of table salt with each 4 cups (1 liter) of dry sand. Simulate an inclined river bed with aluminum foil and pour the sand and salt mixture into it. Place a bowl at one end to catch the river water. Pour some rainwater or distilled water at the top of the river bed. Students can taste the collected water to see if it contains salt or set the collected water out to evaporate. They will realize that in a similar way, rainwater dissolves salts in rocks and soil, carries them from the land by way of rivers, and deposits them in the oceans. Thus, the oceans now contain materials that have drained off the land for millions of centuries. **02 Af Fc**

Removing dissolved materials from ocean water by freezing. Obtain some ocean water or dissolve 1 ounce (28.35 grams) of table salt in a quart (liter) of water. Pour the water into ice cube trays and place them into a freezer. Do the same with an equal amount of plain water. Students can record the appearance of the water in each tray at 10-minute intervals until each is frozen. Next, take several cubes from each tray, melt them in saucers, evaporate the water, and inspect the residue. Students will find that very little residue remains. (When salt water freezes, it forms crystals of nearly pure ice—most of the salt is left behind in unfrozen water.) **03 Ca**

Contributing Idea B. Ocean water exerts more pressure than fresh water.

Comparing equal volumes of ocean water and fresh water by weight. Have students weigh two 1 quart (liter) containers. Fill one with tap water and the other with **04 Cf**

an equal amount of ocean water. If ocean water is not available, dissolve 1 ounce (28.35 grams) of table salt in the quart (liter) of water. Now weigh the containers again. By subtracting the weight of the empty container from the weight of the container with water in it, students will find out how much weight each volume of water has. Similarly, two large balloons can be attached to the ends of two 10 foot (3 m) lengths of rubber tubing. While outdoors, one tube can be filled with tap water and the other with ocean or salt water. Hold the balloons closed so that the water does not enter them. When each tube is filled to capacity, allow the water to flow into the balloons. The tube containing ocean water will expand its balloon to a greater extent. Because of the weight difference brought about by dissolved salt and other materials, water pressure is greater 10 feet (3 m) deep in ocean water than it is 10 feet (3 m) deep in fresh water.

Contributing Idea C. Ocean water has a greater buoyancy than fresh water.

Comparing the weight of an object suspended in air with one suspended in water. Tie a string around a large rock and suspend it from a spring scale. Record the weight of the rock out of the water. Now hang it in a container of tap water and record the weight. Next, hang it in a container of ocean water or water that has 1 ounce (28.35 grams) of table salt dissolved in 1 quart (liter) of water. Let students look for differences in the recorded weights. You might explain that water has a force that pushes up on objects and that this force is stronger with ocean water.

05
Cf

Observing the above-surface and below-surface configurations of floating ice. Place an ice cube in a clear glass of tap water and an identical cube in a glass of ocean or salt water. Students can observe how much of the ice is above and how much is below each surface. (Although the ice floats somewhat higher in the ocean water, about two-thirds remain submerged.) Some students might be interested in researching the tragic story of the Titanic which struck an iceberg in the Atlantic Ocean.

06
Rd
Ca

Contributing Idea D. Ocean water transmits, absorbs, and scatters solar radiation.

Observing the penetration of light through ocean water. Fill an aquarium with ocean water or a solution of salt water—1 ounce (28.35 grams) of table salt per quart (liter) of water. Beam a light from a 35-mm slide projector through the length of the aquarium, and let students measure the diameter of the beam where it shines against a screen 1 foot (30 cm) from the aquarium. Now shine the beam above the aquarium, across the same distance to the screen. Compare this diameter to the previous one. The difference in density between air and ocean water will cause a difference in the refraction of the beam and thus produce images of different diameters. You might tell students that solar radiation in the form of light disseminates in a very short distance after entering water (about 15 meters).

07
Ca
Cg

Observing the effect of water on depth perception. Suspend several corks from strings of different lengths and tie identical weights to them. Arrange the corks in a line in an aquarium. Let students take turns looking straight down through the water at a cork and, while looking, place a mark on the outside of the aquarium with a grease pencil to show the point where they think the top of each cork is. After several students try this, compare the marks with the actual heights

08
Aa

of the corks when viewed through the side. (The marks will tend to be above the actual heights of the corks and the difference in distance will increase toward the lower corks.) Students will realize that water affects depth perception. (Visibility is distorted because of the refraction of light in the water.) Air also affects depth perception, although to a lesser extent. Because we become adjusted to the effect, we find it difficult to judge distances when air is absent. (Astronauts report that they have some difficulty judging distances while on the moon.)

Contributing Idea E. Ocean water conducts electricity better than fresh water.

Measuring the electroconductivity of fresh and salt water. Set up a simple six-volt circuit with a light bulb as shown. Separate the free wires and dip the ends in fresh water, then in a salt water solution. Students will see that the salt water allows the light to light up. Now repeat the activity by beginning with fresh water and adding salt spoonful by spoonful until the light comes on and gets brighter and brighter. Next, let students slowly move the free ends farther apart in the salt solution. You might explain that salinity (the amount of salt) in water is sometimes measured by the brightness of the light. If possible, obtain some ocean water and compare the brightness of the light to the brightness in one of the previous tests. *(Note: In some areas, hard water will conduct electricity; thus distilled water should be used in the initial tests.)*

09
Ca

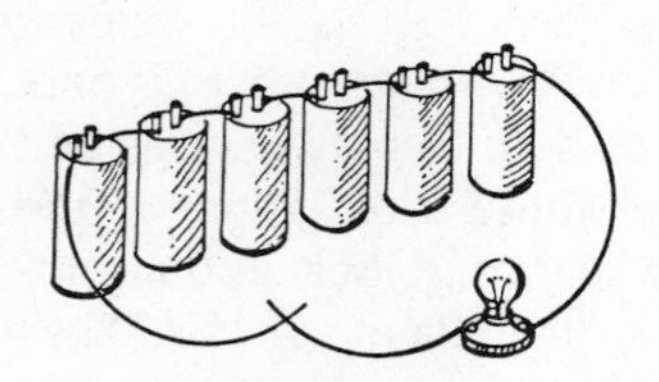

FIGURE 151.09

Contributing Idea F. Ocean water contains living organisms.

Observing small organisms with a microscope. Obtain some ocean water. Use a wet mount slide with a cover slip to let students observe the small organisms with a microscope. (Binocular microscopes are especially good for viewing these organisms.) You might explain that the small organisms are part of a food chain upon which other organisms are dependent. In the ocean, diatoms (algae and plankton) are the primary food producers and the beginnings of all food chains.

10
Aa

Contributing Idea G. The surface of the ocean is made up of waves.

Observing waves at the seashore. If possible, take several field trips to a coastal region and have students record observations of waves at different seasons of the year. They can determine the relationship between waves and a shore's appearance by observing how the waves approach the beach. They can determine approximate wave heights and breaker heights by references to the height of people on the beach and the height of fixed objects such as pilings.

11
Aa
Bc
Cb

Measuring some wave characteristics. Place about 8 inches (25 cm) of water in a large aquarium. Roll a hand towel around the horizontal bar of a wire hanger, then bend the top of the hanger and hook it over one end of the aquarium so that the towel is just at the water level to lessen the bouncing back of waves. Measure 6 inch (18 cm)

12
Cb
Cc
Ch

distances along one side of the aquarium and vertically mark them with a wax pencil or with strips of masking tape. To produce waves, place a flat board or your hand palm down, at the water level and move it up and down about two inches (6 cm) in a rapid but steady rhythm. Students can estimate the wavelength of the waves by referring to the 6 inch (18 cm) marks. They can use a stopwatch to count the number of waves made in 10 seconds. (The number of waves will be the same as the number of downward motions.) They can divide the number of waves into the time (10 seconds) to determine the wave period, then calculate the velocity of the waves in feet (or cm) per second:

$$\text{velocity} = \frac{\text{wave length}}{\text{period}}$$

Determining a relationship between waves and water depth. Place about ½ inch (1 cm) of water in an aquarium or large, flat, deep baking pan. Set up a series of waves through the water by tapping or jarring the container. Students can observe and describe the waves they see, especially the heights and the spacing between them (e.g., the distance between the top of one to the top of the next). Make the water twice as deep and repeat. The students will discover that the waves are now of a different size and wavelength. Explain that within certain limits, the distance between one wave and another decreases as water in oceans, seas, and lakes becomes shallower and increases as water becomes deeper. This experience can also be tried in a bathtub. If possible, let students examine aerial photographs of shorelines for further evidence. Some may wish to see what effect waves have below the surface of the water. To find out, attach a number of small corks to weighted strings so that they float at different depths in a large aquarium. Students can make waves by raising and lowering a flat board on the surface at one end of the aquarium. Have them note any movement of the corks as they observe through the side of the aquarium.

13 Ec Fc

Observing wave action. Add a shallow amount of water to a rectangular heat-resistant baking dish and set it on an overhead projector. Make waves in the pan by tapping the container. Students can observe the patterns of waves by the projection on a screen or wall, and experiment by putting obstructions into the pan to represent wharves or jetties. They will discover that waves seem to move around obstructions and change direction after striking, thus producing a sheltered or calm area on the surface.

14 Aa Ec

Generalization II. Some characteristics of ocean water can be organized.

Contributing Idea A. Fossils in ocean sediments help scientists sequence events of the past.

Making a model of a fossil. Obtain seashells or snail shells. Working individually or in small groups, have students coat one side of each shell with Vaseline, and press the coated side into a block of modeling clay that is about twice as large as the shell. Next, mold a thin rim around the top edge of each block of clay, then remove the shell, leaving a mold. Prepare some plaster of paris. (Follow the directions on the package or add water until the mixture is thick.) Using small brushes, paint Vaseline on the top of the block and inside the mold. (The Vaseline will prevent the plaster from sticking to the clay.) Fill the mold to the top with plaster and let it dry for twenty-four hours. Gently peel off the rim, remove the block of clay from the plaster, and you will have a good cast of one

15 Bb Fc

side of each shell. (The edges of the plaster may have to be trimmed with a dull knife.) You might explain that in a similar way, marine organisms with shells are buried at the bottom of oceans and pressed into the mud, silt, and ooze. When their soft bodies decay and their shells dissolve, a hole or *mold* is left showing the exact shape of the outside of the organism. Tell students that the mold is sometimes filled with other materials that form a *cast* (like the plaster) that is the exact shape of the original organism. Both the mold and the cast are fossils. As an alternate or supplementary activity, students can make a complete fossil shell by pressing the shell between two blocks of clay. (Be sure to coat the entire surface of the shell with Vaseline as well as all surfaces of the blocks.) When pressing the blocks together, try to have only half the shell impression in each block. Separate the blocks, remove the shell, then hold the blocks tightly together with rubber bands. Cut a funnel-shaped hole into the top of the blocks to pour plaster into the mold. Tap the mold gently several times to remove any trapped air, and add more plaster. If any plaster leaks from the mold, press the blocks together where the leak occurs. When the plaster has hardened, open the mold and trim the fossil's edges to produce a good model. Discuss with students how fossils provide knowledge of past ocean life and how fossils in layers of sediments help scientists order sequences of events over geological periods of time.

Contributing Idea B. Some characteristics of ocean water can be seriated.

Making and using a Secchi disk. One of the most widely used instruments for measuring water transparency is the Secchi disk. Such an instrument can be made in several ways.

16
Bb
Da
Gc

a. Students can adhesive tape the edge of a metal coffee can lid, spray white paint on one side and black on the other (or each side can be sprayed half and half or in quarters), tie a string to a horseshoe magnet, and hold the lid with the magnet *(a)*.

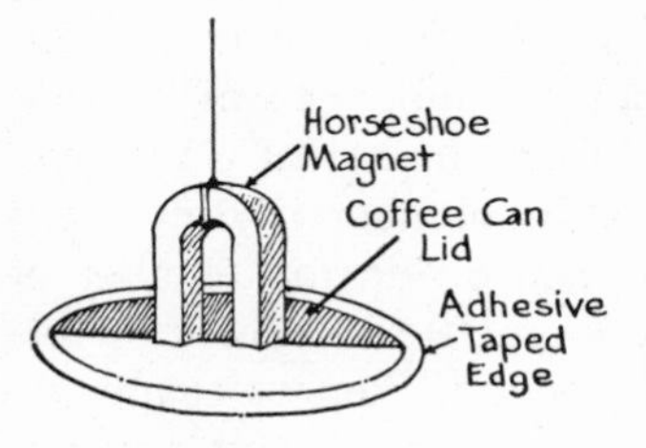

FIGURE 151.16a

b. Students can insert a weighted eyebolt through a metal disk that is about 8 inches (25 cm) in diameter and suspend the disk from a chain or calibrated line *(b)*. Prepare two or three aquariums, one

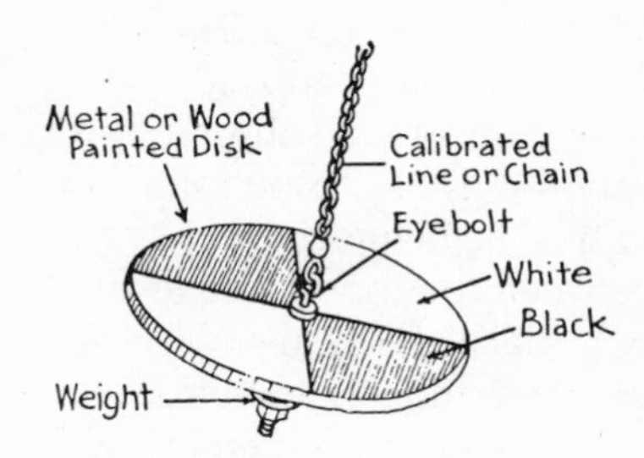

FIGURE 151.16b

with tap water and the others with water and different amounts of powdered milk to make them cloudy. Lower the disk into one aquarium, then another. When the disk disappears from view, the depth of visuality can be recorded by marking the line at the water surface and measuring the length to the disk. Students can order the three aquariums by the measurements they obtain. Tell them that the device they used is called a *Secchi disk,* used to determine the cloudiness or *turbidity* of water. Use it to test other liquids (e.g., baking soda and water, adobe clay and water). *(Note: Rinse out the aquariums when finished—some materials, such as powdered milk, decay quickly and become difficult to remove.)*

Measuring and comparing the transparencies of bodies of water. Secchi disks can measure the transparency of water, expressed as the depth or limit of visibility, of lakes, ponds, estuaries, oceans, or other bodies of water. Lower a Secchi disk on a calibrated line every 3 inches (10 cm) into a body of water until the disk disappears from view. (Be sure the disk remains horizontal with the water surface at all times.) Have a student take a line reading (the distance from the surface of the water to the disk) when the disk disappears from view, then raise the disk until it reappears and take another reading. The limit of transparency can be determined by finding the arithmetical mean between the depth at which the disk disappeared and the depth at which it reappeared. Now let several students take different measurements. (They may find that many variables can influence the readings: the distance the observer's eye is from the surface, the degree of roughness of the water, the time of day, etc.) Students can standardize their procedures to obtain consistent readings. (Consistency of the data can be improved if the disk is viewed with the observer's eye at a fixed distance from the surface and if the observations are made in the shade during the middle of the day.) Have students keep other records such as the date, time, place, and name of observer. For additional activities they can use the disk to (1) determine to what extent water transparency differs for the same body of water from day to day and season to season; (2) investigate what differences there are in the water transparency of local lakes, estuaries, or other bodies of water; (3) find out if water transparency varies significantly with distance from industry or sewage treatment plants. The various bodies of water can be ordered by their degrees of transparency.

17
Cd
Db
Ec

Making and using a cloudiness indicator. Cut six or more rectangular windows in a heavy piece of cardboard. Cover half of each window with strips of plastic from overhead transparencies. The first window might be covered with four strips, the second with eight, and so on. Students can hold collected samples of water in clear glass jars behind windows to find the number of layers of plastic that best matches the cloudiness of each sample (the more plastic sheets, the greater the cloudiness of the sample). This device can be used to identify possible water pollution areas.

18
Db
Gc

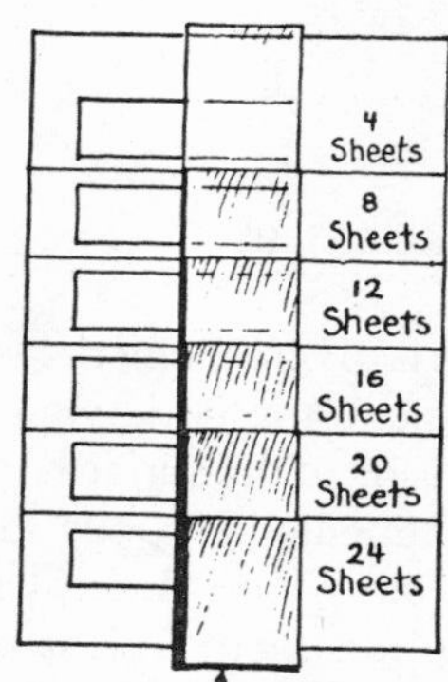

FIGURE 151.18

Contributing Idea C. Some characteristics of ocean water can be classified.

Making and using a water sampler. Obtain water samples from different depths with a water sampler. To make one, let students tape nails or fish weights to the outside of a small bottle to make it heavy, then twist a screw eye into the top of a cork fitted into the bottle. Mark a string or nylon cord at six inch (15 cm) intervals, and tie it to the screw eye. Now tie a one foot (30 cm) length of another string from the screw eye to the neck of the bottle, letting the loop of the string hang loose. Insert the cork just tightly enough so that it will stay in place when the bottle is lifted by the first string. Carefully lower the bottle into the water to a depth from which a sample is to be obtained. Next, jerk the string to remove the cork, wait for the bubbles to stop rising to the surface, then pull up the bottle. Students can pour some of the sample into a baby food jar and label it in terms of location and depth. This sample can be compared with others obtained at different depths and locations. The samples can be classified by color or cloudiness by placing them next to each other to view them from the side against an area of clear blue sky. Students should look carefully for differences in shades and colors.

19
Ca
Dc
Gc

FIGURE 151.19

Contributing Idea D. Some ocean sediments can be classified.

Making and using sorting screens. Students can build a set of sieve screens by constructing three identical square frames using twelve 2 inch by 2 inch by 12 inch (5 cm by 5 cm by 30 cm) pieces of lumber. Use tacks to cover one frame with ¼ inch (.5 cm) mesh hardware cloth, another frame with window screening, and a third with material from an old nylon stocking. Use the screens to separate mud samples by stacking them so that the hardware cloth screen is on top and the stocking screen is on the bottom. Place a mud sample on the top screen, then sprinkle it with a watering can until the soil has been washed through. When the screens are separated, various materials, plants, and animals can be found and studied. Some can be organized by texture, color, and size. Live organisms can be kept in pill vials with some water. Be sure to label samples with location and type.

20
Dc
Gc

Collecting and organizing mud samples. If the water bottom along a shoreline is soft, students can go wading to collect mud samples at low tide with a long handled shovel. A dredge must be used for deep water specimens. Construct a dredge by making holes in the center and around the edge of a coffee can. Find a second can, slightly smaller in diameter, with a plastic lid. Place rocks, nails, or fish weights in the smaller can, then attach the smaller can to the bottom of the coffee can by threading a nylon string through center holes in the cans and tying knots in the string so it won't slip through the holes. Tie a second string (the lifting string) to the open end of the coffee can. Drop the dredge into the water so that the coffee can will embed itself into the mud at the bottom.

21
Dc
Gc
Ge

By pulling on the lifting string, the coffee can will scoop up a mud sample. The dredge should be pulled out of the water by the lifting string. *(Note: If the cord is marked in feet or decimeters, the sample collection depth can be recorded.)* Use a set of sieves, as described in *151.20*, to sort the materials that are found. Collect samples from different locations and keep a record for each (e.g., location, depth, number of specimens, number of each type of specimen).

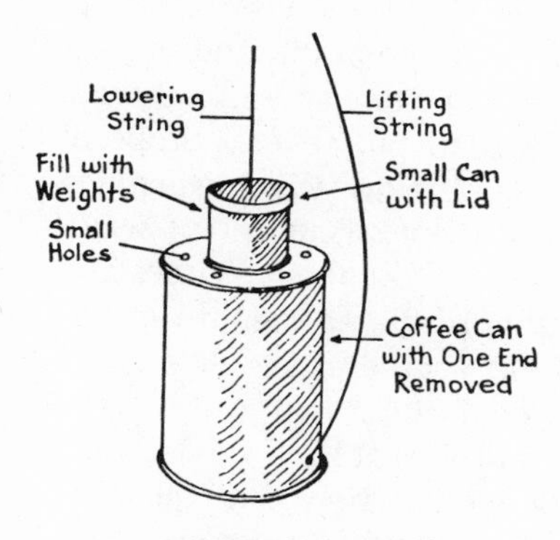

FIGURE 151.21

Contributing Idea E. Ocean life can be classified.

Identifying and classifying some ocean life. If possible, take students on a field trip to study the plant and animal life washed up on the shore. Let them keep a record of where and when particular organisms were found. Reference books can help to identify and classify the organisms.

22 Bc Bd Dc

Collecting and classifying small plant life in ocean water. Students can construct a net to collect small organisms such as plankton. To do this, use string to lace the opening of a nylon stocking to an embroidery hoop or looped coat hanger wire. Place a small bottle in the toe, and use two rubber bands to secure it in place with the mouth toward the hoop. After attaching a towing string or nylon cord to the hoop, tow the device through the water or let a current flow through it. After the device has been in the water for several minutes, pull it out and remove the small bottle. Examine its contents under a microscope. Collect samples from other locations. *(Note: If green material (algae) is collected, place some into a baby food jar. Label jars to show where samples were obtained. Jars can be arranged according to the amount of green material each contains. Generally, the greater the amount of algae, the greater the nutrients in the water.)*

23 Dc Gc Ge

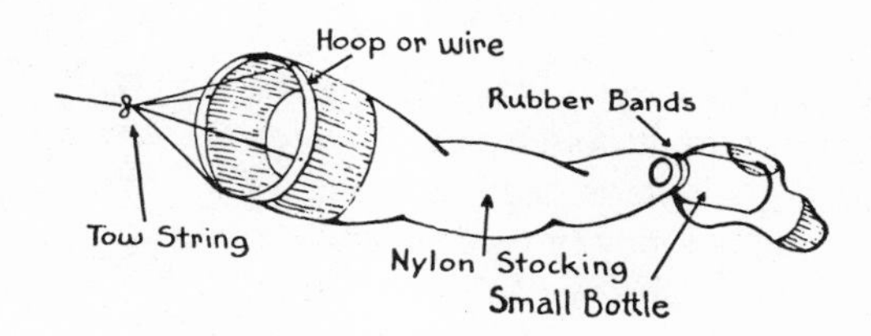

FIGURE 151.23

152 INTERACTIONS

Generalization I. Ocean currents are produced by different sources

Contributing Idea A. Heat can be a source of ocean currents.

Observing that cold water is heavier than warmer water. Put some food coloring into water and freeze it in an ice cube tray. After the water is frozen, place a cube into an aquarium of water. Students can draw a picture of the currents as indicated by the flow of the color (analogous to a cold current) from the cube. From their observations, they can infer that cold water is

01 Aa Be Fc

heavier than warmer water. You might tell them that the temperature of oceans may vary by as much as 65°F (18°C) between the equator and polar regions. They might infer that the cold, polar water constantly descends because it is heavier and tends to move toward the equator, while the warmer surface waters at the equator tend to stay at the surface and flow toward the poles.

Observing that warm water is lighter than colder water. Put a one-hole stopper in a small bottle filled with hot water colored with food coloring. Submerge the bottle into an aquarium of cold water. Students can draw a picture of how the water circulates.

02
Aa
Be
Fc

Observing a current caused by heat. Place a transparent coffee pot of water on a hot plate. As it starts heating, place a few drops of food coloring or washable black ink into the water. Observe the flow patterns. Next, put the pot on only one side of the heating unit and add a few more drops. Students will see a definite current in the water. Explain that what is seen illustrates vertical currents (from ocean depths to the surface) that move north and south toward the earth's poles (cold) from the equatorial region (warm). You might also discuss why the movement takes place. (Heated water is less dense and lighter than cooler water; thus it rises and is replaced by the cooler water.)

03
Aa
Fc

Measuring the temperature of water currents and layers of water. Have students measure and record the temperature of water at various places (top, sides, bottom, corners) in a 5 gallon (20 L) or larger aquarium. They will find that the temperature of the water is much the same throughout the container. Along the same side but at opposite ends, tape two identical thermometers so that the bulb of one is near the surface of the water and the bulb of the other is near the bottom. Now place a row of ice cubes at the end with the thermometer whose bulb is near the surface. (Be sure the cubes are at least 1 inch or 3 cm from the thermometer.) At one-minute intervals, read the temperatures on both thermometers. (The thermometer farther away from the ice will soon show a lower temperature.) If you let students try to explain why this happens, some may suggest that the cold water traveled downward and along the bottom. Let them check such an explanation by using a medicine dropper to put several drops of food coloring just below the melting ice (They will see a colored current of water traveling downward from the ice cubes and along the bottom of the aquarium.) *(Note: The greater the temperature differential between the ice and the water, the faster the current.)* Explain that, in a similar way, temperature differences between equatorial and polar ocean waters and between surface and subsurface waters cause a circulation. (Equatorial waters average about 80°F or 27°C, and polar waters average about 30°F or −2°C; ocean water temperature decreases with depth and is very cold at extreme depths.) A supplementary or additional experience can be set up using an aquarium heater. Attach the heater to one end of the aquarium, and arrange the bulbs of two identical thermometers so that the one nearer the heater is at the bottom of the aquarium and the other is at the top. Students can predict what will happen, then turn on the heater and note the temperatures at one-minute intervals. (The thermometer farther away from the heater will change first.) The movement of the current can be seen by adding food coloring near the heater. *(Note: If the heater is placed near the surface, students will be able to identify*

04
Ce
Fc

thermal layers in the water; if it is placed near the bottom, they can identify movements or currents in the water.) Other variations to this experience can be created. For example, several thermometers and/or ice cubes can be placed at different levels and/or locations; food coloring can be introduced at different locations; and various water temperatures can be used.

Contributing Idea B. Wind can be a source of ocean currents.

Observing and testing the effect of wind upon the surface of water. Fill a long baking pan with water, place it on a flat surface, then wait until the water surface is calm. Without touching the pan, let each student take a turn putting his or her mouth at the water level and blowing gently over the surface of the water using one long breath. Observe any movement in the water and repeat several times. Next, let students experiment with various types of blowing to see what happens (e.g., very strong puffs at regular intervals). They will be able to relate observations to the effect that various winds have upon the ocean's surface.

05
Aa
Bd
Fc

Observing a water current caused by wind. Set up an electric fan so that it blows across a long, wide container of water. If some sawdust or pepper is sprinkled onto the water, students will see that the water moves in a large, circulating path. Explain that in a similar way, wind tends to drive water ahead of it until a continuous current is generated. Tell them that such prevailing winds are the most important causes of ocean currents. Some students might be interested in researching wind and ocean currents in an encyclopedia or other reference source.

06
Aa
Ec

Contributing Idea C. Salinity can be a source of ocean currents.

Observing a water current caused by salinity. Fill four identical bottles, two with salt water (4 spoonsful of salt per cup) and two with tap water. Use food coloring to color one bottle of tap water and one bottle of salt water. Place a card over the mouths of these colored-water bottles and invert them over the bottles containing uncolored water. Carefully remove the cards and let the students observe what happens. (Because there is no temperature difference between the salt and tap waters, students will realize that the mixing is caused by a difference in salinity.) Explain that in a similar way, salinity causes parts of the ocean to move as currents.

07
Aa
Fc

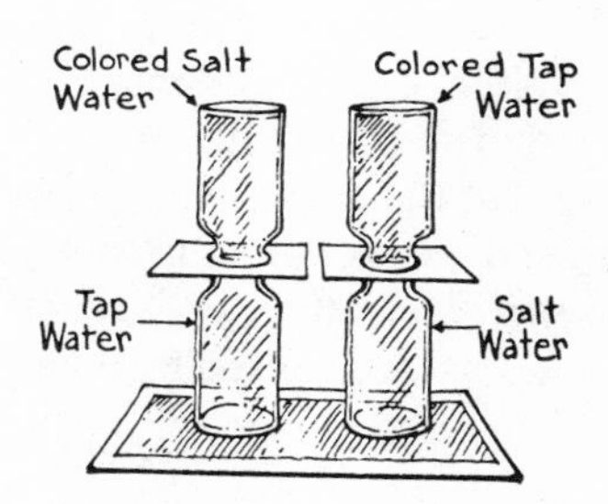

FIGURE 152.07

Generalization II. Waves are produced by different sources.

Contributing Idea A. Wind can produce waves.

Observing water waves. Fill a large, wide-mouth container such as an aquarium or deep flat baking pan with water. When the surface is calm, have a student blow across the water to produce waves. Students can take turns trying to produce different kinds of waves. Some will discover that the

08
Aa
Bb
Ec

height of the waves depends upon the velocity of the wind and the distance over which the wind blows. Tell them that ocean waves behave similarly—those that rise and sink and become regular are *swells;* those that approach the shore, form a steep front, and break are *surf* or *breakers.* Now float a cork on the surface. When the surface is calm, have a student try to move the cork by making waves but without blowing directly on it. After several tries, students will find that they cannot move the cork even though they create waves in the water. (Only the form of the waves moves forward and the cork does not move with a wave.)

Observing waves caused by wind. Fill a long, wide-mouth container, such as a deep baking pan or heat-resistant dish, half way with water and blow a current of air across the water with an electric fan. Students can note the spacing and height of the waves that are produced. Let them experiment by varying the depth of the water and the speed of the fan. From their tests, they will better realize that ocean waves are caused by wind in a similar way. 09 Aa

Contributing Idea B. Earthquakes can produce waves.

Using a model to explain the cause of earthquake waves. Place a brick at one end of a large container. Elevate one end of a second brick to just rest on the first one, then fill the container with water until there are about 2 inches (6 cm) of water above the leaning brick. Explain that the bricks represent two blocks of rock beneath the earth and that movement of the blocks can sometimes cause earthquakes. Now use a small-diameter stick to gently push the leaning brick from the other one. Students will realize how a similar movement under the ocean might cause a wave. You might tell them that such a wave is called a *tsunami* (tsoo nah' me), a Japanese word meaning "big wave in the harbor." 10 Aa Bb Fc

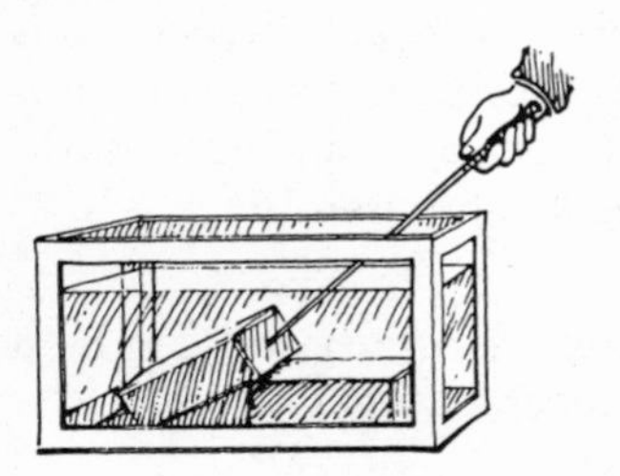

FIGURE 152.10

Generalization III. Tides are produced by the relationship of the moon to the earth.

Contributing Idea A. Tides are indicated by the rise and fall of the water level at shorelines.

Determining changes in water levels at shorelines. If possible, have students place sticks or other markers at the edge of a large body of water such as a very large lake, estuary, or ocean. After several hours, check the markers in relation to the water's edge. Students can compare what they observed with their own seashore experiences. (Sand castles may be washed away as the water level rises.) Tell them that the natural rising and falling of water levels in a large body of water is called *tides.* Some students might be interested in looking up tidal information in the daily newspaper, almanacs, or tide books available at sporting goods stores. 11 Bb Bd Cb

Contributing Idea B. The moon produces tides.

Using a diagram to explain tides. Draw a 20 inch (50 cm) diameter circle on a chart or chalkboard. Tell students that the circle 12 Ca Fc represents the earth, and have them imagine that they are above the north pole looking down upon it. Now draw a 5 inch (12.5 cm) diameter circle 4 feet (125 cm) away to represent the moon. Tell students that the earth and moon attract each other, that the moon's attraction causes the ocean to bulge directly outward toward the moon, and that there is another bulge on the opposite side of the earth. Draw the bulges and letter the high and low positions of the ocean. Put several numerals on the earth and ask students to describe the various tides from those positions as the earth turns. For example, they might say that from position 1, the tide seems to be coming in or rising (because the earth is moving into the bulge) and from position 4, the tide seems to be going out. Students should be able to tell that there are 12 hours between one high tide and the next high tide and 6 hours between changes in tides.

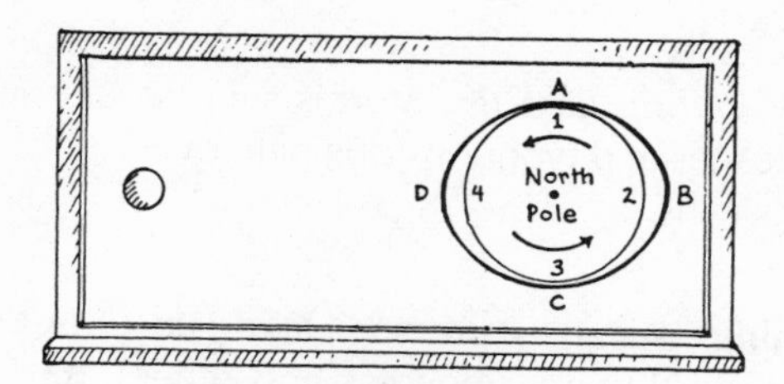

FIGURE 152.12

Using a globe and a smaller sphere to explain tides. Obtain a globe of the earth and a table-tennis ball to represent the moon. Demonstrate that as the earth rotates on its axis, the moon goes around the earth in the same direction, but much more slowly. As you turn the globe and move the ball, show with your hand how the water on the globe is pulled toward the moon. (Although it is pulled almost directly in line with the moon, the tides lag just a bit.) **13 Fc**

Making a model of a tidal basin. Have a student try to carry a shallow pan of water across the room. (This will be very difficult to do without spilling.) The sloshing motion is similar to the motion set up in great ocean basins around the world (e.g., Bay of Fundy). You might use models of the sun, moon, and earth to show that when the sun and moon are in line, their gravitational influences combine to produce the highest possible tides (spring tides); when they are at right angles to each other, their influences divide to produce lower tides (neap tides). **14 Fc**

Generalization IV. Oceans are both constructive and destructive forces as they build up and wear down the earth's surface, creating characteristic shoreline features.

> *Contributing Idea A.* Ocean water physically changes the earth's surface by building up materials.

Observing the relationship between the size of materials and the rate at which they settle in water. Obtain a large test tube that can be corked. Mix at least three sizes of gravel, some sand, and mud. Place the mixture into the tube and fill it with water. Cork and invert the tube. Students will see that as the materials fall through the water, the larger materials settle quickly at the bottom, the medium sizes in the middle, and the fine silt slowly at the top. Explain that rivers, streams, and oceans carry sediments and deposit them on the basis of weight and size. **15 Aa Fc**

> *Contributing Idea B.* Ocean water physically changes the earth's surface by wearing down materials.

Observing the effects of waves on a model beach. Have students make a model beach of sand or dirt at the far end **16 Aa Fc**

of a large rectangular pan or tray. Be sure the edge of the beach is parallel to the end of the pan. Fill the pan about half full of water. Produce a wave by lifting the near end of the pan about 1 inch (3 cm) above the table top and returning it quickly to its original position. Repeat this every five seconds for two minutes. Students can observe and discuss what gradually happens to the model beach. Next, repeat the activity after rebuilding the beach so that its edge is at an angle of 30 degrees to the end of the pan. Discuss the differences in the results.

Collecting and examining seashore sand. If possible, visit a seashore and let students collect samples of sand from different locations. The sand can be placed in small vials or bottles and labeled with its location and date of collection. Upon returning to the classroom, scatter a few grains from each sample onto light green or blue paper. (The color aids viewing of the grains.) Students can examine the grains with a hand lens or binocular microscope if one is available and identify and classify the differences in types of grains that make up each sample. Some students might work out the percentage of each kind and hypothesize where the various grains came from. If possible, return to the seashore at a later date and collect samples from the same locations. Compare these samples with the previous ones to see if changes have taken place. **17 Ca Ge**

Generalization V. The oceans affect climate and weather.

Contributing Idea A. Ocean water evaporates into the atmosphere as moisture; moisture in the atmosphere may condense and return to the ocean.

Making a water cycle model. Inflate a large transparent plastic bag and fit it over a wide-mouth jar or aquarium containing ocean or salt water. Prop up or suspend the bag as shown. Let students observe the bag from time to time. They will see that droplets of moisture appear on the top and sides of the bag, collect, and run down into the aquarium. Explain how what is seen is analogous to the natural water cycle. **18 Fc**

FIGURE 152.18

Contributing Idea B. Storms such as hurricanes and typhoons originate over the oceans.

Researching ocean storms. Have students research in encyclopedias and other sources to find differences and similarities between land (tornadoes) and ocean (hurricanes) storms. They will find that both originate over heated surfaces and blow toward a low pressure center but that the hurricane originates over warm water (the tornado over land in temperate zones), has a path width of hundreds of miles or k (the tornado seldom has a width of more than 2 miles or 3 k), and generally moves at a rate of 12 miles (18 k) per hour (the tornado moves more rapidly). **19 Bd**

Simulating a waterspout. Fill a glass jar three-quarters full with water. Place a lid on the jar. Holding the jar in both hands, have a student move it very rapidly in a circular motion, then quickly stop. Students will see a cyclonic water current that tapers downward like a funnel. Explain that what is seen is analogous to a *waterspout,* a rotating, funnel-shaped column of whirling winds and water extending upward from an ocean or large body of water.

20
Bb
Fc

160 meteorology

161 CHARACTERISTICS

Generalization I. Temperatures on earth vary from time to time and place to place.

Contributing Idea A. The thermometer is an instrument used to measure temperatures.

Making a liquid thermometer. Put a length of glass tubing (about 1 foot or 30 cm long) through a one-hole stopper using a twisting, rotary motion. Fill a bottle or flask to the top with water and add a drop of red ink to make the water more visible. Force the stopper into the bottle so that the water rises into the tube about 3 inches (10 cm). Mark the position of the water, then let a student warm the bottle with his or her hands. Students will note that the colored water rises in the tube. Now cool the bottle with an ice cube or a sponge soaked in cold water and note that the water level drops. To calibrate this instrument, place the bottle into a deep bowl filled with ice cubes. When the liquid in the tube stops descending, tie a string or slip a rubber band around the tube to mark the level of the liquid. Next, place the bottle in a pan of water and heat it. Boil the water in the pan until the level of the colored liquid stops rising. Mark the level with another piece of string or rubber band. The two marks represent the high and low points. A card divided into tenths and hundredths can be placed behind the tube for a scale. This instrument works similarly to commercial liquid thermometers—the liquid expands when heated, contracts when cooled. Commercial liquid thermometers use alcohol or mercury instead of water, since these liquids respond uniformly to temperature changes and do not freeze at temperatures below 32°F (0°C). Other types of thermometers can be compared to this one.

01
Gc

Measuring and graphing air temperatures. Attach a thermometer outside a classroom window. Be sure it is shielded from direct sunlight. (A thermometer will show a higher temperature than the air if it is placed in sunlight.) Have students record the temperature twice each day by checking it at the same time each morning and each afternoon. A record can be kept on a table. The information can be graphed to represent the general directions of temperature change throughout the week. Several interesting variations of this activity can be explored: one student can record the temperature every hour for one day to see how the temperature changes; a group of students can keep records for several months to note temperature changes during the year—weekly or monthly averages can also be computed; individuals or groups can

02
Bf
Ce
Ch

TABLE 161.02. Measuring Temperatures

Week of______________	Morning	Afternoon
Monday	41°F (5°C)	48°F (9°C)
Tuesday	32°F (0°C)	45°F (7°C)
Wednesday	43°F (6°C)	50°F (10°C)
Thursday	45°F (7°C)	59°F (15°C)
Friday	48°F (9°C)	68°F (20°C)

study contrasting temperatures in a) sun—shade, b) moving wind—calm air, c) surface soil—subsoil, d) shallow water—deep water, e) shallow snow—deep snow, and so on.

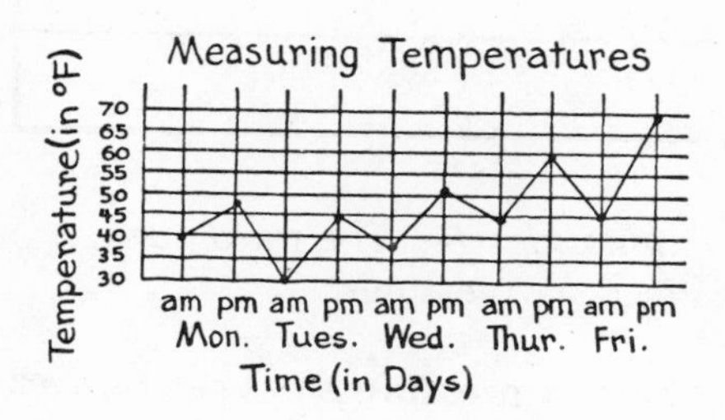

GRAPH 161.02

Measuring the temperature of air at different altitudes. Record the temperature of a thermometer outdoors, then tie the thermometer to a large kite. Let students sail the kite as high as they can. In about 30 minutes, pull the kite in as quickly as possible and record the temperature. Compare the two temperatures. **03 Ce**

Contributing Idea B. The sun's radiant energy is converted to heat energy when it reaches the earth.

Observing that the sun is a source of heat energy. Put equal amounts of ice in two glasses. Place one in the shade and one in the sunlight. Let students observe in which glass the ice melts more quickly. They will realize that the deciding factor is the heat from the sun. **04 Aa**

Measuring the conversion of solar energy to heat energy. Obtain two small identical flasks and coat one with candle soot or cover it with aluminum foil to prevent sunlight from entering. Insert thermometers into one-hole stoppers, and place them into the flasks. Set the flasks in a place that is shaded from direct sunlight until they become cool, then record the thermometer readings. Next, set both flasks in the direct rays of sunlight, and record the thermometer readings every minute until the same reading appears at least three times in a row for each thermometer. The results can be plotted on a graph. Students will readily realize that the sun's radiant energy easily enters the uncovered flask and is converted into heat energy that is measured by the thermometer. The experimental flask remains cool because the radiant energy could not enter. You can make an analogy to the radiant energy that penetrates the earth's atmosphere, strikes the earth, and is converted into heat energy. As a supplementary activity, let students repeat the above procedure after covering the experimental flask with various materials such as colored cellophanes or paint or by filling it with various solid or liquid substances such as soil or water. **05 Ce Ec**

Generalization II. Air pressures vary from time to time and place to place.

Contributing Idea A. The barometer is an instrument used to measure air pressure.

Making an air pressure indicator (barometer). Remove the cork from a small thermos bottle and drill a hole through the cork just large enough to insert a plastic straw or glass tube. Seal the cork in the thermos by dripping candle wax on all connections, then mount the thermos upside down on a stand so that the end of the tube is about ¼ inch (1 cm) from the base. Wrap the thermos in rock wool or some other insulating material, and fit a cardboard box snugly around it. The insulation will help to keep the instrument from acting more like a thermometer than a barometer. Now have students color some mineral oil with food coloring. Set a cup of mineral oil beneath the tube on a day that the air pressure is low (listen to weather forecasts—the low should be below 29.90 inches or 76 cm). Have students observe that whenever the air pressure rises, the mineral oil is pushed up the tube. A scale can be placed behind the tube to note changes. Barometers will work indoors as well as outdoors since the air pressures are about the same. **06 Bb Gc**

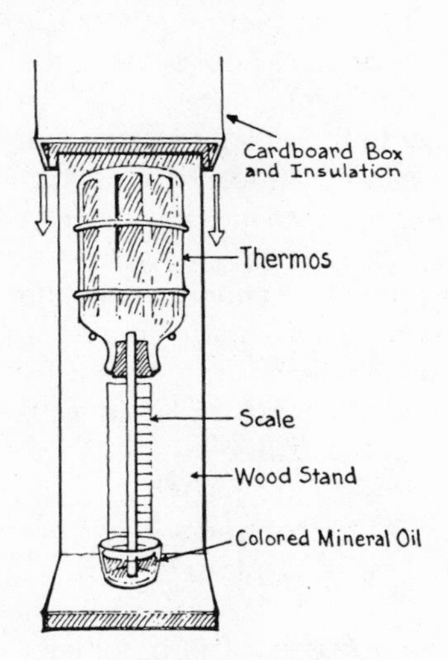

FIGURE 161.06

Measuring and recording the pressure of air. Use a commercial or homemade barometer to keep twice-daily records of changes in air pressure. Note what weather changes take place outdoors with each recording on a table like the one shown. Some weather predictions can be made if wind direction and wind speed information are also obtained. **07 Bc Cd Fb**

TABLE 161.07. Measuring Air Pressure

Week of________	A.M.	P.M.	Observations
Monday	rising	falling	
Tuesday			
Wednesday			
Thursday			
Friday			

Contributing Idea B. The movement of air affects air pressures.

Observing that low pressure areas can be created when air flows swiftly. There are several activities that will help students realize that air pressure can be reduced by rapidly moving air. **08 Aa**

a. Place a table-tennis ball in a funnel, hold it stem downward, and blow up through the stem. Students will be surprised to find that they cannot blow the ball out of the funnel. Next, turn the funnel stem upward while holding the ball inside with one hand. While blowing, release the ball. (The ball should not fall as long as the student is blowing.) Students can realize that the quickly moving air over the ball creates a low pressure area behind it, and the normal air pressure is enough to hold the ball in place. Explain that when air moves rapidly across the surface of the earth, low pressure areas result in a similar way.

b. Put a pin through the center of a 3 inch (8 cm) square of cardboard. Insert the pin into the hole of a spool of thread, and hold the cardboard beneath and flat against the spool. Blow through the spool and remove the hand holding the cardboard. (The card is held in place due to the decreased pressure above the card; the normal air pressure is sufficient to hold the card to the spool.)

Generalization III. Water on earth evaporates into the atmosphere as moisture.

Contributing Idea A. The hygrometer is an instrument used to measure the relative amount of water in the air (humidity).

Making a humidity indicator (hygrometer). There are two basic ways to build an instrument that can measure the amount of moisture in the air.

09 Gc

a. Open two paperclips part way *(a)* and press them about ½ inch (1 cm) apart into a block of soft wood about half-way down one side *(b).* The block of wood should be about 10 inches (25 cm) high

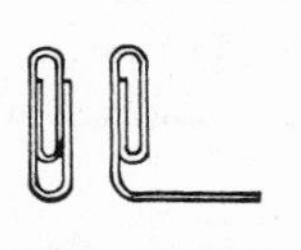

FIGURE 161.09a

FIGURE 161.09b

and sturdy enough so that it will not tip over easily. Glue a fine wire or piece of straw into the eye of a needle *(c),* then set the needle on the two paperclips so that the straw or wire sticks out past the edge of the block *(d).* Wash a long strand

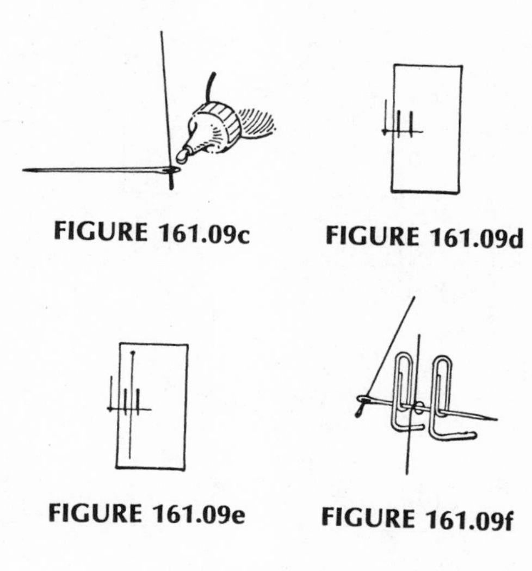

FIGURE 161.09c **FIGURE 161.09d**

FIGURE 161.09e **FIGURE 161.09f**

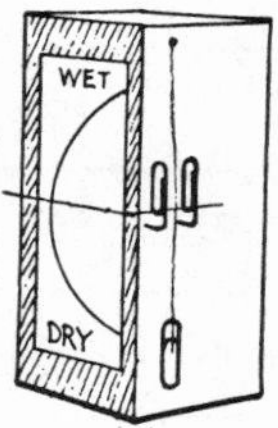

FIGURE 161.09g

of hair in hot soapy water to remove the natural oils. When dry, tie the hair to a pin placed at the top of the block *(e).* Wrap the free end once around the needle in a counterclockwise loop *(f),* then tie a paperclip onto the end of the hair so that it pulls tightly around the needle. Make a measuring scale on a card and mount it on the side of the block *(g).* Set the straw so that it is at the center of the

TABLE 161.09. Relative Humidity Table

Dry-Bulb Temperature Degrees Fahrenheit	*Difference Between Dry Bulb and Wet Bulb Temperatures in Degrees Fahrenheit*																						
	1	2	3	4	5	6	7	8	9	10	11	12	14	16	18	20	22	24	26	28	30	32	34
10	78	56	34	13																			
15	82	64	46	29	11																		
20	85	70	55	40	26	12																	
25	87	74	62	49	37	25	13	1															
30	89	78	67	56	46	36	26	16	6														
35	91	81	72	63	54	45	36	27	19	10	2												
40	92	83	75	68	60	52	45	37	29	22	15	7											
45	93	86	78	71	64	57	51	44	38	31	25	18	6										
50	93	87	80	74	67	61	55	49	43	38	32	27	16	5									
55	94	88	82	76	70	65	59	54	49	43	38	33	23	14	5								
60	94	89	83	78	73	68	63	58	53	48	43	39	30	21	13	5							
65	95	90	85	80	75	70	66	61	56	52	48	44	35	27	20	12	5						
70	95	90	86	81	77	72	68	64	59	55	51	48	40	33	25	19	12	6					
75	96	91	86	82	78	74	70	66	62	58	54	51	44	37	30	24	18	12	7	1			
80	96	91	87	83	79	75	72	68	64	61	57	54	47	41	35	29	23	18	12	7	3		
90	96	92	89	85	81	78	74	71	68	65	61	58	52	47	41	36	31	26	22	17	13	9	5
100	96	93	89	86	83	80	77	73	70	68	65	62	56	51	46	41	37	33	28	24	21	17	13

card. The scale can be calibrated by draping the hygrometer with a cloth soaked in very hot water. Hair stretches when wet and contracts when dry. This action is at an exact rate and is proportional to the amount of water in the air. The humidity under the cloth will cause the hair to stretch and the pointer to rise. When it stops rising, the humidity will be very close to 100 percent. This position can be marked on the card. Other positions can be calibrated by comparing this hygrometer with a commercial one.

b. Mount two identical thermometers on a milk carton or side by side on a board. Cut the tips from a clean white cotton shoestring and slip the loose fibers of one end over the bulb of one thermometer. Tie it in place with a piece of thread, then place the other end into a small container of water. In operation, the two thermometers will produce two different readings. The wet bulb thermometer will give a reading based upon the evaporation of the water from the shoestring. Evaporation has a cooling effect—the reading will be lower than that of the dry bulb thermometer. The rate of evaporation depends upon the amount of water already in the air. When there is a great deal of water in the air, the evaporation is slowed down; thus the temperature reading is higher and closer to that of the dry bulb thermometer. By comparing the readings of the two thermometers, students can determine the relative humidity of the air. Generally a difference in temperature of 15°F (9°C) or more is considered to be an indication of low humidity while a difference of less than 15°F (9°C) indicates high humidity. A table like the one shown can be used to determine the relative humidity more precisely. For example, if the dry bulb thermometer reading is 65°F and the wet bulb thermometer reading is 56°F, the difference in temperature would be 9°F. By reading down the left-hand side of the table to 65°F and across the top to 9°F, the intersection of the two columns is at 56. The numeral in the intersection indicates that the relative humidity is 56 percent.

Measuring and recording the amount of moisture in the air. Have students use hygrometers to measure the amount of moisture in the air. Have them use a table like the one shown on page 108 to keep a weekly record of their measurements. **10 Bc Cg**

Contributing Idea B. Water evaporates.

Observing that water evaporates. Have students observe saucers filled with water over a period of several days. Discuss what happens to the water. **11 Aa**

Measuring the amount of water that evaporates from soil. Fill a flower pot with soil and pour some water into the soil until it begins to drip from the bottom. Weigh the pot, then do not water it again for a week. After a week, weigh the pot again. Students can compare the before and after weights and infer what caused the difference. **12 Cf**

Experimenting to see how different factors affect evaporation. Challenge students to design tests to find out how each of the following factors affects the rate at which water evaporates: temperature, movement of the air, amount of liquid exposed to the air, kind of liquid, amount of moisture in the air (humidity). If needed, help students prepare at least two conditions **13 Ec**

TABLE 161.10. Measuring Humidity (Wet and Dry Bulb Hygrometer)

*Week of*__________	*Wet Bulb*	*Dry Bulb*	*Humidity*
Monday	58°F (15°C)	65°F (19°C)	High
Tuesday	49°F (9°C)	65°F (19°C)	Low
Wednesday			
Thursday			
Friday			

for each test so that results can be compared (e.g., to test the effect of air movement, place some liquid in front of an electric fan and an equal amount in another place where the temperature and humidity are the same, but where there is no wind).

162 INTERACTIONS

Generalization I. As land and water are warmed by the sun, the air above them is heated, becomes lighter, and rises; cooler air, being heavier, moves to replace the warm air.

> *Contributing Idea A.* The wind vane and the nephoscope are instruments used to determine wind direction.

Making a wind direction indicator (wind vane). There are several kinds of instruments that can be made to indicate wind direction. **01 Gc**

a. Cut a notch about 1 inch (2.5 cm) deep in each end of a 1 foot (30 cm) long piece of wood. Cut a small arrow head and a large tail piece from aluminum pie plates. Insert them into the slits, and nail them into place *(a)*. Find the point along the stick where it balances, then drill a hole at that spot just large enough for a small test tube or medicine dropper tubing to fit through. The medicine dropper tubing can be prepared by holding it by the rubber bulb, placing the tip of the dropper

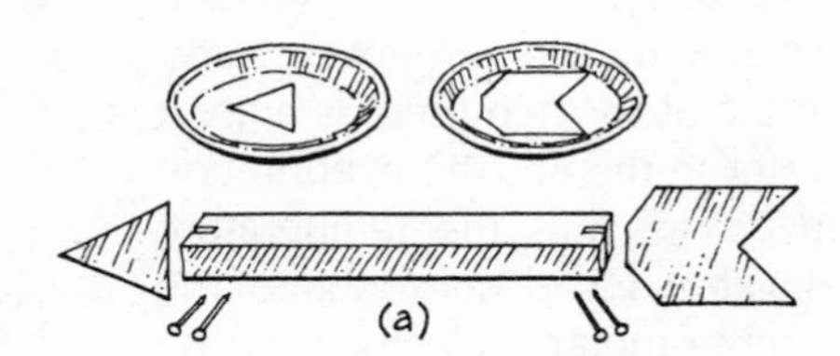

FIGURE 162.01a

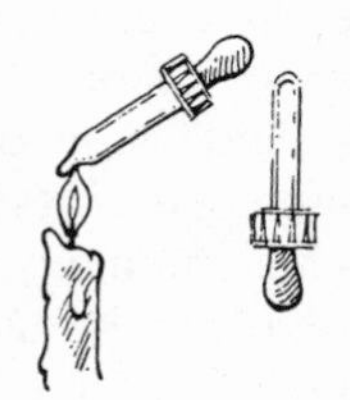

FIGURE 162.01b

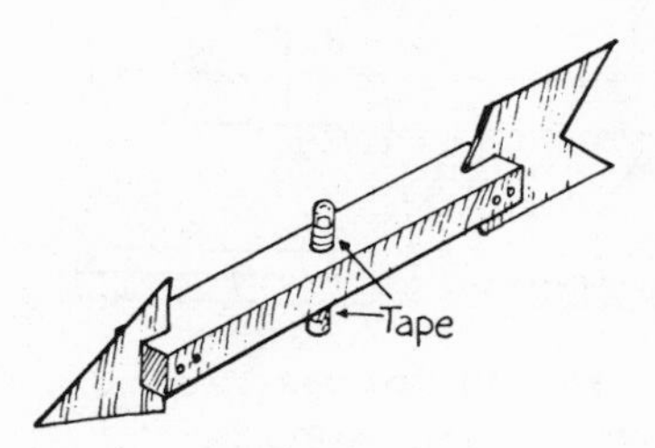

FIGURE 162.01c

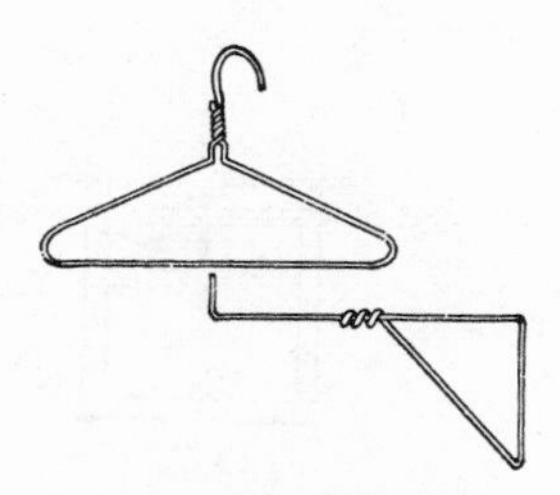

FIGURE 162.01d

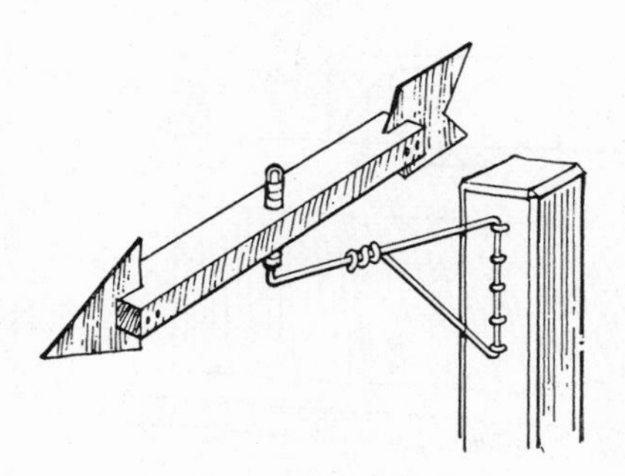

FIGURE 162.01e

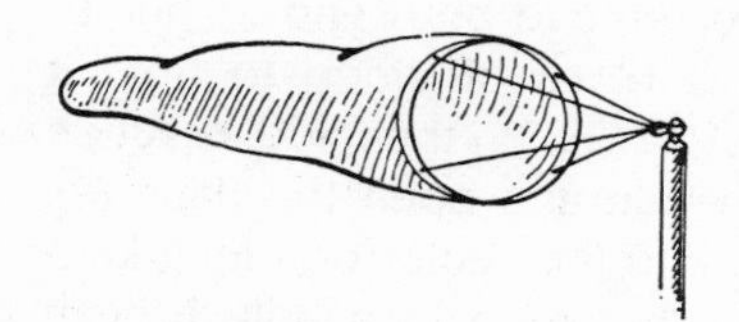

FIGURE 162.01f

into a flame, and rotating it slowly until the opening is completely closed and rounded *(b)*. When the glass is cool, remove the rubber and insert the tubing through the wood. You might need to use some friction tape to keep the wood from slipping off the tubing *(c)*. Now bend a coathanger to form a bracket *(d)*, and mount the wind vane on a post or fence where winds blowing from many directions will strike it *(e)*. Be sure students note that the arrow points in the direction *from* which the wind comes. Tell them that winds are named for the direction *from* which they come (e.g., a north wind comes from the north, and the arrow will point north). Students can use a compass to determine from which direction the winds come.

b. The wind sock is another instrument that is used at airports as a wind direction indicator. It can be made by bending a section of light wire into a circle and attaching some thin cloth to it. It can then be attached to a stick with strings and placed outdoors where the wind will blow freely into it.

02 Gc

Making a cloud direction indicator (nephoscope). The lower part of a moving air mass is usually obstructed and influenced by trees, houses, and other objects; thus, wind vanes (usually near the ground) do not always indicate the true direction of the moving air. To observe movement higher in the atmosphere, have students glue a round mirror to a piece of cardboard, and mark the points of the compass around the mirror. Paste a small paper circle about the size of a dime in the center of the mirror. Set the cardboard on a level spot outdoors with the N pointing north (a compass can be used to orient the mirror). Have students look down into the mirror. When they see a cloud passing over the dime-sized circle, have them follow it with their eyes until it reaches the edge of the mirror. At that point they will see a wind direction indicated on the cardboard. This is the direction *toward* which the wind is blowing. You might remind them that winds are named for the direction *from* which they come.

Recording wind directions. Use a commercial or homemade wind vane or nephoscope and record the wind direction twice a day at the same times each day. Note if there seems to be any relationship between the direction of the wind and the kind of weather that follows.

03
Bc
Fb

TABLE 162.03. Recording Wind Direction

*Week of*__________	*A.M.*	*P.M.*	*Observations*
Monday			
Tuesday			
Wednesday			
Thursday			
Friday			

Contributing Idea B. The anemometer is an instrument used to measure wind speed.

Making a wind speed indicator (anemometer). There are two basic kinds of instrument designs used to measure the speed of winds.

04
Gc

a. The first design requires the student to turn the instrument into the wind. To

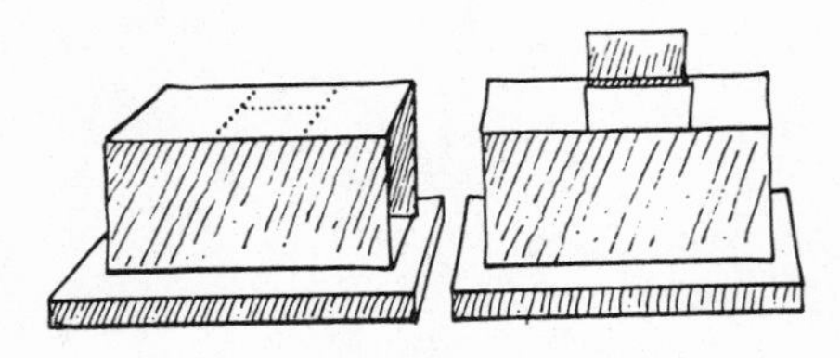

FIGURE 162.04a

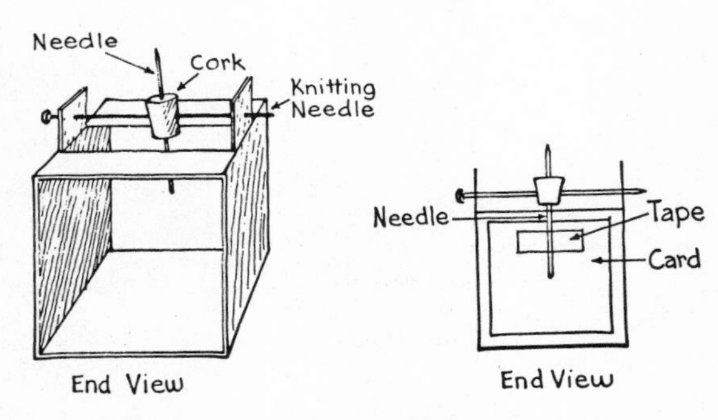

FIGURE 162.04b FIGURE 162.04c

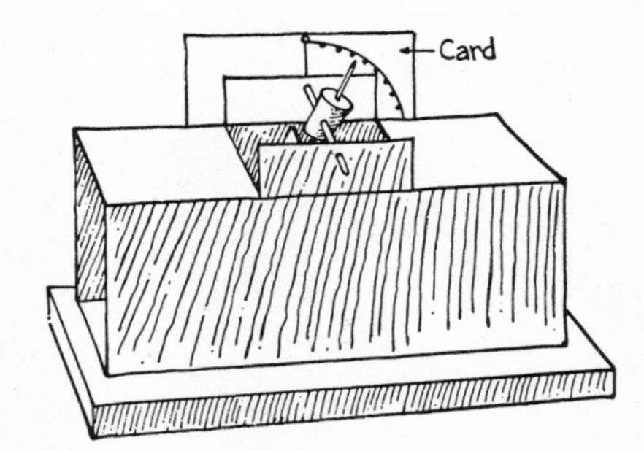

FIGURE 162.04d

make such an instrument, clean a milk carton and remove both ends. Thumbtack it to a block of wood to hold it steady. Cut an "H" in the middle section of the top side and open the flaps *(a)*. Push a cork to the center of a long knitting needle; then push the needle through the flaps *(b)*. Mount a small needle vertically into the cork to serve as a pointer. Cut a square piece of cardboard so that it fits inside the carton. Attach a second, smaller needle to it with tape or glue and stick the needle into the bottom of the cork *(c)*. Students may have to adjust the cork, knitting needle, or the card so that the card swings smoothly within the box. When the wind is blowing, point the indicator into the wind so that it blows through the box. The wind will push against the square card, tilting the cork and moving the needle pointer. The distance the needle tilts indicates how fast the wind flows through the carton. A card can be attached to the carton to make a gauge. When attached, draw a line on it parallel to the needle when it is straight up *(d)*. The gauge can be calibrated using the Beaufort Scale (see *162.06*).

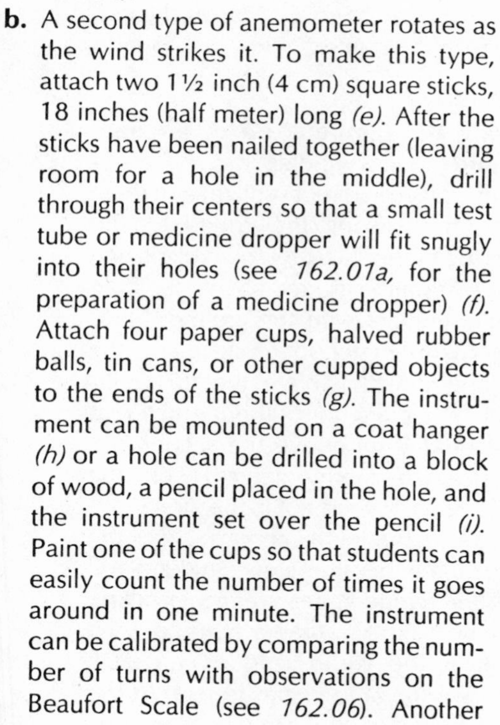

b. A second type of anemometer rotates as the wind strikes it. To make this type, attach two 1½ inch (4 cm) square sticks, 18 inches (half meter) long *(e)*. After the sticks have been nailed together (leaving room for a hole in the middle), drill through their centers so that a small test tube or medicine dropper will fit snugly into their holes (see *162.01a,* for the preparation of a medicine dropper) *(f)*. Attach four paper cups, halved rubber balls, tin cans, or other cupped objects to the ends of the sticks *(g)*. The instrument can be mounted on a coat hanger *(h)* or a hole can be drilled into a block of wood, a pencil placed in the hole, and the instrument set over the pencil *(i)*. Paint one of the cups so that students can easily count the number of times it goes around in one minute. The instrument can be calibrated by comparing the number of turns with observations on the Beaufort Scale (see *162.06*). Another

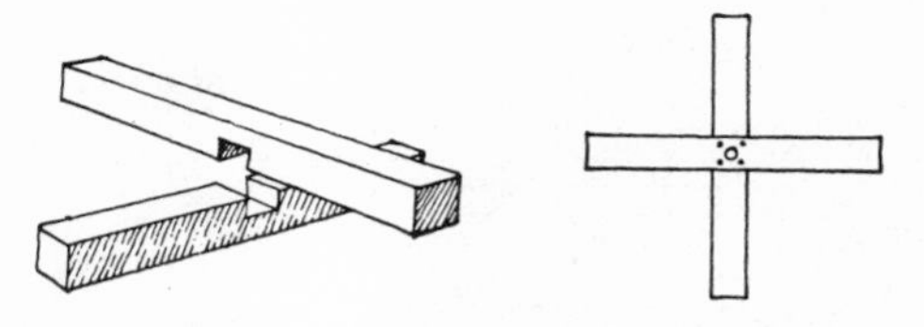

FIGURE 162.04e **FIGURE 162.04f**

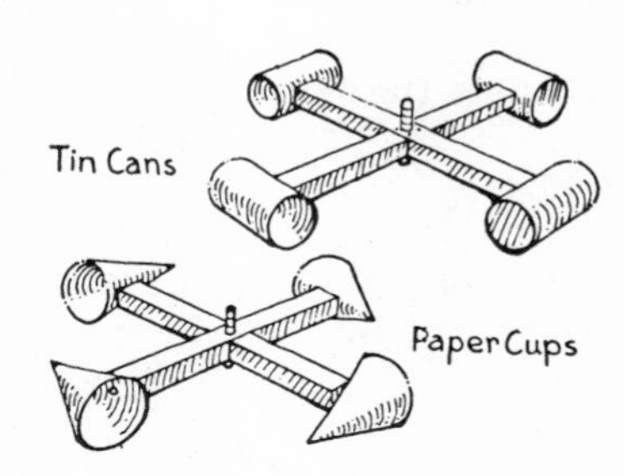

FIGURE 162.04g

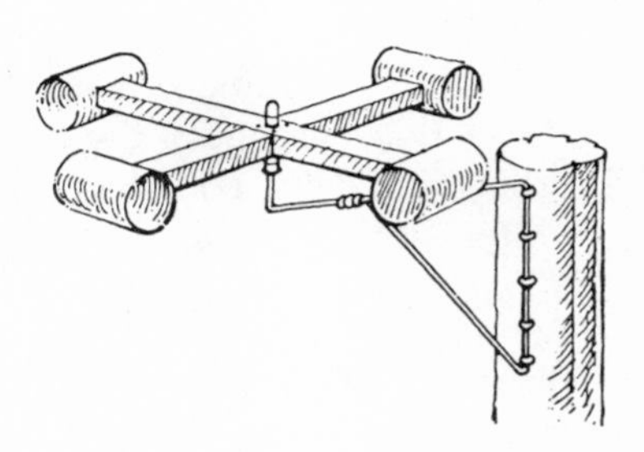

FIGURE 162.04h

FIGURE 162.04i

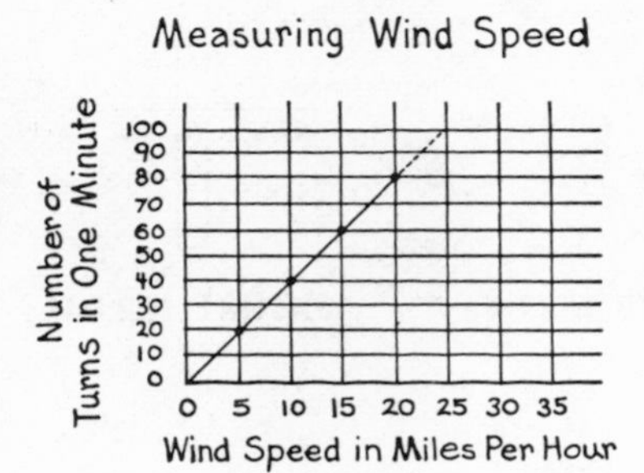

GRAPH 162.04

way to calibrate the instrument is to hold it out the window of a moving car on a calm day. This is done by sitting next to the front right window while the driver drives the car at a steady 5 miles (kilometers) per hour. The number of turns are counted for one minute. (This should be done several times and an average taken.) Repeat for 10 mph (kph), 15 mph (kph), etc. The information can be graphed. When the instrument is mounted on the school grounds, the speed can be determined by counting the number of turns in one minute.

TABLE 162.05. Measuring Wind Speed (Rotary Anemometer)

Week of ______	*A.M.*	*P.M.*	*Observations*
Monday			
Tuesday			
Wednesday			
Thursday			
Friday			

05 Bc Ch Fb

Measuring and recording the speed of wind. Students can keep a daily record of wind speed on a table like the one shown. Measurements should be made at the same time each day, and students should note whether the sky is sunny, cloudy, rainy, etc. Have them study their table after several weeks to see if there is any relationship between the speed and direction of the wind and the kind of weather that follows.

06 Cb

Estimating wind speed without an anemometer. The speed of the wind can be estimated without complex instruments simply by using the Beaufort Scale. The scale allows students to estimate fairly accurately the speed of wind by observing the motions of leaves on trees, chimney smoke, etc.

Contributing Idea C. Air is heated primarily by contact with the ground.

07 Ce

Measuring the temperature of air near the ground. On a windless day, have students find a place in direct sunlight and drive a 6 foot (2 m) stake into the ground. On the shadow side of the stake, let them place one thermometer on the ground so that the bulb is in direct contact with the ground. Place a second thermometer on the stake near the ground and directly above the first thermometer. Attach a third thermometer at least 5 feet (1.5 m) above the second. Be sure that each thermometer is on the shadow side so that it is not in the direct rays of the sun. At 10-minute intervals, check the readings on the thermometers. Students will find that the temperatures are higher nearer the ground. Explain that when cold air moves near the warm land or water, it becomes warmer and begins to rise. As the air rises, it becomes cooler and heavier, then

TABLE 162.06. The Beaufort Scale of the Speed of Wind

Scale Number	*Observation*	*Name of Wind*	*Miles Per Hour*
0	Smoke goes straight up	Calm	Less than 1
1	Direction shown by smoke but not by wind vanes	Light Air	1–3
2	Wind vane moves; leaves rustle	Light Breeze	4–7
3	Flag flutters; leaves move constantly	Gentle Breeze	8–12
4	Raises dirt, paper; flags flap	Moderate Breeze	13–18
5	Small trees sway; flags ripple	Fresh Breeze	19–24
6	Large branches move; flags beat	Strong Breeze	25–31
7	Whole trees sway; flags are extended	Moderate Gale	32–38
8	Twigs break off; hard to walk against	Fresh Gale	39–46
9	Slight damage to buildings	Strong Gale	47–54
10	Trees uprooted; windows break	Full Gale	55–63
11	Widespread damage to buildings	Violent Storm	64–75
12	General destruction	Hurricane	Over 75

moves down toward the land and water again.

Contributing Idea D. Winds are caused by unequal heating of the earth's surface.

Comparing the absorption and release of heat by soil and water. Fill one coffee can with dry soil and another with water. Let each remain in the classroom overnight so that they will be equal in temperature. Students can then put thermometers to the same depth in each and place both cans outdoors in the sunlight. After two hours, check the temperatures to see which is warmer and which is cooler. (The water should be cooler.) Have students share experiences they have had walking barefoot from hot pavement or land to a puddle or

08
Ce

from a sandy beach to the water. They should deduce that soil heats up more rapidly than water. If they next set a can of soil and a can of water of equal temperatures in a refrigerator and check the temperatures every 10 minutes, they will find that the soil cools more quickly than the water. Such data can be easily graphed. Explain that soil or water temperatures influence the air above them. That is, land areas generally warm more rapidly during the day than water areas; thus the air above the land tends to be heated and rise while the cooler, heavier air over the water areas pushes inland to replace it. At night the land areas generally cool more quickly, thus the exchange of air is reversed.

Detecting warm and cool air currents. **09 Aa** Small wind currents in the classroom can be detected using simple wind current detectors. The detector shown can be made by folding a square piece of paper up on the solid lines and down on the dotted lines, then attaching a thread to the center. If held

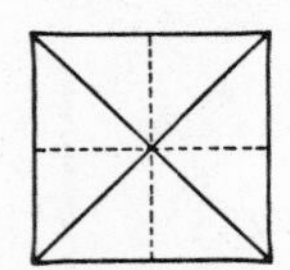

FIGURE 162.09

by the thread over a radiator or light bulb, the device will move and indicate the direction of the current. If temperature readings are taken in different places in the room (e.g., at the ceiling and the floor; at the bottom and top of an open window), students will see that the movement corresponds to the exchange of warm and cool air. Explain that warm air weighs less than cold air, exerts less pressure—thus creating an area of low pressure. The heavier cold air creates an area of high pressure and pushes the warm air upward.

Generalization II. Water on earth evaporates into the atmosphere as moisture.

Contributing Idea A. Temperature change affects evaporation.

Testing to see if an increase in temperature affects evaporation. **10 Ec** Pour ½ cup (100 ml) of water into each of two identical shallow pans. Place one in a warm location and the other in a cool place away from any breeze. Compare the time required for the water to evaporate from each pan. Students can make some judgment about the relationship of heat to evaporation. The test can be repeated and speeded up by placing the experimental pan over a heat source such as a hot plate or radiator. Similarly, two cloths of the same size and same material can be substituted for the pans. Soak each thoroughly in water. Set one in a cool location and one in a warm location. An analogy can be made to clothes on a line on a sunny day and on a cloudy day. In each of the above tests, be sure that the heat factor is the only influence.

Testing to see if a great reduction in temperature affects evaporation. **11 Ec** Thoroughly wet a cloth on a sunny day when the temperature outside is below freezing. Suspend the cloth and observe it every half hour until it is frozen stiff. After several hours, bring the cloth indoors and let students examine it. They will find that the cloth is dry. The time required for the frozen water to leave the cloth will depend upon the relative humidity of the air, the wind, and the temperature. *(Note: The water in* *the cloth actually sublimes from a solid state to a gaseous state.)*

Contributing Idea B. Moving air affects evaporation.

Testing to see if moving air affects evaporation. Wet two identical areas of a chalkboard and fan one with a piece of cardboard or an electric fan. Students will realize that the moving air was the influencing factor in making one area dry faster. Similarly, they can thoroughly wet two cloths of the same size and material, hang one outdoors in a windy location and the other in a sheltered spot. If they touch the cloths periodically, they will find that the water evaporates from the cloth in the wind more quickly. Let them discuss how clothes dry on windy days. **12 Ec**

Testing to see if moving moist air affects temperature. Have students use thermometers to find the temperature of moving air (outside when the wind is blowing or in front of an electric fan). Hang some wet straw, newspaper, or cloth strips in the breeze. Leave a little space between the strips so the air can pass between them. Use thermometers to find out what happens to the temperature of the air after it passes through the wet material. An analogy can be made to air conditioners and how they work. **13 Ec**

Contributing Idea C. Surface area affects evaporation.

Testing to see if surface area affects evaporation. Pour one cup of water into a wide, shallow dish and another cup into a tall, narrow jar. Place both in direct sunlight. Students will see that the one with the wide mouth evaporates faster than the other. Discuss the differences in evaporation from small puddles and large lakes. **14 Ec**

Generalization III. Moisture in the atmosphere can condense into various forms that return to the earth.

Contributing Idea A. The rain gauge is an instrument used to measure rainfall.

Making a rainfall indicator (rain gauge). Any open container with straight sides can be used to measure amounts of rainfall. When the rain is collected, simply have students stick a ruler into the container to see how deep the water is. If the water is 1 inch (3 cm) deep, then 1 inch (3 cm) of rain has fallen. Since most rainfall is less than an inch, a true rain gauge is designed to catch a relatively wide area of rainfall and to funnel it into a narrow area so that it will be deeper and can be measured more easily. *(Note: Measurements must be taken very soon after a rainfall or evaporation will give inaccurate readings.)* Attach a test tube beside a strip of paper on a block of wood. Fill a wide-mouth straight sided jar with 1 inch (3 cm) of water. Pour this water into the test tube and mark the height on the strip of paper. Repeat this procedure using 3/4 inch, ½ inch, and ¼ inch (2.5 cm, 2 cm, 1.5 cm, 1 cm, and .5 cm) of water. Now the jar can be placed outdoors in the open. When rain is collected in the jar, pour it into the test tube to measure how much rain fell. **15 Gc**

Measuring and recording rainfall. Use a rain gauge in conjunction with a wind vane. Information can be recorded on a table. A graph of the data will reveal how much rain **16 Bc Bf Cd**

TABLE 162.16. Measuring Rainfall

Date	*Direction of Wind During Rainfall*	*Amount of Rainfall*

falls in your area in a month and which winds bring rain.

Contributing Idea B. Water condenses.

Observing condensation. Have students observe the condensation of water on glasses of ice water. Let them discuss where the droplets come from. Let them also describe personal experiences that are examples of condensation (e.g., moisture on a mirror in bathroom, moisture on windows inside a closed automobile). **17 Aa Ga**

Contributing Idea C. Temperature change affects condensation.

Observing condensation by cooling through contact. On a very cold day, have students go outside and notice how the moisture from their breath can be seen when it comes in contact with the cold air. Similarly, have them exhale across an ice cube tray or into an open freezer compartment and observe the condensation in the air. They will realize that this phenomena takes place when moist air comes in contact with cooler air or a cooler surface. **18 Aa**

GRAPH 162.16. Relating Rainfall to Wind Direction

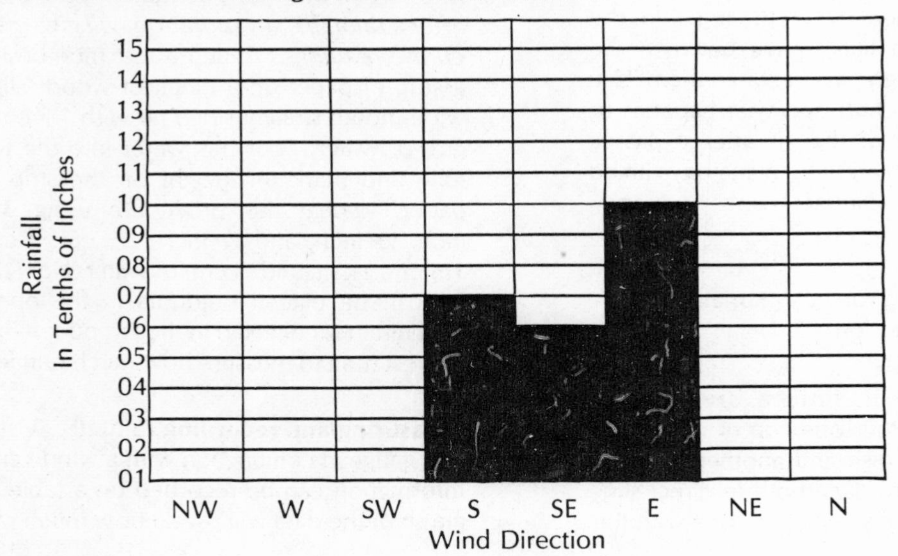

Observing condensation by cooling through rising. Boil some water in a pot, beaker, or tea kettle on a hot plate. When it is boiling, let students describe what they see. **(Caution: Be careful of the steam.)** Students will see clouds of moisture forming as the heated moist air rises into the cooler air of the room. Encourage them to imagine how warm moist air near the earth may rise and cool in the atmosphere in a similar way. *(Note: Rising air cools at the rate of 1°F every 300 feet in the atmosphere.)*

19
Aa
Fa

Observing condensation by cooling through expansion. Pour a cup of hot water into a transparent wide-mouth gallon (4 L) jar. Put a plastic bag inside the jar and fold its edge over the rim, fastening it tightly with rubber bands. Have one student hold the jar firmly on the table, and let another pull upward on the bag quickly. Have them describe what they see inside the jar. (When the bag is pulled upward, the pressure inside the jar is reduced and the warm moist air inside expands. It cools with expansion and a cloud forms inside. When the bag is released, the cloud disappears.) This activity can be repeated several times. The cloud is best seen when the light source is from behind the viewer. Students should be able to generalize that clouds can form in a similar way.

20
Aa
Fa

Contributing Idea D. Dew and frost are formed by the condensation of water vapor on the surface of the earth.

Observing how dew is formed. Remove the label from a tin can. Take some water that has been allowed to stand until it has reached room temperature, and fill the can about halfway. Add a few ice cubes to cool the water. After several minutes, students will see droplets of water forming on the outside of the can. Tell them that the formation is called *dew*. Explain the parallel formation of early morning dew that they may have seen. They will realize that when moist air contacts a cooler surface, such as the side of the can or blades of grass, the moisture in the air condenses on the surface. *(Note: In very dry areas, this activity may not work because there may not be enough water vapor in the air for droplets to form.)*

21
Aa
Bb
Fa

Measuring dew points. Remove the label from a tin can, and half-fill it with water. Place a thermometer in the water, and record the temperature. Add chips of ice, stir, and record the temperature again when droplets form on the outside of the can. Tell students that the temperature reading when the droplets form is called the *dew point* for the temperature of the air in the room. Explain that water droplets or dew on blades of grass or flowers in the morning is formed in a similar way, and the temperature at which it forms is the *dew point.* Let them repeat this activity in different locations and on different days to compare dew points. If other observations are recorded (e.g., room temperature, humidity), factors influencing dew point will be discovered. After several tests, ask students what conclusion they would make if the temperature of the room and the dew point were close together. (There would be a very high relative humidity.)

22
Bb
Ce

Observing how frost is formed. Remove the labels from two tin cans and half-fill one can with water, chips of ice, and a handful of salt to make it very cold. Half-fill the second can with water only. Use two thermometers to record the temperature of each can,

23
Aa
Bb
Fa

then stir the contents of both rapidly. Students can observe the outside of the cans. Tell them that the formation they observe on the first can is called *frost.* Explain the parallel formation of early morning frost on the ground. They can then compare the two temperature readings and realize that when the surface of an object (e.g., the ground) is below the freezing temperature of water, the water vapor in the air condenses on the surface as frost.

Contributing Idea E. Fog is formed by the condensation of water vapor near the surface of the earth.

Observing how fog is formed. Fog is the condensation of moisture in the atmosphere near the surface of the earth. For any of the following experiences, help students realize that real fog is formed in an analogous way through the rapid cooling and condensation of water vapor in the air near the earth's surface.

24
Aa
Fa

a. Heat some water in a container until it boils, then fill a large milk bottle slowly to prevent it from breaking. Next, empty all but 1 inch (3 cm) of water from the bottle and hold it so the light source (e.g., window, electric light) comes from behind the viewers. Now, set the bottle on a tray of ice cubes, and let students describe and discuss what happens inside. They will realize that in a similar way fog forms at night when the earth cools rapidly (represented by the ice tray) and the air next to it cools in turn and comes in contact with the warmer moist air above.

b. Take a large fruit juice can, put some ice in it, and add a handful of salt to make it very cold. Set a smaller can on the ice so that the tops of both cans are even. Pack more ice and salt into the space between the two cans and let a student exhale into the smaller one. The class will see fog form and remain in the smaller can. Point out the parallel to real fog that clings in valleys when the ground is cooled.

Contributing Idea F. Clouds are formed by the condensation of water vapor in the atmosphere.

Observing how clouds are formed. Clouds are condensations of moisture in the atmosphere. For any of the following experiences, help students realize that real clouds are formed similarly through the cooling and condensation brought about by the expansion of rising air.

25
Aa
Fa

a. Pour a cup (200 ml) of warm water into a transparent wide-mouth gallon (4 l) jar. Hold a lighted match in the jar, blow it out, and let it remain briefly. Now place a plastic bag inside the jar, turn its edge down over the rim of the jar, and fasten it securely with rubber bands. Have a student hold the jar firmly on the table, and let another quickly pull upward on the plastic bag. They will see a cloud form inside; when the bag is released, it will disappear. The cloud forms for several reasons: (1) the air in the jar contains invisible water vapor; (2) the air pressure inside the jar is reduced; (3) there are many small particles from the match in the air. As the air inside the jar expands, it cools, and the water vapor condenses as liquid around the smoke particles to form the minute droplets that make up the cloud. Students can test to see if a cloud can be made: (1) without putting water in the jar; (2) with cold water instead of warm water; (3) with water but without smoke, and so on.

b. Obtain two identical wide-mouth jars. Line half the inside of each with soft, black cloth. Add glue to hold the cloths in place, then soak the cloths with water. Cover each jar with a square of glass and set them upright, one in a pan of cold water, the other in a pan of very hot water. Leave the jars for 15 minutes. Remove them from the pans and set the cold jar upside down over the warm jar

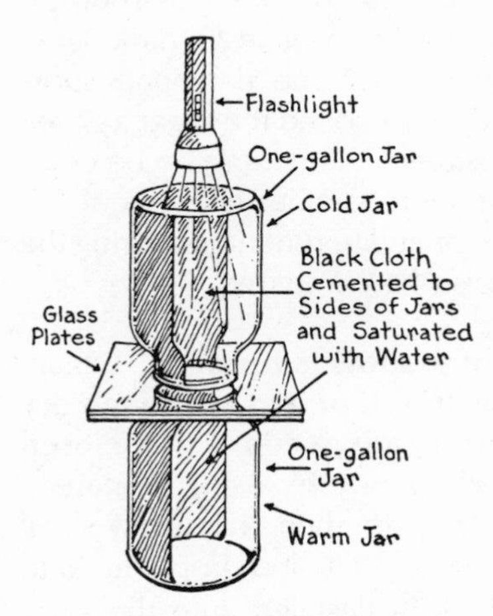

FIGURE 162.25

leaving the glass squares over the openings. Hold a flashlight to shine down through both jars, then carefully remove the glass squares. A cloud will form as the warm moist air rises and comes in contact with the cooler air above. Students can experiment by repeating the activity and reversing the positions of the jars. Tiny tissue paper streamers can be placed inside the jars to indicate the direction the air flows. If possible, observe cloud formations outdoors, and let students judge the relationship of hot and cold air masses based upon their experimental models.

Identifying cloud types. Have students keep records for a week of the kinds of clouds that they see. Help them recognize some of the common types. **26 Ba Bb**

a. *Cirrus, cumulus,* and *stratus* are the basic types of clouds. Cirrus are very high in the atmosphere. They generally look feathery and are composed of tiny ice crystals. They are usually a sign of clear weather. Cumulus clouds are lower than cirrus clouds, and airplanes often fly above them. They look like white puffs in the sky and usually indicate fair weather. Stratus clouds are much lower and foglike. They form gray layers across the sky.

b. *Nimbus* describes basic clouds containing a great amount of water. They are usually very low in the sky and look thick and black. They generally bring rain, snow, sleet, or hail.

c. *Alto* is a prefix meaning high. It is used to describe high forms of basic clouds.

Students can study events that bring about the formation of each type of cloud. They might group cloud types on the basis of the events (clouds that usually follow in sequence with an approaching cold front—altocumulus, cumulonimbus, stratus, stratocumulus; clouds that usually follow in sequence with an approaching warm front—cirrus, cirrostratus, altostratus, nimbostratus, stratus).

Contributing Idea C. When the moisture in clouds continues to condense, it may fall as rain or snow.

Observing how rain is formed. After students have seen how clouds can be formed, have them hold the flat bottom of an ice cube tray slightly tilted but close to steam **27 Aa Fa**

escaping from a boiling teakettle. **(Caution: The steam is very hot.)** Be sure they observe how the water droplets collect, enlarge, and drip off the edge of the tray because of their increasing weight. It should be noted that although the formation of rain is not fully understood, scientists believe that droplets form upon minute particles in the air and continue to increase in size as water vapor cools. Students can try this activity again using two ice cube trays—one empty and one filled with ice cubes. Explain that as the moist air rises (from the teakettle), it cools rapidly against the filled tray and less rapidly against the empty tray. Take care to hold the trays at equal distances from the spout of the teakettle and record the time taken for drops of "rain" to fall from each. They can compare the times. Students can relate their findings to actual events in the atmosphere.

28 Cd Da

Measuring the size of raindrops. On a rainy day, pour some flour through a sifter into a pie pan until it is ½ inch (1 cm) deep. Cover the pan with a large plate and take it outdoors. Hold it in the rain, uncover it, and let the rain fall into the flour for three seconds. Cover the pan with the plate and return indoors. When the pan is uncovered again, students will see wet round lumps where the raindrops fell. Let the lumps dry for about three hours, then use the sifter to separate the lumps from the flour. Measure the lumps with a ruler or seriate them by size to learn something about the relative size of raindrops. *(Note: The lumps are slightly larger than the actual raindrops.)* If records are kept, the sizes of raindrops can be compared during different parts of a storm or between different storms. Another way to measure the size of raindrops is to prepare hoop screens by stretching pieces of discarded nylon stockings over embroidery hoops. When each screen is stretched tightly, staple or tack it in place, then trim away the excess stocking. On a rainy day, press the screens into a large pan with some sprinkled powdered sugar. Tap off any excess sugar, cover the screens, take them outdoors, and hold them horizontally in the rain. When ready, uncover the screens and allow raindrops to fall on them for a timed period (e.g., 30 seconds). Cover the screens again and bring them into the classroom. If the screens are held up against a dark background, students will see darkened spots where each drop removed the sugar as it fell through the screen. Diameters can be compared with a ruler. Samples taken from different locales or at different times during the rainfall can also be compared.

29 Aa Bb Fa

Observing how snow is formed. Obtain some dry ice. **(Caution: Do not touch dry ice with bare hands. Keep it in a closed container when not in use.)** Prepare a cloud as in *162.24b,* then chip a few small pieces from the dry ice, and drop them into the cloud. Hold a flashlight into the small can and exhale gently into the cloud at two-minute intervals. Have students describe what happens. Explain that snow is formed directly from water vapor to ice crystals in a similar way (no liquid state between the two), and that the crystals may fall to the ground as snowflakes. You might explain that if rain freezes when it contacts an object or the ground, the result is called *sleet,* and when frozen particles in the air have additional layers of water frozen to them before they fall to the ground, the result is called *hail.*

30 Aa

Observing how obstacles affect snowdrifts. On a snowy day, prop some card-

TABLE 162.28. Seriating Raindrops By Size

	Small Drops	*Medium Drops*	*Large Drops*
Number of drops in 3 seconds at the beginning of the storm.			
Number of drops in 3 seconds in the middle of the storm.			
Number of drops in 3 seconds near the end of the storm.			

board sheets upright on old snow. Set the sheets so they face in different directions. When it stops snowing, students can examine the distribution of the snow around each sheet. They will be able to describe where the drifting takes place. On the basis of their observations, ask them how they would place a snowfence if they wanted to keep snow from a walkway.

Measuring snowfall. Although snowfall depth is reported on newscasts and is important to skiers, the snowfall measurement by weather bureaus has more meaning when the water content of the snow is determined. To do this, remove the top and bottom from a straight-sided can and push the can straight down into the snow until its rim is even with the earth's surface. Let one student reach under the can and cut the snow even with the bottom edge with a piece of cardboard. Remove the cylinder of snow and dump it into a second can of the same size but which has only one end removed. Melt the snow slowly to avoid too much loss of water by evaporation. Measure the depth of the water that remains. Compare the depth of the water to the depth of the snow that produced it. Since snows are different, students can try this activity with dry, light, fluffy, and packed snows. They will find that some snows are drier than others (the ratio of water to snow depth is less). They can seriate or rank the snows by the amount of water each produces; and if temperature measurements are taken during snowfalls, a relationship between temperature, the type of snow that is formed, and the water content will be discovered.

31
Cd

200-299
Organic Matter

211 CHARACTERISTICS

Generalization I. Animals have identifiable characteristics.

Contributing Idea A. All animals reproduce their kind.

Observing that animals reproduce their kind. Have students find pictures of animals and their young (e.g., deer and fawn, bear and cub, horse and colt, bird and fledgling). They can describe the characteristics of the adults and the young. Have them make a list of the traits that are present in both parents and offspring. 01 Aa Bc

Observing the reproduction of guppies. Guppies reproduce easily under most classroom conditions and can be used to help students realize that subsequent generations maintain the characteristics of the parents. A pair of guppies will live well in a 1 gallon (4 L) aquarium, but the tank will be too small as offspring arrive. The offspring should be placed in other containers. Stock the aquarium with a male and female guppy (the adult female will be larger and less colorful than the male) and some aquatic plants. Keep the temperature from dropping below 72°F (22°C). Feed the guppies daily with commercial food from a pet shop or brine shrimp and daphnia. Students can see the male court the female as he swims around her and turns his body into an S-shape. A pregnant female will produce a brood in twenty-one to twenty-eight days and can continue to do so up to eight months from one mating. The tiny fish are easy prey for the adults, therefore some floating plants should be provided as hiding places for them. Compare the characteristics of the young to those of the adults. Let students tell of experiences they have had in seeing or knowing of the birth of kittens, puppies, hamsters, or other animals. They should realize that in each case, adult animals produce young that carry characteristics of the adults. 02 Aa

Observing inheritance of some physical characteristics. Obtain pictures of the parents (blood relatives) of students and put them on a bulletin board without any identification. Let students study the pictures and guess whose parents are shown. Compare the physical characteristics of the students with those of their parents. Pictures of brothers and sisters can also be used. 03 Aa

Contributing Idea B. All animals grow.

Describing growth in animals. Let students tell of experiences in raising pets, specifically observable indications that the 04 Ba Fa animals being raised were growing. They might describe changes in size, weight, appearance, habits, or behaviors. If they compare a variety of animals (birds, fish, hamsters, snails), students can generalize that all types of animals grow.

Measuring the growth of an animal. Keep a small young mammal as a pet in the classroom (e.g., a mouse or a hamster). Provide it with a good, clean cage and feed it a proper diet. Weigh the animal once a week by placing it in a tall coffee can or cardboard oatmeal carton and setting it on a balance scale. Determine the weight of the animal by subtracting the weight of the container. The weekly measurements can be recorded on a graph to depict the animal's growth over time. Other animals can be measured in the same way. It would be interesting to weigh portions of the food during the study.

05
Bf
Cd
Ch

Contributing Idea C. All animals need food, water, and oxygen to grow.

Observing that animals eat. Place a snail and some green leaves in a jar, and cover it with a piece of nylon stocking. Observe the snail as it locates and eats the leaves. Now place other animals in an appropriate environment (grasshoppers, caterpillars, earthworms, hamsters, and so forth), feed them properly, and observe what, when, and how each feeds.

06
Aa
Fa

Sorting animals by the types of food they eat. Explain that there are three general types of food-eating animals in nature—plant eaters, meat eaters, and those that eat both. Have students collect pictures of animals from magazines and bring them to class. The pictures can be pasted on charts, one for each of the three types of food eaters. To do this, look in encyclopedias or other reference books for physical characteristics that indicate the type of food each animal eats. (In mammals, plant eaters have flat, grinding teeth; meat eaters have sharp, pointed teeth for cutting and tearing; plant/meat eaters have both kinds of teeth; in birds, beaks and feet are indicators.)

07
Db
Fa

Observing that animals breathe. Let students count the number of inhale–exhale movements of their chests in one minute. Now have them exercise for a few minutes by jumping up and down on one foot. Let them sit down and count their breathing rates again. Discuss the effect of exercise upon the body's need to obtain more air (oxygen). Next, see if they can detect breathing in other animals such as dogs, cats, fish, birds, frogs, and lizards. Determine the breathing rates for the different animals and, if possible, do so again after the animal has been active. Students will realize that animals seem to need more air (oxygen) after exercising.

08
Cc
Fa

Generalization II. There is a great variety of physical characteristics among animals without backbones.

Contributing Idea A. Most segmented worms have soft, round, segmented bodies with no legs (although they have appendages), are cold-blooded, and hatch from eggs.

Observing common characteristics of segmented worms. Observe several earthworms to determine similarities among them. Students will find that the animals a) have soft, elongated, round, segmented bodies; b) have a moist skin for body covering; c) have no legs; d) hatch from eggs. When touched, they feel cool. (They are cold-blooded—their body temperature is

09
Aa
Bb

nearly the same as the temperature of their environment.) Animals with these characteristics are *segmented worms.* About ten thousand species have been identified, most living in water.

Collecting earthworms. After a home has been prepared for keeping earthworms, collect them by digging into some rich garden soil covered with dead leaves. At night with a flashlight, in early morning, or after a heavy rain, earthworms can be easily picked up at the surface of the soil.

10
Ge

Housing and caring for earthworms. Fill a small wooden box with about 5 inches (13 cm) of rich garden soil. Dampen the soil and keep it moist but not wet. Place earthworms in the box and set some damp black construction paper or a half dozen sheets of newspaper loosely on the soil. Place the box in a dark location. Earthworms should be fed twice a week by lifting the paper and placing bits of lettuce, moist oatmeal, or moist bread on the surface of the soil. Other foods can be tested to see which ones the worms consume. Another type of container can be built by fastening two pieces of glass on wood strips *(a).* Cover the glass sides with black paper and set the container in a dark place for a few days. When the paper is removed, students will see how the earthworms burrowed underground *(b).* An aquarium or large wide-mouth jar can also be used.

11
Gd

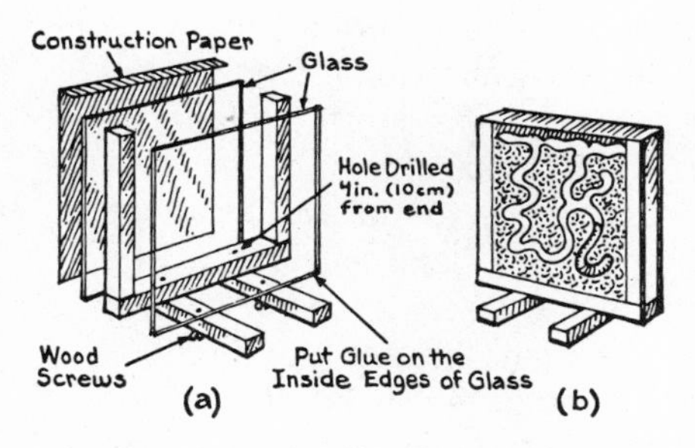

FIGURE 211.11

Contributing Idea B. Most arthropods have segmented bodies (protected by an external skeleton) with segmented legs, are cold-blooded, hatch from eggs, and live on land (some live in water).

Observing characteristics common among arthropods. Examine various animals from this group (phylum) to determine similarities among them. Students might observe a sowbug (crustacean), a spider (arachnid), and an insect. They will find that most of the animals in this group a) live on land (although some live in water); b) have a hard exterior body covering; c) have pairs of jointed, movable appendages for locomotion; d) have pairs of antennae; e) hatch from eggs. Animals with these characteristics are called *arthropods.* About 900,000 species have been identified.

12
Aa
Bb

Observing characteristics common among crustaceans. Crustaceans are a subgroup or class of arthropods. Students can examine various animals from this class to determine similarities among them. They might observe water fleas, a crab, shrimp, and lobster. They will find that most animals in this class a) live in water; b) have two pair of antennae; c) have one pair of legs per body segment; d) hatch from eggs laid in water. Animals with these characteristics are called *crustaceans.* Crustaceans include barnacles, crayfish, and sowbugs.

13
Aa
Bb

Collecting crayfish. Crayfish are fresh-water crustaceans often found under stones or in the mud of fresh-water streams and ponds. They can be caught by dragging a dip net over the muddy bottom or by fishing

14
Ge

with a small piece of raw meat attached to the end of a string. When the crayfish grabs the meat, slowly pull the string from the water. Usually the crayfish will be pulled out with it. Place the crayfish in a container filled with the pond water.

15 Gd

Housing and caring for crayfish. Crayfish are fairly large, and they demand much food. For these reasons they are better suited to a pond habitat than to a classroom aquarium. Plastic wading pools work well with small plant pots for hiding places. For a short period of time they can be kept in large jars or aquariums equipped with aerating pumps. Remove the crayfish to another container when feeding so that the water in which it lives will not become polluted. Feed them small pieces of raw meat or earthworms.

16 Ge

Collecting daphnia. Daphnia are freshwater crustaceans most often found in quiet sections of streams or stagnant pools that are green with algae and free from fish. They are easy to see, but are so small that they can only be caught with a very fine mesh dip net. Place them into a jar filled with the pond water.

17 Gd

Housing and caring for daphnia. Daphnia can be kept in jars filled with water containing algae. Daphnia eat bacteria and algae, but can also be fed once a week with boiled bone meal or mashed hardboiled egg yolk at the ratio of 1 ounce (28 g) of dry material to every 2 quarts (2 L) of water. Adding pulverized manure and placing the jar in sunlight will aid the growth of algae.

18 Ge

Collecting sowbugs. Sowbugs are land crustaceans found in dark, moist environments under boards, logs, and stones. They are related to pill bugs and wood lice. Pill bugs will roll into a ball when touched, but sowbugs will not.

19 Gd

Housing and caring for sowbugs. Sowbugs can be kept in almost any type of container, but are best suited for a woodland environment containing sticks and small rocks under which they can hide. They can be fed very small bits of apple, lettuce, raw potato, bread, or small insects.

20 Aa Bb

Observing characteristics common among arachnids. Arachnids are a subgroup or class of arthropods. Students can examine various animals from this class to determine similarities among them. They might observe spiders. They will find that most of the animals in this class a) live on land; b) have no antennae; c) have four pairs of legs for locomotion; d) hatch from eggs laid on land. Animals with these characteristics are called *arachnids.* Arachnids include scorpions and mites.

21 Ge

Collecting spiders. Spiders are arachnids and can be found almost everywhere—close to or far from structures, on or about shrubbery, smaller plants, and lawns, and near the soil. The easiest way to find them is to first look for a web. It has been estimated that there are thousands of spiders per acre of woodland and meadow and hundreds per house and garden. Different kinds live in different environments. Most can be captured by allowing them to crawl on a stick, then transferring the stick to a container. An orb-web spider (one that constructs a suspended, round web) can be trapped by placing a bag beneath the web then gently tapping the spider so that it drops into the bag. A funnel-web spider (one that constructs its web in the shape of a funnel) can be caught by inverting a jar quickly over the spider when it is at the front of its web and sliding a card under the mouth of the jar. Sweepnetting a grassland area will gather hundreds of spiders. A collection of weeds, rotted leaves, or loose soil in autumn will probably turn up many small spiders. To see them, put the materials in a white pan. The spiders will show up against the light background. In the fall, egg cases can be found. These are often brownish and look like small balls of silk, similar to a round cocoon.

Housing and caring for spiders. Most spiders can be kept in wide-mouth jars for fairly long periods of time. Cover each jar with a piece of cheesecloth or nylon stocking and hold it in place with a canning ring or rubber band. Spiders should be kept in separate jars. They have difficulty climbing on glass, so add some debris to the jar along with a few twigs for them to climb on. Spiders will usually not eat unless their food is alive, so be sure to feed them live meal worms, flies, or other small, soft-bodied insects. You can catch insects in a net swept over a weedy place or trap small flies, beetles, moths, and midges that are attracted to outdoor lamps at night and feed them to spiders. In the winter, you can raise insects such as flies, fruit flies, and mealworms in the classroom for food. Moisture is very important to a spider, so place a small moist piece of sponge or ball of cotton in the jar, and keep it damp.

22
Gd

Observing characteristics common among insects. Insects are a subgroup or class of arthropods. Students can examine various animals from this class to determine similarities among them. They might observe ants, butterflies, grasshoppers, beetles, and flies. They will find that most of the animals in this class a) live on land; b) have one pair of antennae; c) have three pairs of legs for locomotion; d) have one or two pairs of wings; e) hatch from eggs laid on land. If students watch them over time, they will see that most go through several stages to become adults. Ladybug larva are good to work with and easy to collect. Animals with these characteristics are called *insects.* Insects include termites, wasps, moths, crickets, and earwigs.

23
Aa
Bb

Housing and caring for insects. Almost any container can serve as a cage to keep insects for short periods of time. The size of the container should depend upon the size and number of insects collected.

24
Gd

a. *Containers for land-dwelling insects.* Insects can be housed in terrariums or various cages that are quickly and inexpensively made. The following illustrations give several suggestions from which others can be invented. The climate in-

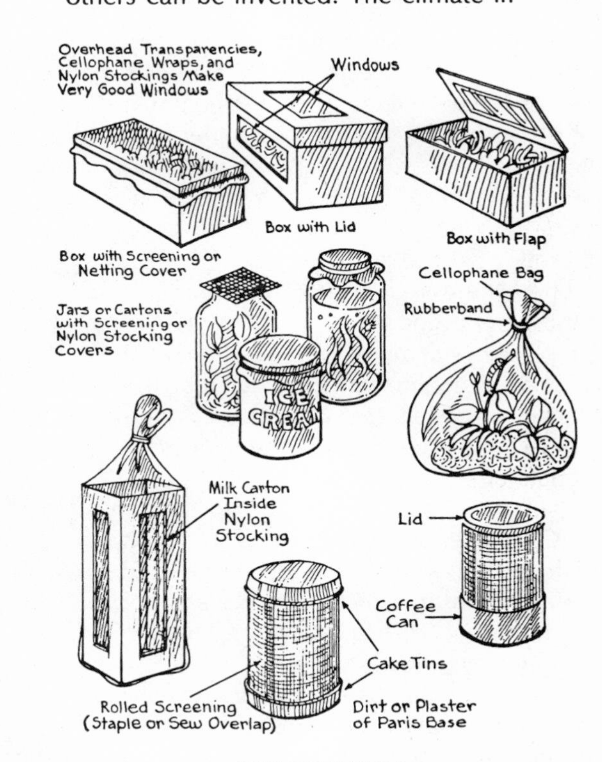

FIGURE 211.24

side a container is important. Try to make the temperature and humidity conditions similar to those where the insect naturally lived. Be sure the container is not placed in direct sunlight or near a heat source. Food can be placed in the bottom of most containers. Leaf-eating insects tend to need fresh food regularly—add fresh leaves every day or so. Other insects might eat fruit, pieces of meat, or fermenting materials. All containers should be kept clean, and any food that has not been eaten after a day should be removed and replaced. Water can be provided in most containers in several ways: a) a jar inverted on a sponge in a small dish or lid; b) a vial full of water plugged with cotton and lying on its side; c) a water-soaked sponge.

b. *Containers for water-dwelling insects.* Almost any glass container can be used to keep water insects. To keep insects such as mosquitoes or water boatmen from escaping, cover the container with a piece of nylon stocking, mosquito netting, or wire screening. The climate inside a container should be kept cool. Keep it from direct sunlight, and set it where there is a free circulation of air. As the water evaporates from the container, replace it with more water from the same pond or stream, or tap water that has been set aside for 24 hours. Water-dwelling insects can also be kept in an aquarium.

Contributing Idea C. Most mollusks have soft, shapeless bodies (protected by a limestone shell) with no legs, are cold-blooded, hatch from eggs laid in water, and live in water.

Observing characteristics common among mollusks. Observe a variety of animals from this group (phylum) to determine similarities among them. Possibilities are limpets, mussels, and clams. Students will find that most animals in this group a) live in water; b) have soft, shapeless, nonsegmented bodies protected by an external shell of limestone; c) have a broad single foot for locomotion; d) hatch from eggs. Animals with these characteristics are called *mollusks.* Mollusks include the common garden snail. About eighty thousand species have been identified.

25 Aa Bb

Collecting snails. Snails are tiny mollusks that can be found in water or on land. Water snails can be found in streams, ponds, or lakes. They can be picked from water plants or dredged from the bottom with a net. Land snails can be found feeding on plants almost anywhere. Look for them in gardens in the early morning or early evening.

26 Aa Ge

Housing and caring for snails. Water snails can be kept in any wide-mouth jar or aquarium filled with pond water and many water plants. Three snails for every gallon (4 L) of water make a balanced aquarium. Water snails will eat algae from the sides of the aquarium and some water plants. Students can watch them move across the glass and see their mouthparts work as they eat algae from the sides. Land snails can be kept on damp peat moss or soil in a glass jar or terrarium. One snail per quart (liter) container is best for balance. Land snails will eat many kinds of leaves.

27 Gd

Generalization III. There is a great variety of physical characteristics among animals with backbones.

Contributing Idea A. Most fish have scales on their bodies, are cold-blooded, hatch from eggs laid in water, and live in water.

Observing characteristics common among fish. Examine various animals from this group to determine similarities among them. Students might observe guppies, goldfish, and various tropical fish in an aquarium. They will find that most of the animals of this group a) live in water; b) have interior skeletons and scales for body covering; c) have fins for locomotion; d) breathe by gills; e) hatch from eggs laid in water. They feel cool to the touch because they are cold-blooded—their body temperature is nearly the same as the temperature of their environment. Animals with these characteristics are called *fish.* Explain that most live in salt water, some in fresh water, and a few in both types of water. About twenty thousand species have been identified. **28 Aa Bb**

Housing and caring for fish. Tropical, fresh-water and salt-water fish each require special aquariums (see *241.01*). Most fish can be fed commercial fish food, small pinches of oatmeal or cornmeal, or live food such as worms and brine shrimp. Do not overfeed the fish (three times a week is usually sufficient). If a fish becomes sick, immediately remove it, and place it in a separate bowl so that it will not contaminate the other fish. **29 Gd**

Contributing Idea B. Most amphibians have smooth or bumpy skin on their bodies, are cold-blooded, hatch from eggs laid in water, and live in water and on land.

Observing characteristics common among amphibians. Examine various animals from this group to determine similarities among them. Students might observe frogs, toads, newts, and salamanders. They will find that most of the animals in this group a) live part of their lives in the water and part on land; b) have an internal skeleton and a moist skin (without scales) for body covering; c) have two pairs of legs for locomotion; d) hatch from eggs laid in water. When touched, they feel cool because they are cold-blooded—their body temperature is nearly the same as the temperature of their environment. Animals with these characteristics are called *amphibians.* About 2,800 species have been identified. **30 Aa Bb**

Collecting salamanders. Salamanders can be found in damp, wooded areas under rotting logs, rocks, and matted leaves. After a rain they can be seen on the rocks and logs. Pick them up by hand on land or scoop them from ponds with a net. Place them in wide-mouthed jars filled with damp leaves and moss. **(Caution: Some salamanders have poison glands on their skins. Wash hands after handling.)** **31 Ge**

Housing and caring for salamanders. Place salamanders in a woodland terrarium made in a large jar or aquarium tank. Plant it with small ferns, moss, and woodland plants. Add rocks for cover and climbing and a container of water at soil level. A weighted wire screen over the top is necessary to keep the salamanders in the terrarium. Salamanders will eat small aquatic life, small insects and their nymphs, earthworms, and meal worms. If some do not eat, return them to where they were found. **32 Gd**

Collecting the eggs and tadpoles of frogs and toads. During the early summer, eggs can be found floating on or near the surface among plants in shallow ponds and marshy places close to shore. Frog's eggs are in jumbled masses of jelly, while toad's eggs are in jellylike strings often wrapped around plants. Use wide-mouth jars or nets to scoop up the eggs. Add several small water plants to the jars. Do not collect too many eggs, for many will die in crowded conditions. In **33 Ge**

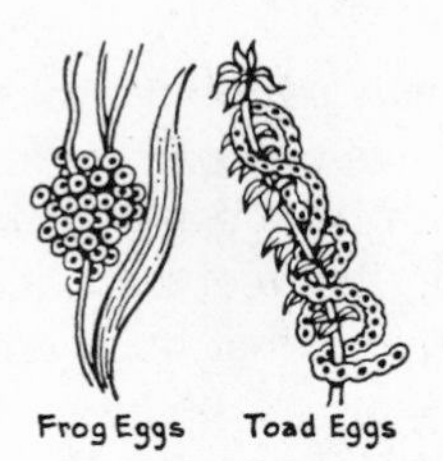

FIGURE 211.33

the spring and summer, frog and toad tadpoles can be found feeding among thick plant growths in calm water. Tadpoles move quickly, thus a large net and a quick hand are needed to scoop them with an upward motion from the water. They can be placed in large jars filled with water and plants taken from the location where they were found.

Housing and caring for the eggs and tadpoles of frogs and toads. Place eggs or tadpoles in large, wide-mouth glass jars or in a fishless aquarium filled with pond water. Include clumps of floating plants and several other water plants. These will serve as places for newly hatched tadpoles to hide. A large rock that extends above the surface of the water or other type of ramp must be available for tadpoles to crawl upon as they develop. Place screening over the top of the aquarium; hold in place with books or rocks. It might be necessary to change the water every few weeks and to clean out the container. Replace the water with more pond water or aged tap water. (Let tap water sit for several days so that the chlorine will evaporate.) Tadpoles will feed on algae in the aquarium or on small bits of lettuce. **34 Gd**

Collecting frogs and toads. Frogs and toads can be caught by hand or with a net at night near plants in shallow, calm waters. They can be located by listening for their calls. If they are in the water, scoop them from underneath with a large wide-mouth net. Both animals can be kept temporarily in a wet burlap sack. **35 Ge**

Housing and caring for frogs and toads. Keep frogs in aquariums that have places for them to climb on dry land. Keep toads in a woodland terrarium. Both will eat live worms, roaches, caterpillars, insects, and small aquatic life. Let students describe the frog as it eats. (Its tongue is hinged at the front of its lower jaw and is flipped out quickly; its eyes retract and bulge because the food is forced down the throat by an inward movement of the roof of its mouth.) **36 Gd**

> *Contributing Idea C.* Most reptiles have dry, scaly skin on their bodies, are cold-blooded, hatch from eggs laid on land, and live on land.

Observing characteristics common among reptiles. Examine a variety of animals from this group to determine similarities among them. Students might observe turtles, tortoises, snakes, and lizards. They will find that most of the animals in this group a) live on land; b) have an internal skeleton and a scaly skin for body covering; c) have two pairs of legs for locomotion; d) have claws on their toes; e) hatch from shelled eggs laid on land (rattlesnakes and garter snakes bear their young alive). When touched, they feel cool because they are cold-blooded—their body temperature is nearly the same as the temperature of their environment. Animals with these characteristics are called *reptiles.* Reptiles include alligators and crocodiles. About seven thousand species have been identified. **37 Aa Bb**

Collecting snakes. Snakes can be found in fields, woods, and near creeks. (**Caution:** **38 Ge**

Some snakes are poisonous so be sure they are properly identified before handling.) Nonpoisonous snakes are easily caught by hand. Those easiest to handle are garter, ringnecked, and hognosed. A snake can be gently, but firmly held just behind the head with one hand while the rest of its body is supported with the other hand. Snakes can be temporarily placed in old pillow cases with the tops tied securely.

39 Gd **Housing and caring for snakes.** Nonpoisonous snakes can be kept in large aquariums covered with weighted wire mesh that folds over the sides. Put a 2 inch (5 cm) layer of sand or gravel at the bottom, and embed a pan of water in the sand. Add some rocks and a sturdy branch on which the snake can climb. Place the container where the snake can receive sunshine, but do not let it get too hot. Snakes will not feed regularly, and some will not eat at all in captivity. If a snake does not eat for several weeks, release it in the area where it was found. Most snakes will eat live insects, worms, grubs, or frogs. Be sure to clean the container frequently to avoid unpleasant odors.

40 Ge **Collecting lizards.** Lizards can be found under logs or among rocks. They are fast runners and a net is helpful in catching them. They can be placed in an old pillow case temporarily.

41 Gd **Housing and caring for lizards.** Lizards can be kept in woodland or desert terrariums. Be sure they contain rocks, twigs, and living plants to climb on, and embed a small container of water in the soil. Lizards will not drink from a dish, but they will take water from the leaves of plants. Spray the leaves of growing plants in the terrarium each day. (Spray the lizard's skin too.) Lizards feed on living flies, moths, and meal worms. Keep the container clean of uneaten food, and set it so that it receives sunlight or artificial light during part of the day and maintains a temperature range between 70°F (21°C) and 80°F (27°C).

42 Ge **Collecting turtles and tortoises.** Turtles can be found in or near ponds, slow streams, and swamps. Tortoises are generally found in more arid regions. Both can be caught by hand. **(Caution: Some turtles and tortoises bite; some turtles carry diseases.)** *(Note: The desert tortoise is protected by law in some areas. Check local restrictions. Turtles purchased in pet stores should be certified as healthy and free from salmonella. Other turtles should be checked by a vet.)* The animals can be placed in a box temporarily.

43 Gd **Housing and caring for turtles and tortoises.** Most turtles can be kept in an aquarium tank filled with 3 inches (7 cm) or 4 inches (10 cm) of water. Include a flat rock or other ramp in the tank so the turtle can crawl onto it. Box turtles and tortoises can also be kept in a terrarium (woodland for some, desert for others). Embed a pan of water in the soil of the desert terrarium. Set the terrarium where it gets some sunshine and some shade. Do not let it get too warm. Feed turtles once a day with commercial turtle food, insects, earthworms, or bits of raw hamburger. Most turtles will eat out of the water, but the painted turtle *must* feed in water. Tortoises are vegetarians and eat less often. They will feed on slices of apple, berries, leafy vegetables, and lettuce. Be sure to change the water twice a week, and clean out uneaten food regularly. **(Caution: Unclean turtle environments can cause**

serious illnesses in humans. Also see caution in 211.42.)

Contributing Idea D. Most birds have feathers on their bodies, are warm-blooded, hatch from eggs laid on land, and live on land.

Observing characteristics common among birds. Observe a variety of animals from this group to determine similarities among them. Students might observe local birds as they feed. They will find that most of the animals in this group a) live on land and can fly; b) have an internal skeleton and feathers for body covering; c) have one pair of legs and one pair of wings for locomotion; d) have bills (without teeth); e) hatch from shelled eggs laid in nests. If live birds can be handled, they will feel warm because they are warm-blooded—they maintain a steady body temperature different from their environment. Animals with these characteristics are called *birds.* About 8,600 species have been identified.

44
Aa
Bb

Housing and caring for birds. It is not advisable to confine birds in the classroom, however, if a cage is temporarily necessary, the size is of great importance. For two small birds such as canaries, budgies, or finches, the minimum size should be about 20 inches by 10 inches by 15 inches (50 cm by 25 cm by 40 cm). For larger birds such as mynas, the minimum should be 25 inches by 20 inches by 15 inches (65 cm by 50 cm by 40 cm). Tame birds will eat commercially prepared bird foods, lettuce, carrots, apples, and pieces of bread. Provide fresh water daily. Caged birds also need a separate container for mixed grit and cuttlefish bone for roughage. Remove all perishable food at the end of each day. The cage floor should be covered with fine sand or paper and changed frequently. *(Note: It is illegal to disturb wild birds or their nests.)*

45
Gd

Making bird feeders to attract wild birds. Bird feeders are easy to build and offer students many opportunities to observe and study wild birds. Stock the feeders with commercially prepared wild-bird food, various seeds, grains, apples, bread, raisins, cracked corn, oats, peanuts, sunflower seeds, or suet. A bread pudding can be prepared by heating suet until it liquifies, stirring in raisins, unsalted peanuts, and various kinds of seeds as it cools, then pouring it into paper cups or over pine cones as it thickens. The cups or cones can be suspended from trees to attract birds.

46
Ec
Gc

a. *Simple feeders.* Hang pine cones, coffee cans, gourds, shallow boxes, half a coconut, or a whole coconut. Imagination will enable students to create other simple feeders.

b. *Tray feeders.* Place food on open trays by windows or on stands. Tray feeders should have upturned edges to keep the food from falling off. Water containers can be placed on tray feeders. A covered tray on a pulley clothesline makes a feeder that can be brought in for restocking.

c. *Log feeders.* Drill holes in an old log or block of wood to contain food. Add pegs on which birds can perch.

d. *Basket feeders.* Tape two slotted soap

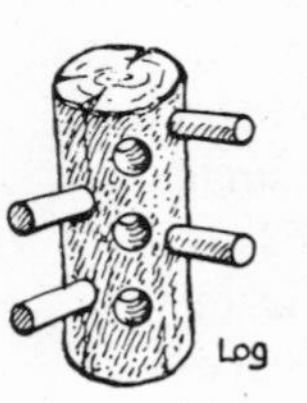

FIGURE 211.46

dishes together or fold wire mesh into a box shape and fill the interior with suet.

Students can keep records of the number and types of birds feeding each day and/or the time of day each type feeds. They can also keep longitudinal records to see which kinds of birds are attracted to the feeders during different seasons of the year. Experiment by varying the food to see which kinds of birds are attracted to particular types of food (e.g., seed-eating birds such as sparrows are attracted to chicken feed, weed seeds, and bread crumbs; fruit-eating birds such as robins and cedar wax wings are attracted to berries and various fruits; insect-eating birds such as chickadees, jays, and creepers are attracted to suet, nuts, peanut butter, and sunflower seeds). Also experiment by coloring the food used to see if birds show a preference for certain colors. *(Note: If you start a feeder in the fall, continue through the winter because birds will depend upon it.)*

Contributing Idea E. Most mammals have some fur or hair on their bodies, are warm-blooded, bear their young alive, provide milk for their young, and live on land.

Observing characteristics common among mammals. Observe a variety of animals from this group to determine similarities among them. Students might observe mice, hamsters, dogs, cats, and rabbits. They will find that most of the animals in this group a) live on land; b) have internal skeletons and hair or fur on their skin for body covering; c) have one or two pairs of legs for locomotion; d) have glands that produce milk to feed their young; e) bear their young alive. Animals with these characteristics are called *mammals.* Mammals include deer, horses, elephants, monkeys, and people.

47
Aa
Bb

Housing and caring for small mammals. Guinea pigs, hamsters, white rats, rabbits, and similar small animals can be raised in the classroom. It is best to obtain such animals from a reliable source (e.g., laboratory) to be sure they are not diseased. Also be sure to check any local restrictions on keeping such animals in the classroom. For most, a large wooden box with a wire mesh cover is sufficient. (Hamsters will gnaw their way out of a wooden box and are better kept in a wire cage.) Cedar shavings or shredded newspapers should be placed in the bottom of the cage and replaced daily. Water should be placed in a heavy dish to keep it from being overturned or a water-drip bottle can be made or purchased from a pet store. Most small mammals will eat commercial pellets, lettuce greens, and carrots. Hamsters will also eat grains such as corn, oats, wheat, and fresh fruit. Provide straw, hay, or cotton for animals that like to nest.

48
Gd

Generalization IV. Animals can be organized by their characteristics.

Contributing Idea A. Animals can be seriated.

Seriating animals. Obtain a number of seashells of the same kind. Have students arrange the shells from smallest to largest or by some sequence in their color patterns. Some students might research factors about animals to seriate them. For example, animals can be ordered from slow moving (snail or slug) to fast moving (cheetah or swift), from small land animal (shrew) to large land animal (elephant), small water animal (gobie) to large water animal (whale or

49
Da

shark), and small flying animal (hummingbird) to large flying animal (eagle or condor).

Contributing Idea B. Animals can be classified.

Classifying animals by observable characteristics. Prepare a set of ten pictures of birds or sets of other animals. Ask students to note differences in coloring, marking, size, shape, and other features. Next, ask them to note at least three ways in which the animals are alike. Now ask each of them to bring one picture of some animal to school. Divide a bulletin board into four sections with a picture of one animal in each (selected for diverse characteristics). Ask the students to place their pictures in the section where the posted animal is most like theirs. As pictures are placed, have students give their rationales. They can discuss the characteristics of the animals in each group. When finished, have them divide one group of pictures into two or more groups. The subdividing process can continue as long as the characteristics of the animals suggest further groupings. **50 Ca Dc**

212 INTERACTIONS

Generalization I. Reproduction is a process by which animals produce offspring.

Contributing Idea A. Some animals hatch from eggs.

Observing the eggs of animals. The eggs of many animals such as frogs, toads, snails, and insects can be kept in aquariums or terrariums in the classroom and watched over time. Under the proper conditions, the eggs will hatch, and students will be able to see the development of the young animal. **01 Aa**

Contributing Idea B. Some animals do not hatch from eggs.

Researching animal reproduction. Have students tell of experiences they might have had observing or hearing about the birth of kittens, puppies, rabbits, or other mammals. It is possible for students to observe the birth of guinea pigs, gerbils, or white rats kept in cages in the classroom. You might wish to discuss human birth, and let students research information about various animal live births. Guppies are among the few fish that give birth to their young alive. Although it is difficult to see the actual live birth of guppies, students can realize that the new babies in an aquarium were born alive. **02 Bd**

Generalization II. Animals respond to certain environmental conditions.

Contributing Idea A. Animals respond to touch.

Observing that earthworms respond to touch. Tap the head of an earthworm gently with a finger or pencil eraser. (The worm will contract and move backward.) Students will realize that the worm seems to be sensitive to things that touch it. Let them try touching other parts of the worm to see if it is equally sensitive in all places. They can also set an alarm clock with its face side against the soil where worms are located. When the alarm is set off, worms will come from their burrows. Explain that earthworms do not hear sound as we do—they feel the sound vibrations from the ringing. (Students can touch the clock lightly to feel the vibrations.) Explain that the vibrations make the **03 Aa**

earthworm feel uncomfortable, and it attempts to get away from the feeling.

Observing that snails respond to touch. With a pencil, gently touch the eyes at the ends of the extended stalks of a snail. (The stalks will retract.) Touch the snail in various places and observe what happens. If the snail is overly disturbed, it will retreat into its shell, Now prepare a board by cutting 3 or 4 holes in it. The largest hole should be able to accommodate the snail's shell easily. The smaller holes should accommodate the snail's body, but not the shell. Place the board in the path of a moving snail, and let students see how the snail is able to select the right-sized hole to travel through.

04
Aa

Contributing Idea B. Animals respond to sound.

Observing that some animals respond to sound. Set two snails near each other and make a loud noise (e.g., fire a cap pistol, suspend an alarm clock in the air and ring it). Observe the behavior of the snails. This can be repeated using other animals such as fish (use a cricket-clicker in and out of the water), insects, amphibians, and so on. *(Note: Some animals will "feel" the sound vibrations rather than "hear" the sound as we do. See 212.03.)*

05
Aa
Fa

Contributing Idea C. Animals respond to smell.

Observing that earthworms respond to smell. Dip a piece of paper into vinegar or household ammonia. Bring the paper to within an inch (2 cm) of an earthworm's head but do not touch the worm. Students will see an immediate reaction and can realize that the worm seems to sense odors. Try bringing the paper near other parts of the worm to see what happens. Also, try other aromatic substances.

06
Aa

Observing how well people detect smells. Hide an aromatic substance, such as an open perfume bottle, in the classroom. Have students use their sense of smell to track it down. Next, have them pretend they are bloodhounds. Sprinkle a trail of 1 inch (2 cm) squares of blotter paper, soaked in clove oil, across a field. Blindfolded students can take turns trying to follow the trail on their hands and knees. Others can watch, and discuss the problems the "bloodhound" student has. You can vary this activity by letting the trail circle across itself or by crossing it with a different scent. Make analogies between this experience and the tracking done by animals such as dogs.

07
Ad

Contributing Idea D. Animals respond to gravity.

Observing that snails respond to gravity. Let snails climb a board, and observe whether or not the snails always climb upward. When a snail reaches the top of the board, turn the board over, and observe again. Keep the board vertical, and test by rotating it 45° or 90° each time. Next, tilt the board various degrees to see what effect the angle of tilt has upon the direction the snail travels. For each test, students can record observations and draw conclusions.

08
Aa

Contributing Idea E. Animals respond to light.

Observing that some animals respond to light. If students look carefully, they will find that worms do not have eyes, yet they seem to be sensitive to light. For example, if worms are kept in a damp box of soil and covered with moist newspapers for several

09
Aa
Fa

hours, some worms will be found on the surface of the soil when the newspapers are removed. Students will see them move rapidly into burrows or begin burrowing when the light strikes them. Try placing a worm on a sheet of damp newspaper and cover it with a box so that it is in complete darkness for an hour. Darken the room, remove the box, and turn a flashlight with a small beam on the head end. To create a small beam, place a cone-shaped piece of paper over the flashlight, snipping off the pointed end. Students will see an immediate reaction and a movement away from the light. Let them test the light beam against other parts of the worm's body. Test other animals in a similar way to see which are most sensitive to light.

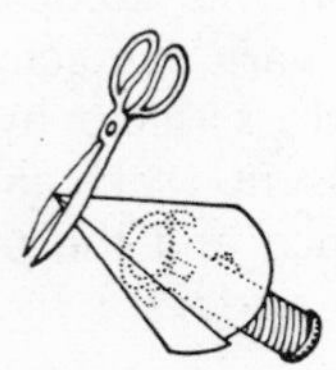

FIGURE 212.09

Contributing Idea F. Animals respond to temperature changes.

Observing that some animals respond to cold. Chip a piece of ice and use a pointed tip to touch an earthworm in various places to see which parts of its body seem most sensitive and least sensitive to cold. Next, chill the worm in a container of soil in a refrigerator for ten minutes and observe how it moves on the surface of the soil. Students can determine how cold influences the worm's movement. Challenge them to create ways to test the effect of cold on other animals such as grasshoppers or crickets.

10
Aa
Fa

Observing that some animals respond to heat. Near one end of a terrarium, place a heating unit or lamp with the bulb turned away from the interior. Put two snails near the heat source, two midway in the terrarium, and two as far from the heat as possible. Observe whether the snails move toward or away from the heat. Other animals can be observed in a terrarium in a similar way. Water snails can be tested by placing one in a can of water, lighting a candle, and holding it beneath the snail. Observe the snail's reaction. Do not leave the candle in one position too long for the heat might injure the snail.

11
Aa
Fa

Experimenting to see the effect of temperature changes upon fish. Place a goldfish into a wide-mouth jar half-filled with water. Place a thermometer in the jar so that it can be read without disturbing the fish. Record the temperature of the water and the number of gill beats of the fish in one minute. Add crushed ice to the water very slowly to avoid exciting the fish. Continue adding ice until the temperature is near freezing (32°F or 0°C). Again count the number of gill beats in one minute. Next, add warm water to the jar slowly to avoid exciting the fish until the temperature is raised by 5°. Record the gill beats per minute. Continue adding warm water by 5° intervals until the temperature reaches 86°F or 30°C. By studying the gathered data, students will be able to tell what happens to the fish's respiration rate as temperatures increase or decrease.

12
Cc
Ce
Ec

Observing hibernation. Place a frog in a large jar of water. When the frog floats at the surface, count the number of moves its throat makes in a minute to determine its rate of breathing. Now add ice to the water,

13
Aa
Bb
Fa

one piece at a time. As the water cools, students will see the frog move to the bottom and become less active. When this happens, determine the rate of throat movements again. When nearly all motion stops, tell students that the frog is in a state of *hibernation.* (The heartbeat and respiration are very slow.) Explain that, to avoid being frozen, frogs hibernate during the cold winter by digging beneath the soft mud of a pond or stream. They breathe under the water by taking in small amounts of oxygen through their skin. For an additional experience with hibernation, put a different frog in a jar containing a small amount of water. Set the jar in a refrigerator (not the freezer) overnight. When the frog is removed, it can be turned onto its back and its movements observed. Keep track of the time it takes the frog to become active again. Some students might be interested in researching reference books to find out what other animals hibernate during winter.

220 plants

221 CHARACTERISTICS

Generalization I. Plants have identifiable characteristics.

Contributing Idea A. All plants reproduce their kind.

Observing seeds produced by plants grown from seeds. Let students examine several different kinds of seeds (e.g., bean seeds, tomato seeds) and draw pictures of them. Plant the seeds in potting soil. After germination, continue growing them until they flower and produce new seeds. Examine the new seeds and compare them with the pictures drawn previously. Students will realize that the plants reproduced their own kind. Plant the new seeds and observe their growth. *(Note: Direct sunlight or a special artificial lighting system is necessary if plants are to be grown in the classroom.)*

01
Aa
Be
Ca

Observing plants grown from spores. Fern spores can be planted on the outside of a moist flowerpot to let students see how new ferns develop. Begin by letting a fern frond that has spores on it dry out for several days. Obtain a clay flower pot and wash it thoroughly. Next, soak it for several hours, changing the water several times to remove minerals in the clay that might keep the fern spores from developing. Fill the pot with wet newspapers or paper toweling, and invert it in a saucer of water. (The newspapers will keep the flowerpot moist.) Shake the dry leaf over the pot so that some of the spores drop off. Cover the pot and saucer with a glass jar and set it in a dimly lit place at room temperature. Add a little water to the saucer whenever the surface of the pot begins to dry out. In three to four weeks, the spores will form small green spots. Tell students that these are the sexual plants and are called the *prothallia.* The larger *prothallus* will be about ¼ inch (1 cm) in diameter. These structures contain the egg-producing and sperm-producing parts of the fern. As each egg is fertilized by a sperm cell, it begins to grow into a new fern plant. In about eight weeks, the prothallia will shrivel, leaving a small fern in their place. The fern can be transplanted to a pot of rich soil if care is taken not to damage its fragile roots.

02
Aa
Bb

Contributing Idea B. All plants grow.

Germinating seeds and measuring growth. Sprinkle grass seeds in a small box of potting soil. Be sure the seeds are spread sparely. Soon after they germinate, have students note that the plants are thinly spaced. Continue observations for several weeks. Students will find that the grass not only grows taller but also grows more blades over time. Challenge them to devise a way

03
Aa
Cc

to measure the increase in density of the grass (e.g., blades of grass might be counted periodically within a small defined area).

Observing growth from spores. Obtain a fern frond that has spores on it. Let it dry out for several days. Place some rich soil into four pots, and cover it with ½ inch (1 cm) of fine, washed sand. Pour boiling water over the sand and pots to sterilize them. (Bacteria and mold spores are difficult to keep out of the developing fern spores—sterilization and four pots increase the probability of success.) Shake spores from the fern frond over the soil, and cover each pot with a sheet of glass. Set the pots in saucers of water and keep them moist in a shady location. Students will see the plants develop through two stages in the following two to eight weeks. **04 Aa**

Contributing Idea C. All plants need food, water, and oxygen to grow.

Experimenting to see the effect of nutrients in soil upon plant growth. Prepare five soil containers. Into the first, put only garden soil. Into the second, put soil plus fertilizer in the quantity recommended by the manufacturer. Fill the third, fourth, and fifth with soil and one-fourth, one-half, and three-fourths the amount of fertilizer respectively. Plant three pea seeds in each container and water them equally. If the heights are measured daily, the data can be graphed. Continue taking measurements past the seedling stage. Students will find that the amount of nutrients in the soil influences the growth of the plants. Many variations of this activity can be tried. For example, students can compare the addition of nutrients to sand, vermiculite, or sterilized soil (soil that has been baked in an oven), compare soils from different locales, or compare different types of fertilizers. **05 Bf Ec**

Comparing the growth of watered and unwatered plants. Plant equal amounts of radish and lima bean seeds in potting soil in each of 6 pots. Place all but one pot in quart (liter) milk cartons. Prepare one pot so that it gets no water and increase the amount of water in several increments in each of the four other pots so that the soil in the last pot is standing underwater. For the last pot, seal the drain holes in the bottom of the pot, or set the pot into a larger container filled with water. Cover the tops of each carton with a plastic bag held in place with rubber bands. (This will keep the water from escaping and reduce the need for watering during the test.) Place all the containers where light can enter them. Set the uncovered plant near the others for comparative purposes. Give this plant normal watering. Observe the plants on a regular basis. Students will find that water is necessary for growth, but that too much water is harmful. Explain that too much water blocks the flow of air to the plant's roots, and plants need oxygen in the soil for growth. **06 Ca Ec**

Contributing Idea D. All plants are sensitive to their surroundings.

Experimenting to determine the environmental factors to which plants are sensitive. Grow seed or seedless plants in a terrarium. Let students devise ways to find out the effect of different factors on the growth of the plants. Some factors they might test are temperature, light, water, soil nutrients, and wind. As they devise their tests, help them to control all factors except the one they are testing. If numerical data are kept regularly (e.g., height of growth **07 Ea Eb Ec**

each day), the results can be placed on a graph.

Generalization II. There is a great variety of physical characteristics among seedless plants.

Contributing Idea A. Fungi live off other plants, animals, and dead organic matter.

Observing mushrooms. Mushrooms are the best known fungi and can be found in woods, fields, and orchards. They appear quickly after warm spring and autumn rains. Students can see that a mature mushroom consists of a stalk and a cap, the latter opening after the fungi pushes through the soil or decaying log. If they examine the underside of the cap, they will see numerous plates, called *gills,* radiating from the center. On the outside of each gill are hundreds of cases containing *spores.* These spores can be examined with a microscope. It has been estimated that a single mushroom produces as many as ten billion spores. **(Caution: Explain that some mushrooms are poisonous and that there is no certain rule to tell which are edible.)** Students might compare mushrooms with other types of fungi, such as mold and mildew.

08
Aa
Bb

Making mushroom spore prints. Students can make prints from mature, fresh mushrooms. After picking, a mushroom should be placed in a plastic bag and printed as soon as possible. To print, carefully remove the stem from the cap and place the cap with its spore-bearing side down over a piece of paper. Since spores are usually the same color as the underside of the cap, it is best to use white paper if the underside is pink or brown and dark paper if it is light-colored. For the best results, cover the cap and paper with a glass bowl, cake cover, or jar to keep the spores from being blown off by air currents. Let it stand overnight. When the cap is removed from the paper, a spore print should remain. Spray immediately with clear hair spray or other spray fixative to preserve it. *(Note: Do not hold the spray can too close to the spores or the force will disturb the print.)*

09
Gd

Contributing Idea B. Ferns and mosses are simple plants.

Collecting ferns and mosses. Ferns and mosses are primitive seedless plants found in wooded areas, near streams, and in other damp, shady places. When collecting mosses, it is important to dig out their root-like structures along with some soil. These structures, called *rhizoids,* absorb moisture and nutrients from the soil. Wrap them carefully in damp paper toweling to prevent drying, and keep them from direct sunlight until they can be transplanted.

10
Bb
Gd

Housing and caring for ferns and mosses. A variety of ferns and mosses will grow with a minimum of attention in a classroom terrarium. Prepare the terrarium in a large jar or aquarium tank. Place 1 inch (3 cm) layer of gravel or crushed rock on the bottom to provide proper drainage. Add ½ inch (2 cm) of sand, then a generous layer of garden soil or humus. Arrange the soil to provide a higher terrace at one end. Rocks can be placed on top of the soil and ferns and mosses planted. When transplanting, firmly embed the plants into the soil. Let students create their own landscaping patterns. Now add water to the low end of the terrarium until the gravel is nearly covered. (This should provide adequate moisture for the soil.) Next, cover the terrarium with a sheet of glass to prevent evaporation and to main-

11
Gd

tain a humid condition inside. Check occasionally to be sure the terrarium remains moist. If it becomes too wet or if molds appear, remove the cover to let it dry out for a while. (Powdered sulfur sprinkled into the container will help prevent the growth of molds.) Set the terrarium where it will receive a medium amount of light; avoid direct sunlight and overheating. A closed terrarium will maintain itself for a very long time.

Generalization III. There is a great variety of physical characteristics among seed plants.

Contributing Idea A. Seeds have similar and dissimilar characteristics.

Examining the external parts of seeds. Remove lima beans or peas from their pods, and notice the rough portion on the seeds where they were attached to the pod. Tell students that this marking is called the seed *scar*. Have them find the seed scars on other seeds. Now place the seeds in warm water, and note the air bubbles coming from a tiny hole near the seed scar. You might tell them that this tiny hole is called the *micropyle* and that it is the main point through which water can enter the seed to enable it to begin its growth. Other seeds can be tested in a similar way to locate their micropyles. From their observations, students should infer that seeds contain air. Let the seeds continue to soak, and observe them at one-hour intervals. As the water enters the micropyle, the seed swells. A lima bean seed will be well soaked in about three hours. Compare the soaked seeds with dry ones in terms of size, color, shape, and texture.

12
Aa
Bb
Fa

Examining the internal parts of seeds. Soak various seeds overnight, then let students dissect them. They will find two main classes of seeds. Most seeds divide into two parts and are called dicotyledons (e.g., beans, peas, peanuts). Those that do not divide are called monocotyledons (e.g., corn, oats, coconuts).

13
Bb
Ca

a. *Dicotyledons.* Distribute soaked lima bean seeds to each student. Place the seeds on paper towels or plates so that the parts will not get lost. With a pin, toothpick, or dissecting needle have each student carefully peel the cover *(seed coat)* from the seed. The protective-covering purpose of the coat can be discussed. Now open the seeds into two portions. (Each portion is called a *cotyledon.*) Find the tiny plant *(embryo)* inside. Hand lenses can be used to observe the leaves, stem, and root of the embryo. Similar seeds can also be soaked and compared with the lima bean seed. For example, the familiar unshelled peanut can be dissected to find the same parts—the reddish brown seed cover, the two seed halves, and the plant.

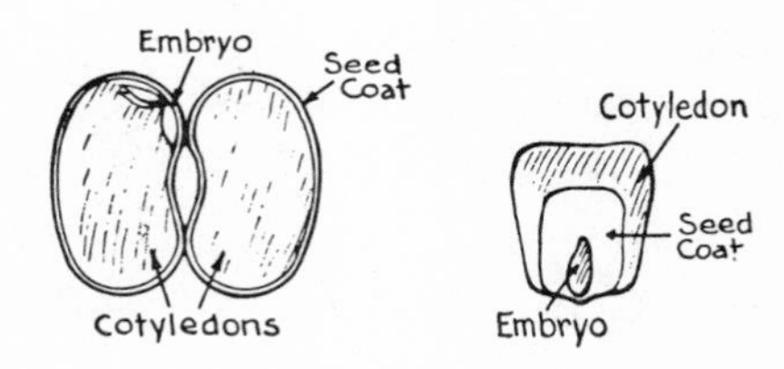

FIGURE 221.13

b. *Monocotyledons.* Distribute soaked corn seeds. The seed coat can be removed with a sharp implement. Beneath the coat, students can see the embryo.

Contributing Idea B. In most seed plants, the root grows at the base of the stem.

Determining the location of root growth on stems. Germinate various seeds in a soilless garden (see *441.01*). Observe the root growth, and have students note that it is at the base of the plant stem. By examining many kinds of plants, they can infer this commonality among most plants. In their observations they may note one root that appears larger than the surrounding or branching roots. Tell them that this is the main or *primary root,* but that not all plants have primary roots. Have them describe how the primary root differs from the other roots in location, appearance, and texture. The three basic groupings for root systems can be used as a guide for students as they identify and match roots in the classroom.

14
Aa
Bb
Fa

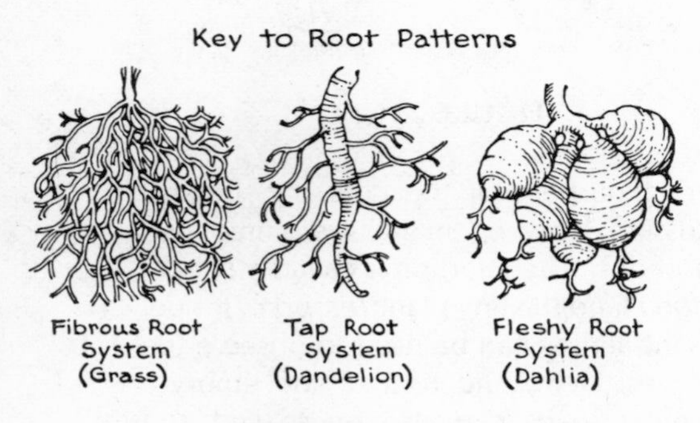

KEY 221.14

As a supplemental activity break open a pot in which a plant has been growing for some time. The mass of roots can be studied. Also, seeds can be planted in soil next to the side of a glass jar. The jar can be wrapped in black paper to keep sunlight from the roots. After a week, the paper can be removed and the roots observed.

Contributing Idea C. The external characteristics of a stem imply an order to its parts and appendages.

Observing stems. With careful observation, much can be inferred from the details of the external characteristics of a woody stem. Distribute a woody stem, such as a willow or horse chestnut, to students. Let them observe and describe all the characteristics they see. Although the important characteristics are labeled in the illustration, the

15
Aa
Ba
Fa

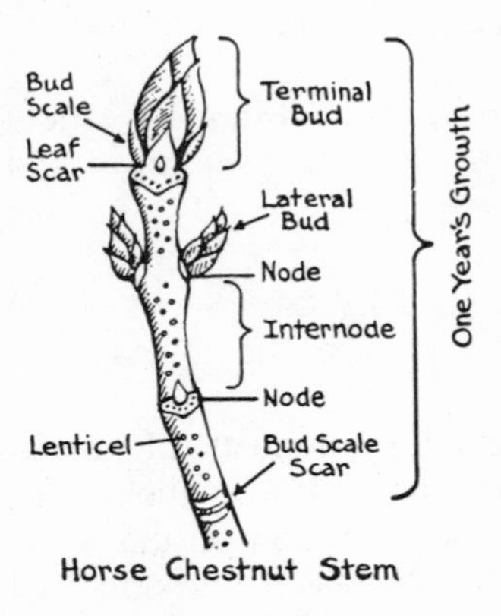

FIGURE 221.15

students' discoveries of the parts and the descriptive words they use are of utmost importance. Among the significant points that can be realized from the examination of a number of different stems might be: the roots are found at the base of the stem, stems usually grow upward, and a large bud *(terminal bud)* is found at the tip; there are usually enlargements on the stem *(nodes)* where leaves occur; there are spaces between leaves *(internodes)* of various lengths; where leaves have fallen, leaf scars remain on the stem (students can remove leaves to observe the place where the scar will form); there is a tiny bud *(lateral bud)* above each leaf or leaf scar (the bud remains when a leaf falls); pores *(lenticels)* can be seen on the internodes of some stems such as birch, horse chestnut, and cherry; the bud scale indicates where new growth began (students can tell the age of a branch by counting the growth areas between the bud scales). From observing these characteristics of various stems, students might be able to draw some inferences and generalizations about the parts of stems: between the base and tip of most stems are enlargements (nodes) separated by smooth stem sections (internodes); between the base and tip are leaves and buds with the bud always preceded or followed by a leaf or leaf scar; a leaf scar always denotes the place where a leaf has been; a bud scale marks the place where new growth began.

16 Fb

Determining order in stem characteristics. By observing various stems, students will realize that all plants have an ordered relation of parts. A base, denoted by roots *(R)* precedes a leaf or leaf scar *(L),* which precedes a bud *(B),* which precedes a tip or terminal bud *(T).* Using this formula *(RLBT),* students will be able to determine the direction of any stem. To check their understanding of the relationship of the parts, bring to class a willow or other woody stem with the leaves removed. See if students can determine the direction of the stem and provide evidence for their judgment.

Contributing Idea D. Stems have definite external characteristics.

17 Bf Ca Fa

Comparing the external characteristics of stems. Have students remove new plants from the soil and observe them to determine where a stem begins (above the ground or below the ground, etc.). Then bring to class a variety of stems to compare for similarities and differences (size, covering, etc.). The various stems can be grouped by tallying characteristics on a histogram. Encourage students to use their senses of touch and smell as they observe. When the tallies are completed, they can interpret the results to see which shape is most common, which texture is least common, which color is the rarest, and so on.

Contributing Idea E. Stems have definite internal characteristics.

18 Aa Fa

Observing the annual rings in woody stems. A Christmas tree stem can be cut into segments and distributed to students, or a freshly cut tree stump can be used for this purpose. From observations and inferences, students can discover that: woody stems increase in diameter; growth slows down or stops during cold and dry seasons (variance in spacings between rings indicates climatic effects upon the growth of the plant); rings that are not concentric (wider on one side than the other) indicate that the plant grew on a slope or hill with the wider rings being

HISTOGRAM 221.17. Comparing the External Characteristics of Stems

Count	*Color*			*Texture*				*Length*		*Shape*					*Support*			*Straight*	
	green	brown	other	smooth	rough	fuzzy	other	short	long	round	flat	ovular	other	leaves	flowers	branches	other	yes	no
25																			
20																			
15																			
10																			
5																			

on the downhill side; light and dark portions within the same ring indicate seasonal differences.

Comparing grains in different woody stems. Gather samples of various kinds of wood such as pine, ash, birch, and redwood. Cut some of them crosswise and some lengthwise to let students observe and compare the grains and color differences. Cut ends can be shellacked for preservation.

19
Ca

Contributing Idea F. Leaves are appendages of the stem and are arranged in different ways upon the stem.

Determining the location of leaf growth on stems. To help students realize the positional relationships between the leaf and stem parts, have them observe potted plants, plants growing in the school area, or cut stems with leaves attached. Many different kinds of plants should be observed so that patterns can be realized. A few suggestions concerning possible observations are given here. By carefully observing, students will see that:

20
Aa
Ca
Fa

a. The leaf is usually supported by a stalk *(petiole)*. In a few plants (all grasses) the leaf may be attached directly to the stem.

b. At the base of the petiole, on either side of some plants, there are two appendages *(stipules)*. In most plants these are absent.

c. Leaves occur at the nodes of a stem. If there is one leaf at each node, the leaves form a spiral or alternate pattern. If two leaves are at a node, they are always opposite. If there are more than two leaves, they are whorled. It is not unusual for plants to have combinations of the different leaf arrangements.

d. Buds are always above where the leaf meets the stem, but some buds may be so underdeveloped that they cannot be seen. Although botanists use certain terms for the various observable parts and arrangements, it is best to let students describe what they see in their own terms.

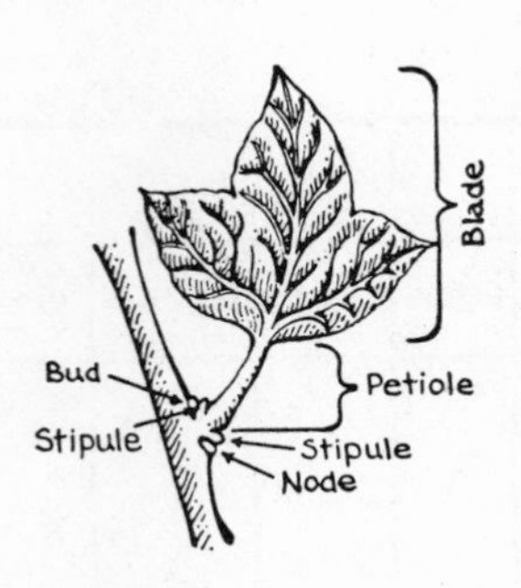

FIGURE 221.20

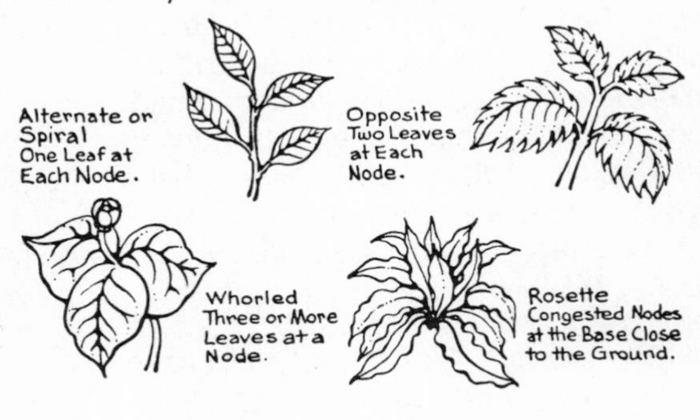

KEY 221.20

Contributing Idea G. Flowers are appendages of the stem and are arranged in different ways on the stem.

Observing the structures of flowering plants. Students can discover that leaves and flowers each appear at a node on a stem; thus the two can be related as both being appendages. Closer examination will

21
Aa
Cd

reveal, however, that the flower appendages differ from leaf appendages in having no buds in their axils. If observations and measurements can be made periodically, students will realize another difference—that the internodes of leaf appendages elongate over time, while the internodes of flower appendages do not.

Comparing the arrangements of flowers on stems. Bring to class different kinds of flowers from the eight types shown. Often local dry weeds will best show the various structures. Students can work in small groups for this activity. Although they generally realize that some flowers are produced only at the tip of a flower stalk (e.g., poppy), they will readily discover other common arrangements (types of inflorescences) from observations of the many examples.

22
Ca
Dc
Fa

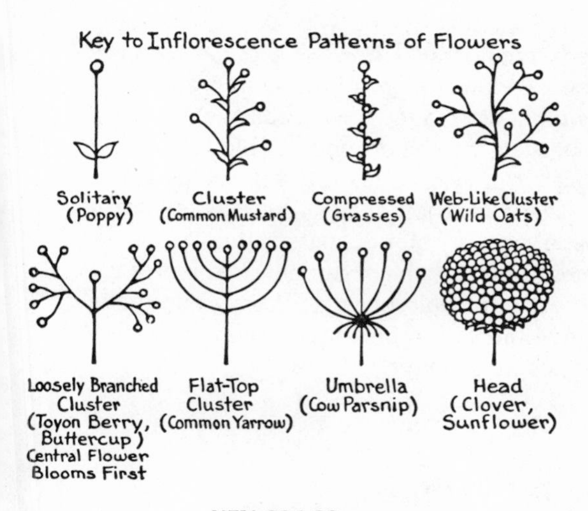

KEY 221.22

The inflorescence types can be used as a basis for grouping or classifying other flowers that are brought to class. You might explain that inflorescence, or flower arrangement on a stem, is thought to be an advantage to a plant, for the grouping together of many tiny flowers increases the coloration or brightness (e.g., geranium). The bright coloring tends to attract animals, aiding in pollination.

Generalization IV. Plants can be organized by their characteristics.

Contributing Idea A. Plants can be seriated.

Seriating seeds. Have students collect fruit of different types. Count the seeds contained in each, and arrange the findings in some numerical order. In some fruit there may be only one or two seeds (e.g., carrots produce only two seeds per fruit), in some there may be several (e.g., peas, beans), while in others there may be many seeds (e.g., tobacco produces more than 40,000 seeds per fruit). Students can also determine whether all fruit from the same plant produce the same number of seeds.

23
Cc
Da

Making a histogram of the numbers of seeds in pods. Obtain a large bag of pea pods from a grocery store. Have each student remove two pods from the bag. Open each pod and count the number of peas in each. Paper plates are helpful in keeping pieces of the pods off the desks. Ask who found the least number of peas in a pod and write the number on the chalkboard. Ask who found the most and write the number. Draw a horizontal base line between the two numbers and divide the space into equal segments. For example, if the range for the number of peas in the pods was from 3 to 12, prepare a base line with equal segments labeled 3, 4, 5, 6, 7, 8, 9, 10, 11, 12. After the line has been prepared, open a pea pod and count the number of peas in it. Put an X on the board above the numeral that

24
Bf
Da

represents the number of peas in that pod. Repeat for a second pea pod. After this demonstration, let students come to the board and mark the number of peas found in each of their pods. The result will form a histogram similar to the one shown. From

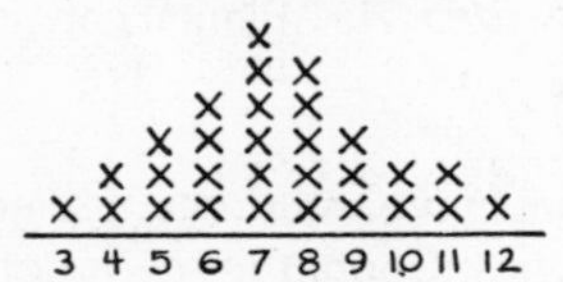

HISTOGRAM 221.24

the results, questions that will lead to inductive thinking can be asked: "If you were to pull one more pea pod from the bag, what is the most likely number of peas in the pod?"; "Is there a relationship between the length of a pod and the number of peas in the pod?" The activity can be continued with string bean pods or other types of seed pods. If fresh pods are not in season, two #303 cans of whole beans will provide enough bean pods for a class of thirty students. Apples in season can be cut in half between the stem and blossom ends, the seeds can be counted, and a histogram made of the count. *(Note: Seeds from plants in the same family, produce identical numerical seed patterns.)*

Contributing Idea B. Plants can be classified.

Classifying seeds. Let students bring seeds that they find to school. Have them classify the seeds in as many ways as they can. Classifications will depend upon the seeds that are collected and are best if they are invented by the children. They need not agree with scientific classifications; however, students should be able to provide rationales for their groupings. For example, they might group seeds by shapes (triangular, circular, egg-shaped, curved, coiled), by colors (orange, purple, striped, spotted, black, white), by sizes or weights (small or large), or by textures (horns, wings, tails). If fairly large numbers of seeds from the various classifications are obtained, students can set them outside, observe, and keep records concerning the animals that eat the seeds. Judgments about which animals eat which seeds can be made. Students can hypothesize why some seeds are not eaten by any animals.

25 Dc Fa

Classifying seeds by types of dispersion. Divide the class into several small groups. Pass one envelope containing various seeds to each group. Ask students to divide the seeds into sets by the way they think the seeds might travel from one place to another. Encourage group discussion and experimentation to see how seeds might travel (e.g., seeds can be dropped or blown to see what air or wind does to them). Reclassify the seeds on another basis such as texture, shape, size, or color. Reclassify them several times. This will help students realize that there are multiple ways to classify and that some objects can belong to several classifications at one time.

26 Dc

222 INTERACTIONS

Generalization I. Germination is the process by which an embryo starts to grow into a new plant.

Contributing Idea A. Some plants grow from seeds.

01 Aa Cd Fb

Observing that some plants grow from seeds. Have students take a shallow wooden box about 1 foot (30 cm) square and about 4 inches (10 cm) deep and fill it with top soil from a place where many plants have been growing, such as a vacant lot, garden, or nature area. Keep the soil in the classroom for several weeks or more, moistening it frequently. Students can observe and record the different kinds of plants that grow. Similarly, dig up a square foot (25 cm) of soil in the winter. The soil should be about 4 inches (10 cm) deep. It can be placed in a container in the classroom and kept moist. Observe the emergence of plants and make some judgments about seeds in winter.

Contributing Idea B. Not all seeds germinate.

02 Cc Fb

Recognizing that not all seeds germinate. Let each student plant two dozen radish seeds in a flower pot. Keep the soil moist. In about one week, they can count the plants to determine what fraction of the planted seeds germinated.

03 Cc

Comparing the germination ratios of different kinds of seeds. Students can make seed testers to help them determine how many seeds germinate in a given example.

a. Fold in half a long piece of paper toweling or cloth such as muslin or gauze. Mark equal-sized squares on one side. Number the squares and, using different seeds for each square, place ten within each square. Fold the other half of the towel over the seeds and carefully roll it up. Tie it loosely with string, moisten it, and set it in a saucer of water. Keep it moist and in a warm place for three days, then carefully unroll it and tally how many of each kind of seed germinated. The tester can be rerolled and opened after another three-day interval. A histogram showing the number that germinated can be tallied.

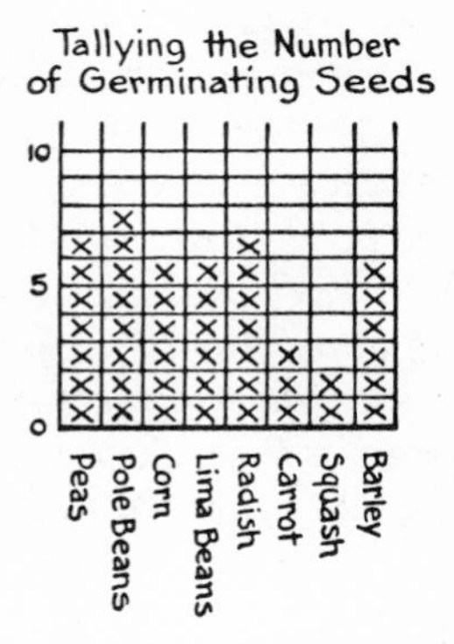

HISTOGRAM 222.03

b. Percentages of germinating seeds can be obtained by dividing a 10 inch by 10 inch (25 cm by 25 cm) cloth into one inch (2 cm) squares. Place a seed in the center of each square. Place a second piece of cloth over the seeds and sprinkle it with water. Set the cloths and seeds on several sheets of newspaper. Carefully roll them up. The newspapers will help keep the cloths moist. Fasten the ends loosely with string, and keep the papers moist. After one week, carefully unroll the tester, and count the seeds that germinated. A percentage or ratio can be obtained easily. For example, if 78 seeds germinated, the sample of seeds would be rated at 78 percent.

Contributing Idea C. Different seeds germinate in different lengths of time.

04 Ch Da

Measuring the germination times of different seeds. Use small containers filled with clean sand and plant various seeds such as orange, lemon, apple, pumpkin, barley, oat, corn, or bean in different containers. Plant two or three of a kind in a container, and make sure that all seeds are planted at the same depth and receive the same amount of water. Keep a record of the germination times. The selected seeds can be ordered by the sequence in which they germinate. *(Note: Some seeds can be defective and will never germinate; others can take several months or more. For example, apples, peaches, plums, and many other kinds of fruit seeds will not grow immediately after taking them from a fruit. Such seeds generally need time to ripen in a cool, moist location.)*

Contributing Idea D. Germinating seeds can be observed and measured.

Drawing pictures of germinating seeds. Have students draw pictures of seeds during stages of germination. To aid their observa- **05 Be Ca**

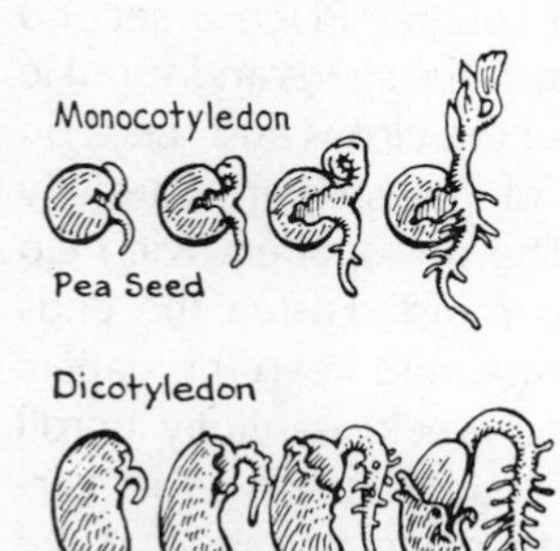

FIGURE 222.05

tions and drawings, place a sheet of ¼ inch (1 cm) graph paper behind the seeds in a soilless garden. Observations can be copied on another sheet of graph paper. Pictures of the germination stages of monocotyledons and dicotyledons can be compared.

Measuring the rate at which various parts of a germinating seed grow. Select a seed, such as a pea or bean, that has just broken through its seed coat. Place an ink mark at the tip of the root and at the tip of the stem. Let the seed continue to germinate **06 Bf Cd Fa**

FIGURE 222.06

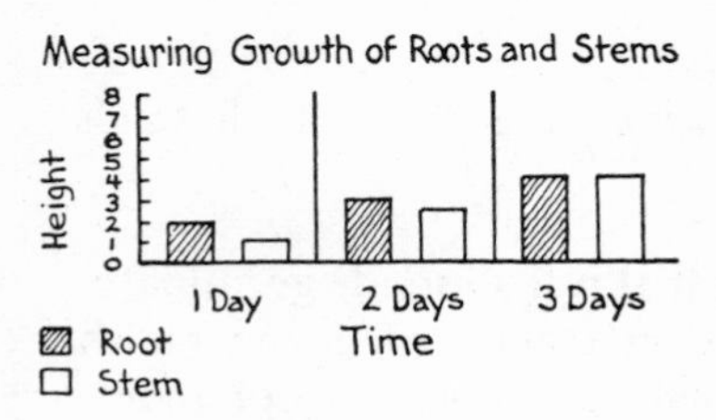

GRAPH 222.06

in a soilless garden. Mark the root and stem daily or every other day in the same manner. A simple bar graph can be prepared to depict the growth of the root and stem side by side. Students can be asked to make some judgments about the results (e.g., "Which part grew the most after one week?").

Observing and measuring the strength of a germinating seed. Plant seeds two inches (5 cm) apart in a large flat box. Place two equal-sized cigar boxes or trays over the seeded areas. Weigh one of the boxes down with a few light objects. Observe the results after several days. Similarly, weigh several kinds of large seeds and record their weights. Plant each seed at the same depth. Weigh several quarters (they are about 1/5 **07 Aa Cf**

ounce or 6.25 grams each) and place one over each seed. Students can observe what happens to the quarters after several days and compute the ratio of the seed's weight to the weight it could lift. Now find out the maximum amount of weight a seedling will lift by weighing fifty-cent pieces, silver dollars, or other similar objects. Over each seed, place a differently weighted object, observe, and compute ratios. The strengths of other seeds can be measured in the same way, then ordered by their relative strengths.

Generalization II. Somewhat specific conditions are needed for seeds to germinate.

Contributing Idea A. Seeds need some water in order to germinate.

Experimenting to see the effect of water on germination. Plant four bean seeds in wet soil and four in dry soil. Make sure that one container is kept moist and the other dry. Other factors such as temperature, light, and ventilation should be normal and equal for both containers. Let students observe what happens. Similarly, place folded paper towels in the bottoms of two glasses. Sprinkle some seed on each, cover them with another paper towel, and label the glasses *A* and *B.* Soak the towels in glass *A* and keep them moist. Keep the towels in glass *B* dry. The other conditions should be the same for both containers. **08 Ec**

Measuring how much water is absorbed by a seed. Use a balance scale to weigh twelve lima bean seeds and record their weights. Soak them in water overnight, then weigh them again. Compute the percentage of increase. The results can be compared with similar measurements taken of corn, beans, peas, oats, etc. Kinds of seeds can be ordered by their average percentages. **09 Bf Cf Db**

Contributing Idea B. Seeds need some warmth in order to germinate.

Experimenting to see the effect of temperature upon germination. Prepare two identical soilless gardens with four identical seeds in each. Place one garden in a very cold place and the other in a relatively warm place. If a refrigerator can be used, the second garden should be placed in a warm dark location because the interior of the refrigerator is dark. Students can observe which seeds germinate first. **10 Ec**

Experimenting to see the effect of extreme temperatures upon the germination of seeds. Boil ten seeds for ten minutes or heat them in an oven at 150°F (65°C) for fifteen minutes. Place the same number of seeds in dry ice for the same amount of time, or place them in the freezer of a refrigerator overnight. Do not alter a third set of ten seeds. Allow all the seeds to return to room temperature, then place them in a soilless garden to let them germinate. Have students observe the results. **11 Ec**

Contributing Idea C. Light is not needed for seeds to germinate.

Experimenting to see if light is necessary for germination. Plant equal numbers of several kinds of seeds in two identical soilless containers. Place one container in the dark (e.g., a closet, cabinet, or drawer) and one in indirect sunlight. Keep the temperature normal and equal for both. Have students compare the growths. As an alternate approach, a cover can be placed over one container to keep out light. **12 Ec**

Experimenting to see if the amount of light has an effect on germination. Have students keep all conditions the same for different soilless seed containers except for the amount of sunlight. A light meter can be used to check the amount of light in different locations. Patterns in germination can be compared among the different locations. 13 Ec

Contributing Idea D. Soil is not needed for seeds to germinate.

Experimenting to see if soil is necessary for germination. Obtain some sand and wash it clean to remove all the soil. Plant seeds in several containers of the sand, keeping the sand moist and warm. Plant identical seeds in soil for comparisons. The seeds in the sand will grow up through the sand and continue growing for a short period of time (as long as the cotyledons provide food for the plant). Other material such as vermiculite can be used in place of the washed sand, or seeds can be grown in soilless gardens. Students will realize that soil is not necessary for germination to take place but it is necessary for growth to continue. 14 Ec

Contributing Idea E. The depth of a seed in soil will not affect germination.

Experimenting to see the effect of depth on germination. Fill a straight-sided, glass container with good garden soil. Plant seeds at different depths next to the glass. Plant two seeds one inch (2.5 cm) deep, two seeds two inches (5 cm) deep, two at three inches (7.5 cm), and two at four inches (10 cm). Keep the soil moist and fasten a sheet of black paper around the jar. Remove the paper daily to observe which seeds germinate. Students can keep a record of which seeds sprout and how long it took each to sprout. They can also record how many plants grew above the surface of the soil and from which depths they came. They can realize that deeply planted seeds will germinate, but will have used their food material before reaching the surface. *(Note: The black paper in this activity is to keep light from the seeds until they break through the soil's surface, thus simulating the lack of light underground.)* 15 Ec

Contributing Idea F. The position of a seed in soil will not affect germination.

Experimenting to see if the position of a seed affects germination. Soak some seeds overnight, then mount them on needles so that they point in various directions. Mount the needles on a cork, and set it in a shallow dish of water. Cover the cork with a glass and observe. Similarly, students can place seeds sideways, upsidedown, right side up in a soilless container. They will find that seeds will germinate from any position. 16 Ec

Generalization III. Plants respond to certain environmental conditions.

Contributing Idea A. Gravity influences roots to grow down and stems to grow up *(geotropism).*

Experimenting to determine the effect of gravity upon roots and stems. Let students germinate seeds (e.g., lima bean, radish) between a blotter and glass. Several days after germination turn the apparatus on its side, and observe the roots and stems for several more days. (The roots will change direction and grow downward; the stem will change direction and grow upward.) Repeat the turning every several days or try various partial turnings. For example, move the ap- 17 Ec Fa paratus one-eighth turn each day in the same direction or move it one-fourth turn every four days, alternating the direction at each turn. The roots and stems will always respond to gravity in the same way. To compare the effect of gravity upon different plants, prepare six soilless containers with three bean, three radish, and three corn seeds in each. As the seeds germinate, set two containers on their sides and invert two containers. Keep the remaining two unchanged. Let students compare the roots and observe the direction of growth. They can infer that the roots of plants will always grow downward, and the stems will always grow upward. Similarly, gravity's effect upon just the stems of plants can be seen by growing some seeds in a soil container and turning it on its side after the plants have grown to sufficient heights.

Observing the effect of artificial gravity on the growth of plants. Attach soilless containers (see *441.01*) to the turntable of a phonograph, and observe the directions of growth of the roots and stems as they are rotated for several days. Different speeds can be tested. 18 Aa Ec

Observing that twining plants only grow upward. Grow a twining plant along an upright string or wooden support. After it has grown sufficiently, let students gently remove the support from the soil without detaching the plant. Set the support horizontally, and observe the plant for several days. (It will stop twining.) Turn the support upside down and observe again. (The last coils will unwind, and the plant will start to climb upward again.) 19 Aa

Contributing Idea B. Roots grow in the presence of water.

Observing that roots grow in the presence of water. Several activities indicate that roots develop and grow in relation to the amount of water present. 20 Aa

a. Half-fill a shallow box with soil. Using adhesive tape or an oil-base clay, plug the holes in two porous flower pots. Set the pots in diagonal corners of the box. Plant seeds evenly throughout the box. Fill one of the pots regularly with water but not the other. Do not water the soil. Over several weeks, have students note the order of germination and the appearance of the plants nearest the watered pot. (The plants will be taller and healthier looking.) After about two weeks, carefully remove both pots, and notice the mat of roots around the water-filled pot. Remove several plants from various sections of the box and observe the direction of root growth. As a variation, one flower pot can be placed in the center of a box with seeds evenly planted around it. Fill the pot with water and observe.

b. Plant soaked seeds evenly in soil next to the glass side of an aquarium. Set a small, plugged, porous flower pot in one corner of the aquarium next to the glass. Add water to the pot regularly, but do not water the soil. Over several weeks, students can observe the roots growing next to the glass and see that they extend within the moist area.

Contributing Idea C. Stems above the ground, leaves, and flowers are affected by light *(phototropism).*

Observing that some stems bend toward light. Plant a number of bean seeds in a box and water them as needed. When the plants have reached a height of about 5 inches (15 cm), students can change the po- 21 Aa sition of the box in relation to the angle of the sun. In a short time, the stems of the plants will bend back toward the sun's rays, but the bending will be in a different place on the stem. Students can change the box's position in relation to the sun any number of times, observing the change in direction of stems. (They will always bend toward the light.)

Observing that some leaves and flowers turn toward the light. 22 Aa Place a well-grown geranium or coleus plant in a sunny window, and observe that the leaves seem to follow the light throughout the day. (Actually sunlight inhibits growth—the shaded side of the plant continues to grow and causes it to elongate on one side and bend away from the shade.) After watching the plant for several days, turn it so that the leaves point away from the light at midday. Observe what happens. Students can keep a record of how long such turning takes. Several kinds of plants can be compared by their turning times. The blossoms on plants can be tested in the same way.

Observing that some leaves open and close in response to light. 23 Aa Have students observe clover leaves to see how they are affected by sunlight. During the day they can see that the three clover leaflets are spread out and change their positions. At night the leaflets close. (The two side leaflets move together until they are facing, while the top leaflet folds over them.) The effect of night can be simulated by covering a clover plant with anything that will keep out the light such as a large can or box. In about half an hour, the leaves will start to close. In an hour, the leaves will be folded in their nighttime positions.

Experimenting to see which colors influence the growth of plants. 24 Bf Ec Have students grow a patch of grass in the classroom or use a patch of lawn in the schoolyard. Divide a fairly thick section of the grass into four sections. This can be done by placing cardboard walls around the sections to keep light from leaking from one section to another. Cover one section with blue cellophane, another with red, and a third with green. Leave the fourth section uncovered, or cover it with clear cellophane *(a)*. Let students record their observations as the days pass. Note the length of the stems, color of the plants, and other general appearances. Students should be able to make some judgments as to which colors of light seem to be the most important for plant growth. (Red is best.) As an alternative, the open ends of several shoeboxes can be covered with different colors of cellophane. Several holes should be punched in the sides of each box near the bottom for adequate air circulation. Unless the plants are large, they will fit easily

FIGURE 222.24a

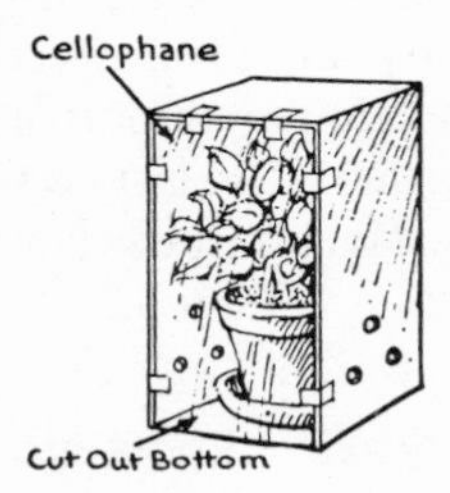

FIGURE 222.24b

into the shoeboxes when placed on end. The boxes make it easy to water the plant and to take daily measurements *(b)*. A graph can be kept to show any changes in the growth rates of plants. Comparisons can be made, and students will find that not all colors affect growth in the same way.

Contributing Idea D. Some plants respond to touch.

Observing that roots react to obstacles in their path. Plant some seeds in glass containers filled with soil. Make sure the seeds are placed next to the glass for observation. Add pieces of wood, rocks, or other obstacles to the soil beneath the seeds and under the root tip. Students will see that the roots grow around the objects. Similarly, germinate seeds on blotting paper or a sponge, and place small pieces of wood or other obstacles at the tip of the roots. Observe on a daily basis how they react. **25 Aa**

Observing that stems react to obstacles in their path. Set a plant directly under some blockage, such as a low shelf, and observe it for several days. **26 Aa**

Observing that some plants react dramatically when touched. Obtain a mimosa plant from a nursery. Let students touch the narrow leaves with a pencil. (The leaves will fold together and droop rapidly.) The plant will not be harmed, but it may take several hours for it to regain its original shape. **27 Aa**

Observing that twining stems respond to touch. Observe a climbing plant such as a sweet pea, morning glory, pole bean, or moon flower. Note that the narrow tendrils on such plants precede the rest of the plant and curl around sticks or strings. Also note the direction in which the tendrils turn. (The mentioned plants all turn in a counter-clockwise direction.) Have a student carefully unwrap a tendril from another plant, and gently wrap it several times around a support in a clockwise direction. In a short time, the tendril will unwind completely. (Tendrils of a plant turn only in one direction.) **28 Aa**

Contributing Idea E. Some plants respond to heat and cold.

Observing that some plants respond to heat. Bring a lighted match near a terminal leaflet of a mimosa plant. The entire leaf will respond immediately. After a few hours, the original shape will be regained. Other plants can be tested in the same way. Students can also notice that the leaves on many plants become limp on hot days. **29 Aa**

Observing that some plants respond to cold. Place a mimosa plant in a cold place for about ten minutes. Have a student gently touch a terminal leaflet and note that the plant no longer responds to touch rapidly. When the plant returns to room temperature, its sensitivity to touch also returns. Other plants can be tested in the same way. **30 Aa**

Contributing Idea F. Some plants respond to seasonal changes.

Experimenting to see what causes leaves to change color. Set twigs with leaves (e.g., red maple, sugar maple, poplar, oak) in glasses filled with different solutions such as salt water, sugar water, vinegar water, or combinations of these. Place some in a refrigerator at night and others near a heater. Be sure to keep identical twigs in plain water and under normal conditions as controls. Keep notes on what happens and try other tests. **31 Ec**

230 microorganisms

231 CHARACTERISTICS

Generalization I. Microorganisms have identifiable characteristics.

Contributing Idea A. All microorganisms reproduce their kind.

Observing that microorganisms reproduce their kind. Have students culture protozoa, bacteria, mold, yeast, and algae. As each culture develops, they will easily see that the increases in populations involve new organisms that are like the parent. **01 Aa Fa**

Contributing Idea B. All microorganisms grow.

Measuring the growth of microorganisms. Microorganisms are too small for the growth of an individual organism to be measured in the classroom; however, growth of populations can be measured. **02 Cd Cg**

a. *Protozoa.* Obtain some pond water (see *231.11*) and develop a culture. The number of protozoa in a drop of pond water can be counted and compared with a later sample. Students can keep count of each type of organism that they see. A few drops of gelatin or methyl cellulose added to the water will slow the animals down for better viewing.

b. *Bacteria.* Inoculate the culture medium in a petri dish (see *231.14–16*). Cover and seal the dish. Students can measure changes in the size and shape of the growths inside.

c. *Molds.* Grow mold on bread (see *231.18*). Students can cut paper strips each day to represent the distance each mold grows. *Aspergillis elegans* is a relatively slow growing mold that grows on bread. It leaves a ring of spores with each day's growth, making it easy to measure on a daily basis.

d. *Yeasts.* Culture some yeast (see *231.20*). Students can estimate the number of plants in one-fourth of a viewed area, then multiply the number by four to obtain an estimate of how many there are. Recount later, after the yeast has risen.

e. *Algae.* Students can see the increase in green algae in an aquarium kept in strong, but not direct sunlight.

Contributing Idea C. All microorganisms need food and water to grow.

Observing that microorganisms need food in order to grow. As students culture various microorganisms, they will realize the importance of providing the proper food source for each kind. For example, they can culture yeast in one container, but not add molasses to the recipe for a second container (see *231.20*). Similarly, they can **03 Aa Fa**

try to grow bacteria in a petri dish that does not contain a culture medium.

Observing that yeasts require food for growth. Put 1 teaspoon (5 ml spoonsful) of cornstarch in a pint (500 ml) of water. Remove a spoonful of the mixture and add a little dilute iodine. (It will turn blue indicating the presence of starch.) Add a yeast cake to the jar. Repeat the iodine test several times a day. (The color will show less and less starch until none is left.) If students test for the presence of sugar during this activity, they will realize that the yeast uses the starch for food and turns it into sugar. **04 Aa Fb**

Observing the effect of water upon the growth of bacteria. Obtain some powdered milk. Prepare a cup (250 ml) in a jar by adding water to it according to directions. Place an equal amount of the powder into a second jar, but keep it dry. Cover the jars and place them on a window sill. Tell students that some scientists have estimated that a drop of milk might contain as many as 100 million bacteria. Open the jars after two days to see if they smell the same. Students will realize that the lack of water had an effect on the growth of bacteria in the milk. **05 Aa Fa**

Experimenting to see the effect of moisture upon the growth of molds. Place a piece of white bread in each of two coffee cans. Moisten the bread in one can, but not in the other. Leave both uncovered for three hours, then cover and place both in a warm dark place for a week. Compare the growth of molds on each piece of bread. Students will find that the dry bread shows much less growth than the moistened bread and conclude that moisture is important for the growth of molds. **06 Ec Fa**

Contributing Idea D. All microorganisms are sensitive to their surroundings.

Observing that some microorganisms are sensitive to light. Place a culture of euglena in a long test tube sealed with a cork or rubber stopper. Hold the test tube in a horizontal position. Using hand lenses, students will see that the animals become evenly distributed in a short time. Now hang a piece of dark cloth over half the test tube. Do this gently so as not to disturb the animals. After an hour, remove the cloth and immediately observe again. Students will find that the euglena tend to be in the lighted portion of the tube, and realize that the animals sense light and seem to have a preference for it. They can test the influence of light on one-celled plants by placing cultures of them in locations with different lighting conditions. **07 Aa Ec Fa**

Observing that microorganisms are sensitive to temperature changes. Place a culture of protozoa in a long test tube sealed with a cork or rubber stopper. Hold the test tube vertically. Using hand lenses, students will see that the animals become evenly distributed in a short time. Now place the base of the tube against an ice cube. Students will find that the protozoa tend to move to the top of the tube, and realize that the animals sense cold and seem to avoid it. Students can test the influence of heat and cold upon one-celled plants by placing cultures of them in locations of different temperatures. **08 Aa Ec Fa**

Generalization II. There is a great variety of physical characteristics among microorganisms.

Contributing Idea A. The microscope is an instrument used for viewing objects that are very small.

Using a compound microscope. Show students that microscopes have at least two curved lenses that work together. The lens **09 Bb Ga**

nearest the object being viewed is the *objective lens.* The lens nearest the viewer's eye is the *eyepiece.* Have students locate the *platform* or *stage* beneath the objective lens. The *clips* hold glass slides on the platform. Beneath the platform is a *mirror* that can be turned to reflect light from a lamp or window onto the viewed object. **(Caution: Never turn the mirror directly toward the sun.)** The mirror can be adjusted until a uniform circle of light without shadows is seen. This circle is called a *field.* Place a small drop of water in the center of a clean, dry glass slide. Cut a small letter "e" from a newspaper and lay it right-side up in the drop of water. Slowly place the cover glass over the water by placing one edge down first (to prevent air bubbles from being caught between the glasses). Now place the mounted slide on the platform, clip it into place, and move it with thumbs and forefingers until the letter "e" is in the center of the platform opening. Turn the *course adjustment knob* to move the objective lens down as far as it will go, but be careful not to touch the slide. (Look at the slide from the side to be sure the objective lens does not touch or break the slide.) Students can take turns looking through the eyepiece, watching the field as they turn the coarse adjustment knob slowly toward them. This will raise the body of the tube and the lenses, and students will see the letter "e" appear. The adjustment can be moved slowly up and down to focus the letter. Let students move the slide to the right, left, up, and down. They will find that everything seen through the eyepiece is reversed. Some students might be interested in looking up the work of Anton von Leeuwenhoek (1632–1723), a Dutch merchant—the first person to perfect the microscope.

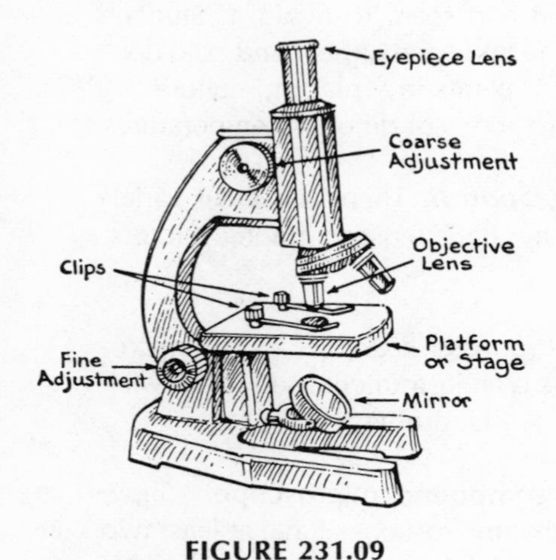

FIGURE 231.09

Preparing microscopic slides. Place a culture or specimen in the center of a clean, dry glass slide. If the specimen is liquid, a cover glass can be placed over the top by placing one edge down first to prevent bubbles from becoming trapped beneath the glass *(a).* If the water begins to dry up while viewing, use a medicine dropper to place a

10
Ga

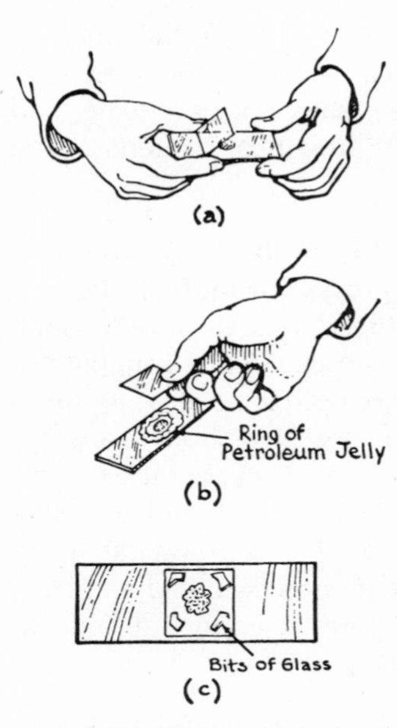

FIGURE 231.10

little more water along the edge of the cover glass. Bacterial cultures can be fixed to slides by spreading a drop so that it forms a thin film on the glass. Allow it to dry, then pass it through a flame three or four times to kill the culture and fix it to the slide. Now put several drops of methylene-blue stain on the culture to improve viewing. Leave it for two minutes, then dip the slide in water to wash off excess stain. Blot gently with a paper towel and view under a microscope. Sometimes it is best to not press the object being viewed with the cover glass. Special slides with indentations in them can be purchased or the cover glass can be raised by making a ring of petroleum jelly slightly smaller than the cover on the slide. Place a drop of a culture in the ring and add the cover glass *(b).* An alternative method is to break a cover glass into several pieces, place them on a slide, cover them with an unbroken cover glass, and glue the pieces in place *(c).* A culture can be dropped between the slide and the raised cover.

Contributing Idea B. Protozoa are simple one-celled animals that are too small to be seen with the unaided eye.

Collecting protozoa. Protozoa can be found primarily in water. Although they can be collected from puddles of standing water almost anywhere, the best source is unpolluted ponds. Students can collect jars of pond water, including some mud from the bottom. Let the water settle. After two days, stir the water without disturbing the mud, fill a medicine dropper with the water, and place a few drops in a watch glass or test tube. Students can use hand lenses to observe the contents. They will see some of the larger protozoa such as the *paramecium.* Tell them that some protozoa cause diseases in people—one causes malaria, carried by the anopheles mosquito; another causes sleeping sickness, carried by the tsetse fly. Some students might be interested in researching information on the prevention of such diseases.

11
Bb
Ge

Culturing protozoa. Place plant material such as dry grass, hay, or leaves in a quart (liter) wide-mouthed jar. Fill it three-fourths full with tap water that has been standing for two days to allow the chlorine to evaporate. Cover the jar and set it in a well-lighted place, but away from direct sunlight. In two days, students can observe drops of the water under a microscope. The population and varieties of protozoa will increase for several weeks.

12
Gd

Observing the physical characteristics of protozoa. Although there are about fifteen thousand species of protozoa, most are too small to be seen without a microscope. To see some protozoa, prepare a microscope slide by placing a large drop of pond water on it and carefully covering it with a cover glass. Set the slide on a microscope or a microprojector platform. When focused, move the slide slowly until an animal is viewed. The protozoan most likely to be observed is the *paramecium.* A diagram of this animal can guide students' observations *(a).* They will see that it has a definite, permanent shape and moves rapidly by the coordinated beating of a great many fine, hair-like structures on its body. You might tell them that the structures are *cilia* and that the dark spots in the animal are *nuclei.* (The smaller nucleus controls reproduction and the larger controls most of the animal's other functions.) As the animal swims, food is swept into the oral groove and eventually is passed out through the anal pore. Students might also observe an *amoeba*—a slow moving, clear animal with a constantly

13
Aa
Bb

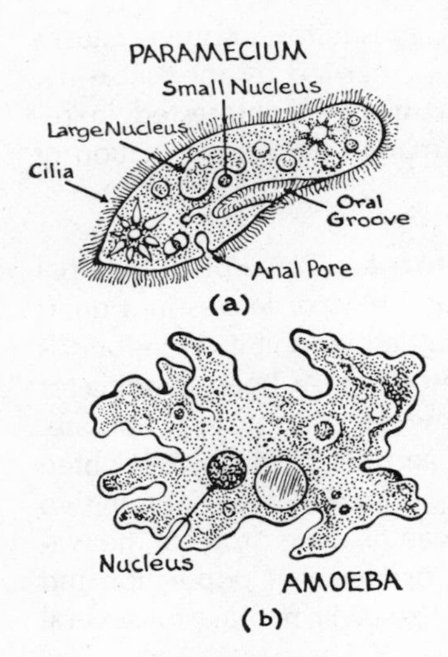

FIGURE 231.13

changing body shape *(b).* It has one nucleus and obtains food by engulfing it. Now obtain several samples of water from a pond, stream, puddle, and faucet. Make a slide for each type, label the slides so they will not get mixed up, then look for similarities and differences in the kinds and numbers of protozoa in the samples.

> *Contributing Idea C.* Bacteria are simple one-celled plants that are too small to be seen with the unaided eye.

Culturing bacteria. Bacteria are non-green plant microorganisms that can be found almost everywhere. They can be cultured in the classroom by placing plant material (such as dry grass, beans, peas) or various meats into a small amount of water in an open jar for several days. Cover the jar for several more days. Observable changes in the contents and the production of bubbles and odors will indicate the growth of bacteria. Cultures can be transferred to a nutrient medium in petri dishes or test tubes to observe the growth of colonies. To inoculate a medium, dip a sterile cotton swab or glass rod into a culture and streak it across a medium. (**Caution: Under most conditions bacteria are harmless; however, disease-producing strains can appear in a culture, thus students must be careful not to spill a culture and must always wash hands and materials thoroughly after working with one.**)

14
Gd

Observing bacteria. Bacteria are the smallest living things known. Some are so tiny that it takes 25,000 of them to make a line 1 inch (2.5 cm) long. Because individual bacteria are too small to be seen, they are best studied in colonies. Growth of a colony can be seen by transferring a culture to a nutrient medium in a petri dish or test tube or a microscope slide. A sterilized cotton swab or glass rod can be used to streak a culture across a glass slide. When the slide has dried, cover the streak with red ink, and let it stand for several minutes. Gently rinse the red ink off with running water. Examine the slide under a microscope. (The red ink will help students see the culture more clearly.) You might tell them that there are about two thousand known kinds of bacteria. Some cause disease in people, but others produce good food flavors in cheeses, cream, meats and other foods.

15
Aa

Preparing culture mediums. A study of bacteria requires some nutrient medium for bacteria growth. Test tubes and petri dishes containing sterile nutrient can be purchased, or students can prepare their own in several ways. Different mediums will allow different bacterial colonies to develop. For each of the following suggestions, be certain that containers are clean and sterilized. (**Caution: Before discarding any medium on which bacteria have grown, be sure to**

16
Gc

sterilize the medium by heating or by soaking it overnight in a disinfectant [e.g., 3 percent Lysol].) Glassware and dishes can be sterilized in an oven at 110°F (45°C) for one hour or in a pressure cooker at fifteen pounds for thirty minutes. Hands should be thoroughly washed before and after handling materials. *(Note: Shallow glass dishes or glass coasters with glass covers can be substituted for petri dishes. Baby food jars and test tubes also work well.)*

a. *Agar medium.* Put 6 level teaspoons (5 ml spoonsful) of nutrient agar into 1 cup (250 ml) of water. Shake vigorously to get as much into solution as possible. Add another cup of water, cover, then sterilize in a pressure cooker. Do this by placing the agar container in a basket just above the water level. (A small can or rubber stoppers can be used to raise the basket.) Put a lid on the cooker and clamp it tight. Allow steam to flow freely at a moderate rate for ten minutes. Place the pressure control on the vent pipe, raise pressure to fifteen pounds, then lower the heat just enough to maintain the pressure for thirty minutes. At the end of this time, turn off the heat, and allow the pressure to return to zero of its own accord. Let the cooker cool for two minutes, remove cover, allow agar to cool for ten minutes, then pour it into sterilized petri dishes, test tubes, or other containers. As a variation, clear beef broth can be substituted for the water.

b. *Gelatin medium.* Dissolve one beef bouillon cube in eight ounces (250 ml) of boiling water. Add a spoonful of sugar. Prepare unflavored gelatin according to directions on the package (use Knox gelatin or a bacteriological gelatin from a scientific supply house), but substitute the beef broth for the boiling water. Cool the solution and pour it into sterilized petri dishes or test tubes. Cover and let containers cool.

c. *Potato medium.* Cut wedges from a large, peeled, uncooked potato. Place one wedge into each test tube to be used. Add 1 teaspoon (5 ml spoonful) of water to each tube, plug with cotton, and sterilize them in a pressure cooker at 15 pounds for thirty minutes or in a double boiler for one hour. Similarly, thin slices of white potato can be placed into petri dishes, sealed, and sterilized. Wedges or slices of carrot, sweet potato, apple, pear, or orange can be substituted for the white potato.

17 Ec

Determining the presence of bacteria. Prepare five or more sterilized petri dishes or test tubes containing a nutrient medium. Do not open them until students decide what to do with each. They might let a fly walk across one, place a hair on one, touch one with the tip of a lead pencil, finger, or a coin, cough or sneeze into one, place food such as a drop of milk, piece of meat or cheese into another, or expose several to the air for different lengths of time. Soil samples can also be tested. Cover, label, and seal each container with adhesive tape. **(Caution: Once sealed, do not open a container again. This will prevent exposure to pathogens that could develop and be dangerous. When study is completed, carefully destroy and dispose of the colonies by boiling them in a large quantity of water or by soaking them in a strong disinfectant for twenty-four hours.)** For alternative experiences students might compare (1) a five-minute exposure of a dish in a quiet room without people to a five-minute exposure of another dish when people are moving about; (2) a washed finger with an unwashed finger. For each experience, be sure students maintain an unopened container for comparative purposes.

Contributing Idea D. Fungi are simple plants, some of which are too small to be seen with the unaided eye.

Culturing molds. Fungi vary in size from microscopic plants to larger forms such as mushrooms; molds are a common form of microscopic fungus that cannot make their own food (they do not contain chlorophyll). To grow some, place a moist blotter or paper towel on the bottom of a jar. Set a piece of white bread on the blotter and leave the jar open for thirty minutes. Put a lid on the jar, label it, and store it in a dark place for a week. At the end of that time, students will see how the mold has grown. You might tell them that there are more than eighty-five thousand known kinds. Some are beneficial in the manufacturing of beverages and cheese and in the production of penicillin. Some are a nuisance, such as those that grow on leather; others are harmful and cause disease in plants (mildew on grapes, rot on peaches) or in people (ringworm, athlete's foot). Now have students compare cultures grown on different kinds of bread—homemade, commercial white, whole wheat, rye, etc. (Commercial breads include preservatives that will affect the growth of molds.) **18 Gd**

Observing the physical characteristics of some molds. Culture some white-bread mold. After a week, students will see a variety of molds growing on the bread. Let them count the different colors. Explain that each color is a different colony. Black mold is a common variety that can be examined with a hand lens. Tell students that the small round balls they see on the end of the slender stalks are cases containing *spores.* Explain that when the cases ripen, they release the spores into the air, and some fall upon the bread to form new mold. Roquefort, blue, and Gorgonzola cheeses grow penicillin molds. These can be removed from the cheeses with a fine needle and placed on a microscope slide. Their structures can be seen under a microscope. **19 Aa Bb**

Culturing yeasts. Yeast plants are a common form of microscopic fungi that cannot make their own food, thus do not contain chlorophyll. More than fifty-five thousand kinds are known. Yeast plants can be grown by mixing baker's yeast with warm water and a little molasses. Allow the mixture to stand for one hour. Place a drop of it on a glass slide and observe under a microscope or on a microprojector. Students can easily see the yeast plants. Ask them to estimate the number they see. One way to estimate is to divide the visual area into fourths, count the number in one fourth of the area, then multiply the number by four. Now let the mixture stand overnight. Look at it the next day and determine the number. Tell students that these tiny yeast plants are in the air almost everywhere. **20 Cb Gc**

Experimenting to see the effect of temperature changes on the growth of yeast. Make some bread dough with yeast. Divide the dough into three parts. Put the first in a refrigerator, the second in a warm place, and the third in a hot place. Students can examine the dough each hour and discuss the effect of the different temperatures. **21 Ec**

Making bread. Students can make bread by mixing 1.2 cups (300 ml) of flour with 3 teaspoons (5 ml spoonsful) of salt and 3 of sugar. Add 4 cups (1,000 ml) of water and 2 packages of dry yeast. While making bread, be sure all equipment is clean and **22 Ec Gc**

that hands are washed thoroughly. As students work with the bread dough, let them describe how it feels. Let the dough rise for 1½ hours, put it into small greased pans, and let it rise again. Bake for ten minutes at 450°F (235°C). For a comparative experience, prepare two batches of dough—one using dry yeast and one without yeast. Compare the breads before and after baking in terms of color, smell, taste, and texture.

Contributing Idea E. Algae are simple plants, some of which are too small to be seen with the unaided eye.

Culturing algae. Algae vary in color and size from microscopic plants to larger forms such as seaweeds and kelps. Green algae is a common microscopic form that is widely distributed in ponds, lakes, streams, and on land in damp places. More than sixty-seven hundred kinds have been identified. Students can collect some green scum that forms on stagnant pools and transfer it to a large aquarium or jar. Algae can be cultured in an aquarium placed in strong, but not direct sunlight. A stocked aquarium usually provides enough nutrients for the algae to grow. **23 Gd**

Observing the physical characteristics of algae. Place a drop of algae on a microscope slide and examine it under a microscope. Individual plants will be visible. Students can count the number of different forms they see and use reference books to identify them. They can observe a single plant for a while and perhaps see it reproduce by fission. **24 Aa Bd**

Generalization III. Microorganisms can be organized by characteristics.

Contributing Idea A. Some microorganisms can be seriated.

Seriating bacteria and molds. Culture some bacteria and mold. Students can measure the sizes of the different colonies on a daily basis for a week, then order the organisms by their growth rate. **25 Da**

Contributing Idea B. Some microorganisms can be classified.

Classifying microorganisms. Some microorganisms can be classified by the substances they grow on, their colors, or the conditions under which they grow best. Let students research to identify each type in reference books. Molds are generally classified by color. Bacteria are classified by three basic shapes—rodlike *(bacilli),* spherical *(cocci),* and spiral *(pirilla).* **26 Bd Dc**

232 INTERACTIONS

Generalization I. Reproduction is a process by which microorganisms give rise to new generations.

Contributing Idea A. Some microorganisms reproduce by fission or conjugation.

Observing reproduction in protozoa and algae. Place a large drop from a rich culture of protozoa on a microscope slide and place it under a microscope or on a microprojector. Move the slide slowly until a large paramecium is found, then follow it. If the sample begins to dry, add more water. Rarely, students will see the paramecium divide to form two paramecia. Tell them that this type of reproduction is called *fission* (simple cell division). Explain the protozoa can also reproduce by joining together, then dividing (a process called *conjugation*), but the chances of seeing this in the classroom **01 Aa Bb**

are slight. Algae can similarly be observed in hopes of seeing reproduction by fission or conjugation.

Contributing Idea B. Some microorganisms reproduce by means of spores.

Observing reproduction by molds. Allow some moist white bread to become moldy in a coffee can. Students will soon see a black mold on the bread. This is a very common variety of mold called *rhizopus.* It **02 Aa Bb**

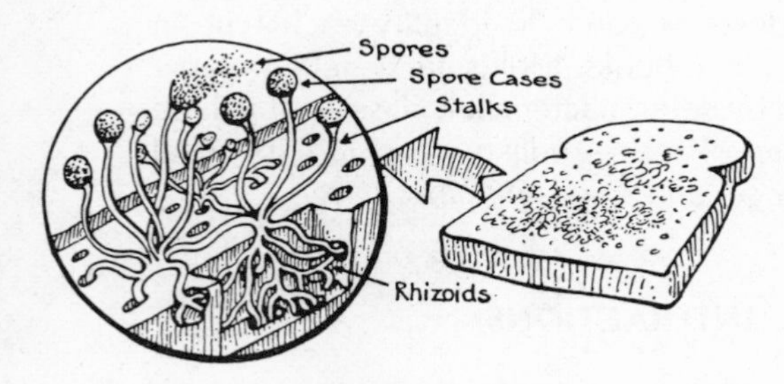

FIGURE 232.02

can be examined with a hand lens to clearly see the stalks topped by black balls. Explain that the balls are *spore cases.* The cases can be broken open and the *spores* placed on a glass slide for observation under a microscope.

Contributing Idea C. Some microorganisms reproduce by budding.

Observing reproduction by yeasts. Place a piece of baker's yeast into a small bottle and add a little molasses and warm water. Shake the mixture. Observe a drop of the mixture under a microscope. Students will see a small bulge grow out of a larger cell, remain attached for a short period of time, then eventually separate from it as an independent plant. Explain that this kind of reproduction is called *budding.* **03 Aa Bb**

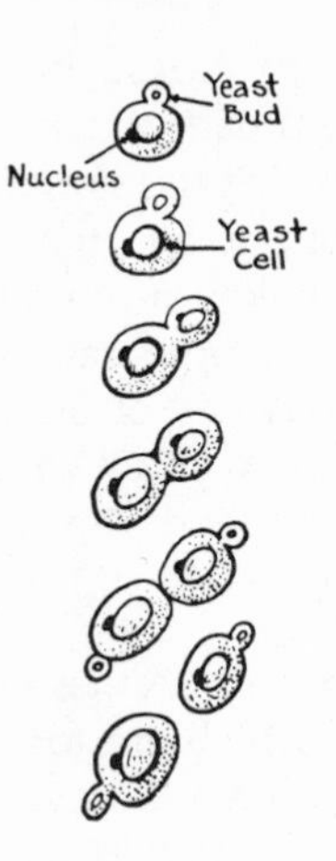

FIGURE 232.03

Generalization II. Various microorganisms operate in special ways that keep the organism alive and growing.

Contributing Idea A. Certain bacteria break down proteins, carbohydrates, or other substances for their life functions.

Observing the growth of bacteria on different foods. Protein is an important part of food, present in cheese, meat, milk, gelatin. Mix 1 ounce (25 g) of powdered, unflavored gelatin with a cup (250 ml) of garden soil, pour the mixture into a quart (liter) wide-mouthed jar, and label it "Protein." Carbohydrate is an energy food, present in bread, flour, starch, sugar. Mix 1 ounce (25 g) of sugar with one cup (250 ml) of garden soil, pour the mixture into an identical jar, and label it "Carbohydrate." To a quart (liter) bottle of water, add 9 teaspoons (5 ml spoonsful) of mineral fertilizer, shake the mixture, and allow it to settle. When settled, fill each of the jars with the liquid, cover the jars, and let them stand for two or three weeks. Keep daily records of observations. Students will smell odors and see different colors appear. Tell them that some bacteria **04 Aa Bb**

break down protein food in the absence of oxygen, a process call *putrefaction,* while others break down carbohydrates in the absence of oxygen, a process called *fermentation.* Some students might be interested in researching: (1) the uses people have made of fermentation; (2) the work of Louis Pasteur (1822–1895), the French chemist who studied fermentation.

Contributing Idea B. Certain molds grow only on certain foods.

Experimenting to see on what foods molds will grow. Bring to class several samples of moldy food such as an orange with green or blue mold covering part of it, an apple with a brown spot, a sweet or white potato with black mold on it, and a carrot with a soft, moldy area. Now scrape some mold from each sample onto healthy specimens of the same kind. Put the healthy specimens in sealed jars. Next, jab a sterilized needle or stiff wire into the moldy portion of one sample, then pierce the skin of healthy samples of different fruits and vegetables. Repeat this for each of the moldy samples, sterilizing the needle each time. Put all the inoculated, healthy speciments in sealed jars. Students can compare the results after several days. After examining the healthy specimens that had mold scraped onto them, discuss how fruits and vegetables are structured to resist infections. Students will also realize that certain molds only grow on certain foods.

05
Ec
Fa

Contributing Idea C. Not all microorganisms need oxygen to live and grow.

Experimenting to see if air is necessary for the growth of bacteria. To a quart (liter) bottle of water, add 9 teaspoons (5 ml spoonsful) of mineral fertilizer, shake the mixture, and allow it to settle. Now add ½ inch (1 cm) of soil and ½ teaspoon (5 ml spoonsful) of sulfur to one of three identical jars with lids. Pour the fertilizer mixture into the jar to a level of 1 inch (3 cm) above the soil. Add the same amount of soil and mixture to the second jar, but do not add sulfur. Add the mixture and sulfur to the third jar, but do not add the soil. Swirl the contents around and seal each jar with its lid. Let the jars remain undisturbed for two weeks, but have students record observations daily. They will see bubbles appear and smell odors when the jars are opened. Since all the jars were covered to prevent exposure to air, students will realize that the changes within must be due to bacteria that can live and grow when sealed off from air. From the results of the three conditions, they should also realize that the bacteria inside must use sulfur instead of oxygen for growth. Tell them that some bacteria, called *aerobic bacteria,* need air to grow while others, called *anaerobic bacteria,* grow without air. Explain that septic tanks make use of the latter type to change waste products into harmless material. Sometimes anaerobic bacteria poison foods by growing in vacuum-sealed cans.

06
Bb
Ec
Fa

Contributing Idea D. Some microorganisms produce carbon dioxide as they live and grow.

Testing the gas given off by some microorganisms. Drop a piece of baker's yeast into a small bottle, add a little molasses and water, then shake the bottle. Insert a piece of glass tubing through a one-hole rubber stopper and seal the bottle with the stopper so that the tubing does not touch the mixture. Pour some limewater into a second small bottle. Insert two pieces of glass tubing into a two-hole rubber stopper and seal the

07
Fb

bottle with it. One piece of tubing should be long enough so it will be in the limewater, and the other should not touch the limewater. Connect a rubber tubing from the first bottle to the longer tube of the second bottle. Now set up a second set of two bottles in the same way, but place moldy bread or another type of fungus (e.g., mushrooms) in place of the yeast mixture in the first bottle. Now set up a third set for control purposes by putting only molasses and water in the first bottle. After twenty-four hours, students will find that nothing happens in the control bottle; thus, the yeast or moldy bread must be responsible for making the limewater cloudy. Cloudy limewater indicates that carbon dioxide is being given off by the mixtures. Explain that yeasts and molds are non-green plants, and since they cannot manufacture their own food, they must live on other food sources. As an alternative experience, place a solution of yeast and water in small amounts in the bottom of a soda pop bottle. Add a few drops of syrup to one bottle, flour, gelatin, or grape juice to others. Firmly attach a balloon to each bottle and observe what happens. Students can see which substances mixed with yeast produce the most active results and test the gas that inflates the balloon. (It will be carbon dioxide.)

Observing fermentation by yeast. Follow a recipe to make bread dough (a mixture of flour, sugar, and water). Divide the dough into two parts. Mix some yeast into one part and set both in glass jars in a warm place for a few hours. Students will observe that the dough with yeast rises. By examining the dough, they will see that there are holes in it caused by gas bubbles. Explain that the gas (carbon dioxide) is produced by the action of the yeast on the sugar and that this process is called *fermentation*. **08 Aa Bb**

Generalization III. Natural and artificial conditions affect the growth of microorganisms.

Contributing Idea A. Temperature changes affect the growth of microorganisms.

Experimenting to see the effect of different temperatures upon the growth of protozoa. Pour equal amounts of pond water (or other water containing a protozoa culture) into three clean jars. Store one in a refrigerator, one in the classroom at room temperature, and one in a warmer location. After twenty-four hours, use a medicine dropper to place a large drop from each jar on different microscope slides. Place cover slips on them, and observe under a microscope or with a microprojector to see what effect the different temperatures had. **09 Ec Fa**

Experimenting to see the effect of different temperatures upon the growth of bacteria. Place two drops each of different milk products such as cottage cheese, half-and-half, ice cream, and buttermilk, in different petri dishes. Make three sets and place one in a refrigerator, one in a hot place, and leave one at room temperature. After two days, smell them and observe samples under a microscope. Students will note marked differences in the products kept in a refrigerator. **10 Ec Fa**

Experimenting to see the effect of different temperatures upon the growth of molds. Moisten two pieces of white bread and place them in two coffee cans. Leave them uncovered for three hours, then cover both. Place one in a refrigerator and the other in a dark, warm place. Compare the growth after a week. Students will find that the bread in the refrigerator shows much **11 Ec Fa**

less growth than the bread in the warm location. They should conclude that warmth is important for the growth of molds. A similar activity can be done with yeasts.

Contributing Idea B. The growth of microorganisms can be controlled.

Using high temperatures to kill bacteria. Pour half a culture of bacteria into each of two beakers. Boil one for twenty minutes and immediately seal it with a piece of plastic wrap and a rubber band. Seal the unboiled water in the same way. After two days, open and smell the contents and observe differences. Students will realize that high temperatures kill bacteria.

12
Fa
Ga

Experimenting to see what effect toothpastes have upon the growth of bacteria. Place a small amount of four or five different brands of toothpaste into separate sterile petri dishes containing a culture medium. Expose the dishes to the air for thirty minutes. Seal each lid with adhesive tape, and label it with the date and brand of toothpaste inside. Place the dishes in a warm, dark place. In three days students will see which toothpaste contains chemicals that have the greatest effect upon bacterial growth. Similarly, they can prepare the dishes with toothpaste samples, then use a toothpick to scrape some material from between their teeth and make a line on each dish with the toothpick.

13
Ec

Experimenting to see what effect bandages and disinfectants have upon the growth of bacteria. Obtain two apples of the same variety, size, and condition. Cut a small slit into each of them with a knife. Place an adhesive bandage on one, leaving the other exposed to the air. After three days students will find that the bandage slows down bacterial growth. This activity can be repeated by putting some iodine or other disinfectant on the cut of one apple instead of the bandage.

14
Ec

240 ecology

241 ENVIRONMENTAL CHARACTERISTICS

Generalization I. Plants and animals can be found in a variety of environments; each kind exists only in certain environments.

Contributing Idea A. Some plants and animals live in water. The aquarium is a simulated environment designed for observing marine plants and animals.

Building aquariums. Aquariums provide first-hand observations of marine plants, animals, and their relationships. The most useful aquarium simulates a particular environment and includes natural balances and ecological relationships. When properly set up, freshwater and saltwater aquariums need little attention.

01
Gc

a. *Containers.* Commercially prepared tanks, preferably with slate or glass bottoms, are most satisfactory; however, almost any sturdy glass container from a drinking glass to a large container will work. Small-mouthed containers are not good, for they do not provide enough oxygen for most fish. It should be kept in mind that a commercial aquarium is fragile and tends to leak once it has been bumped, moved, or allowed to remain empty for a long time. It is also good to remember that the tank should never be lifted when it is filled with sand and water and that an empty tank should not be lifted by the top frame but should be supported by the bottom frame. Although cool freshwater aquariums might be covered or not covered, tanks used for tropical fish should include a cover that is slightly raised to allow air to circulate over the water. A cover will help keep the temperature constant and retain moisture. All-glass containers or containers with reinforced corners of stainless steel are best for saltwater environments. Be sure to spread several layers of waxed paper under the tank to protect the surface where it rests.

b. *Sand.* If sand is placed in the aquarium first, water clouding can usually be avoided. The sand should be coarse and well washed before being placed into the container. To wash the sand, simply place it in a clean can, fill the can with water, stir, and pour out the dirty water, holding the sand in the can with your hand. It might be necessary to repeat this process several times. The layer of sand used in the aquarium might vary in depth, but 1 inch (3 cm) is sufficient. If it is piled higher toward one end, refuse will settle toward the other end, making it easier to remove. Ornaments such as castles and large stones should be avoided because they provide crevices in which refuse will collect. Refuse, if allowed to remain, will foul the water. Sea shells should not be used in freshwater aquariums. They dissolve slowly, releasing too much lime into the water.

c. *Water.* If you do not use the water where the fish are found, ordinary tap water is fine. Remember, however, that tap water contains chemicals such as chlorine that will kill fish as well as the disease germs it was meant to destroy. To make tap water usable, let it stand (not in the tank to be used) for several days to allow the chlorine to evaporate. In order not to disturb the sand or gravel base (which might create a clouding that will take several days to clear), place a saucer or large piece of paper over the sand and pour the water directly over it. Fill the tank to half its capacity and then add plants. Anchor them securely, complete filling the tank, and remove the paper. Allow the water to stand for several hours to adjust to room temperature. The water temperature should be kept fairly constant and should be the same as the habitat that the aquarium represents. When the climate is extremely warm, an aerator will help to cool the water. If the climate is too cold, various heating devices can be obtained. The water temperatures for most freshwater fish are best between 65°F and 70°F (18° and 21°C), but they can stand extremes from 50°F to 75°F (10° to 24°C). Tropicals survive best between 73°F and 78°F (23° and 26°C), but will stand the narrow limits of 70°F to 85°F (21° to 30°C). For saltwater aquariums, fresh seawater is best. Artificial seawater, however, can be made by dissolving 10½ ounces (300 g) of pure table salt, 1½ ounces (42 g) of magnesium chloride, 1 ounce (28 g) Epsom salts, and ½ ounce (14 g) plaster of paris in every 3 gallons (11 L) of water. The salts can be of a technical grade. It is possible to buy prepared mixes at some pet stores. Water that evaporates from saltwater aquariums should be replaced, otherwise the salt content that does not evaporate will become excessive and will lessen the oxygen content. A mark at the original water line will help to keep track. A hydrometer can be floated in the container to check the salinity of the solution. The desired reading is 1025, but any point between 1020 and 1030 is satisfactory. Add more salt if the reading is too low. The hydrometer should be left in the tank so that a reading can be taken each day. The temperature for saltwater aquariums should be kept around 65°F (18°C). The water in a well-planned freshwater or seawater aquarium will seldom need to be changed, but if the need arises, remove the fish and place them in a container filled with water from the tank. Remove the remaining water with a siphon, clean the tank, and wash the sand. Rebuild the tank as previously described.

d. *Animals.* It is best to allow aquarium water to sit several days before introducing any animal life. When the aquarium is ready, place the fish and their water in a jar and float it in the aquarium for about an hour. The temperature of the aquarium and the water in the jar will gradually match, and the fish can be placed into the aquarium without shock. This precaution is necessary for the health and survival of the fish. Know your animals and plan for their proper food in advance. Foods will vary depending on the type of fish. Local pet stores carry products for all types. Feed the fish small amounts three times per week. The more varieties of animals in the same tank, the more interesting and varied will be the observations. A cool freshwater aquarium can be stocked with goldfish, minnows, dace, and sunfish. There is a great variety of compatible fish appropriate for tropical aquariums, and saltwater aquariums might include such fascinating creatures as barnacles, corals, small crabs, sea anemones, and mussels. To create a balance in a freshwater tank, add scavengers such as snails and catfish. These housekeepers are active at night and clean up any refuse they find on the bottom, thus helping to keep the water from becoming foul. The snails will also eat algae off the glass. One medium-sized snail for every gallon (4 L) of water is sufficient. Watch for dead snails, because their decaying bodies rapidly poison the water. Other animal forms, such as water insects and other invertebrates (beetles, hydra, copepods, ostracods) can also be placed in aquariums.

e. *Plants.* Plants can transform a plain-looking aquarium into an attractive one. They are also beneficial to the environment in practical ways, producing oxygen and consuming carbon dioxide. They use refuse as fertilizer, produce shade, and provide hiding places and egg-laying spots for the fish. There are many varieties of plants that might be chosen. Cool, freshwater aquariums might include Sagittaria, Ludwigia, water poppy, and Nitella. Tropical tanks are colorful when arranged with Anacharis, Ludwigia, Vallisneria, and sword plants. Saltwater containers can use Enteromorpha, sea lettuce, rockweed, and Ceramium.

f. *Equipment.* A great deal of equipment can be obtained to improve the efficiency of an aquarium: filters, bubblers, aerators, controlled feeders, and so forth. Most are unnecessary, but a few slightly improve the operation of the aquarium. A dip tube is essential for removing sediments that collect in the tank. Decaying matter, which soon fouls the water, should never remain in the container. Mechanical filtering of the water through glass wool and bone charcoal makes cleaning the tank a rare necessity. These filters are usually attached to the back of the tank. Subsand filters are also available and depend upon bacterial action to purify the water. Aerators have several advantages. They cause bubbles to rise through the water. The bursting bubbles break the surface tension, increase the air-to-water surface, and maintain more uniform chemical conditions. Saltwater tanks should be provided with aerators. Tropical fish tanks require heating units and thermometers.

g. *Problems.* All plants need light in order to manufacture their own food. Too much light, however, will cause green algae to form on the sides of the tank; and if the water is not fresh, too little light will cause brown algae to form. They will not hurt the fish, but algae are often unsightly. The condition usually clears by changing the amount of light on the tank. When the edge of the water becomes surrounded by bubbles or when the fish swim at the surface sticking their mouths out into the air, something is wrong. Usually the trouble is pollution or a lack of oxygen in the water. Check for dead snails, dead fish, or uneaten food.

After an aquarium is established, students can test the effect of changes in light or temperature upon the environment. They might observe that one or two animals dominate or become the "bosses" for other animals. Have them introduce a new animal into the aquarium to see how the others act toward it. They might also train fish to respond at feeding time to a tap, whistle, or other signal.

Contributing Idea B. Some plants and animals live on land, or in water and on land. The terrarium is a simulated environment designed for observing land plants and animals.

Building terrariums. Terrariums provide first-hand observations of land plants, animals, and their relationships. Basically there are three types of land environments, each with a particular set of environmental characteristics.

02
Gc

a. *Woodland environment.* To create a woodland environment, cover the bottom of a container with a layer of gravel for good drainage, then add several inches of fertile soil. Bury pieces of charcoal into the soil to absorb gases and prevent the soil from souring. Add a few rocks and a dish of water to provide the necessary moisture for this environment. With a glass top, the terrarium will maintain its own water cycle. Mosses, ferns, and other small woodland plants can be placed in the container and watered infrequently. If mold appears, remove the moldy plants and leave the top of the container open to allow some of the excess moisture to evaporate. Small animals such as frogs, turtles, toads, salamanders, newts, and snakes can be added. *(Note: The animals must be compatible and you should not put in too many. Be sure to remove any excess food that these animals do not eat.)*

b. *Swamp environment.* To create a swamp environment, use bog soil or make a base of sand, peat, and gravel in equal parts about 2 inches (5 cm) deep. Place a pan of water on a gentle slope of earth to make it easy for animals to climb in and out. Mosses, ferns, Venus's-flytraps, and pitcher plants can be placed in the container. Set the mosses in shallow soil and the other plants in deeper soil. To make an attractive display, place taller plants in the back and on the sides. Avoid overcrowding, but arrange everything so that no bare earth shows. After planting, sprinkle the terrarium with water. A glass top will help this terrarium maintain its own water cycle. Frogs, newts, baby alligators, tadpoles, salamanders, and turtles will live in this moist, marshy habitat if they are limited in number and type. Most of these animals will prefer live food such as insects, mealworms, caterpillars, worms, moths, and slugs.

c. *Desert environment.* To create a desert environment, build a base with 2 inches (5 cm) of sand. Add rocks and a small dish of water. Plant desert plants such as cactus. Small desert snakes, lizards, and horned toads can be kept in this habitat. *(Note: Desert plants decay from too much water, and molds might grow.)*

After a terrarium is established, students can test the influence of changes in light or temperature on the environment. For example, they might turn it part way around for a few days to see what happens to the plants and animals. They might also try setting it under a 60-watt light all night for several weeks to see what happens.

242 INTERACTIONS

Generalization I. Plants are adapted to their environment for survival. Various structures of plants operate in special ways which keep the plants alive and growing.

Contributing Idea A. Roots anchor plants and hold the soil in place.

Observing the anchoring power of roots. Have students pull up several different kinds of weeds from a lot near the school. Note and discuss the root structures of those with the strongest anchorage. Now find the same kind of weed growing in both a damp place and a dry place, noting the root structure and anchorage power of the weed in each place. If possible, obtain corn plants for observing their anchoring root systems. (Corn has *prop roots* that extend from the stem above the ground to the soil.) Students can look for other plants with prop root systems: mangrove trees grow in saltwater swamps and support the tree from both its trunk, and its branches; banyan trees grow in the tropics and have very visible prop root systems. **01** **Aa** **Bb** **Ca**

Generalizing that roots hold the plant in the soil and keep the soil from washing away. Grow some grass in moist soil on a board, in a shallow baking pan, or in a sand table. After the grass has grown to about ½ inch (1.5 cm) in height, clip the blades so that only the roots remain. Prepare a second pan with identical soil, but without grass. Slightly tilt the pans over a sink, and sprinkle water on them. Make comparisons to see which pan best retains the soil. Draw inferences about local surroundings where the soil is easily washed away during storms. **02** **Fa**

Contributing Idea B. Roots absorb water and dissolved substances from the soil.

Observing the importance of roots to plants. Obtain two identical leafy plants. Cut the roots from one and leave them on the other. Replant both, and observe their growth at daily intervals. Students will realize that roots provide an essential function for the growth of a plant. **03** **Aa**

Observing that roots are the primary means by which a plant absorbs water. Obtain two closely matched plants. Water the soil of one normally, and keep track of how much water is used. Using the same amount of water, water only the leaves and stem of the second plant. To prevent the water from dripping onto the soil, waterproof the pot opening by carefully wrapping aluminum foil or plastic wrap around the base of the plant. Let the foil extend over the sides of the pot. A tray underneath the pot can collect any runoff. Within a few days, observable results will be fairly conclusive: The plant whose leaves and stem were watered will be drooping; the soil watered plant will be thriving. **04** **Aa** **Ec**

Observing the path of the conduction tubes through a root. Stand a carrot in a glass of red-colored water. After two days, have students first cut the carrot crosswise to observe the colored tubes, then cut it again lengthwise to trace the path the red water has taken. It can be observed that the tubes go from the branch roots to the main root, then to the stem and leaves. **05** **Aa**

Generalizing that roots absorb dissolved substances, but not suspended solids. Insert three similar plants into test tubes containing respectively—water, red ink, a suspension of red congo. After a few days, the plant with ink will have visibly absorbed the color, while the others will have absorbed only the water. This activity can be done with other solutions and suspensions—e.g., red dye (solution) and tempera paint (sus- **06** **Ec** **Fa**

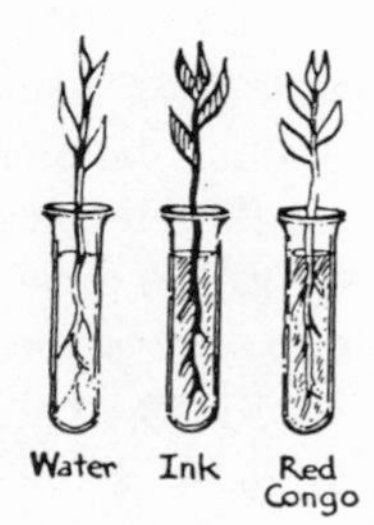

FIGURE 242.06

pension). Students will be able to infer from several tests that plants can absorb soluble materials but not nonsoluble materials.

Contributing Idea C. The stem is the part of the plant that supports other organs.

Comparing differences in stems. Let students tour the school grounds or neighborhood to see (1) if all plants have stems; (2) which portion of the plants are stems; and (3) what differences there are in the way stems grow and support appendages. They can observe that not all stems are rigid or upright (e.g., ivy, morning glory, water hyacinth, water lily), that some creep or twine, and that some that float on the surface of water are hollow and air-filled, while others beneath the surface are not (e.g., bladderwort). **07 Ca**

Contributing Idea D. Stems conduct liquids and minerals from the roots to other organs.

Observing that stems transport liquids to the various parts of the plant. Obtain a white carnation and submerge its stem in water. While under water, cut off the end of the stem. (Holding it under water prevents air from entering and blocking the conducting tubes.) Place the carnation in a container **08 Aa Ca**

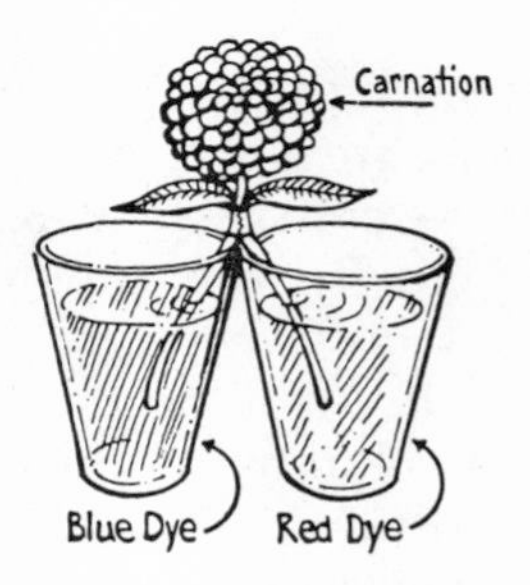

FIGURE 242.08

half-filled with water with food coloring, vegetable dye, or washable ink. Set the container in sunlight for several hours. Observe the carnation at half-hour intervals. Students will see that the petals become colored and infer that the colored water rose up the stem through the conducting tubes. Several variations can be used as supplementary activities.

a. White lily or aster stems can be used in place of carnation stems to show the transportation of a colored liquid to the flower. Celery stalks with leaves attached can be used to show the transportation to the leaves.
b. Split the stem of a white carnation into two or more parts with a razor blade. Extend the split several inches up the stem and wrap the stem with tape above the cut to avoid further splitting. Gently spread the sections out and place them into containers, each filled with differently colored water. Observe the plant at half-hour intervals.

Experimenting to see if stems carry materials dissolved in water. Obtain three healthy stalks of celery. Place one in a jar of sand and water, the second in a jar of sand and sugar water, and the third in sand and colored water. From observations over several hours, students will see that in **09 Ec**

the third stalk, dissolved material had been transported through the stem. By cutting the first and second stalks crosswise, students can taste the liquid droplets which form at the cut to see if any dissolved sugar was transported through the stem. Jar number one is used as a control to make sure that the dissolved sugar in jar number two was transported by the stem and not manufactured by the plant.

Contributing Idea E. Some leaves trap insects.

Observing leaves that trap insects. 10 Aa Many insect-trapping plants can be obtained from local nurseries. A hand lens is useful in observing them. Insect-trapping plants secrete liquids that dissolve the soft body parts of insects and small animals and absorb them as food. These plants, however, also manufacture food in sunlight as other plants do.

a. *Sundew Plant.* Students can see that the leaves of the sundew plant are covered with hundreds of tiny red hairs that look like pins on a pincushion. They can note that a drop of red liquid forms at the tip of each hair. Have them touch a droplet with a toothpick. The droplets are sticky and can be pulled away like a piece of gum. Students can infer how the sticky droplets capture small insects.

b. *Butterworts.* Students can observe that the oval yellow leaves of butterworts are covered with a shiny, sticky substance that traps small insects. These leaves generally lie flat on the ground and the edges of the leaf roll inward to hold the victim.

c. *Pitcher Plant.* This plant resembles a pitcher. Students can see that it retains rainwater and that many bristly hairs surround the mouth of the pitcher. They can note that these hairs all point down and inward. From these observations they can infer that insects attracted into the flower cannot leave and are drowned.

d. *Venus's-flytrap.* The leaves of this plant have two lobes, each fringed with bristles. On the surface of each lobe, students will see three hairs. A touch of a hair with a pencil point or toothpick will close the two lobes like a trap. Students can feed small insects to this plant.

Contributing Idea F. Reproduction in flowering plants is accomplished through pollination.

Generalizing that both male and female organs are necessary for reproduction in flowering plants. 11 Bb Fa Let students examine the different parts of a simple, complete flower (one which has both female and male organs) such as the tulip or Easter lily. They can easily find the male organs *(stamens)* that produce a powdery substance *(pollen)*. Let them touch the large tops of the stamen *(anthers)* and feel the yellow dust that sticks to their fingers. Let them blow the pollen from their fingers. Have them move the stamens of the flower with a pencil tip, and they will see the female organ *(pistil)* which looks like a slender vase. Touch the top of this organ *(stigma)* and feel its sticky surface. Describe the shape of its lower part *(ovary)*. After blowing pollen from their fingers and feeling the sticky stigma, students will be able to infer one way by which pollen can reach a stigma (by being wind blown). They will also be able to infer that the sticky stigma is built to receive the pollen.

Observing that pollen must fall upon the stigma of a pistil before seeds can form. 12 Aa Keep a flower under unpollinated conditions for control purposes. This can be done by putting plastic wrap around the pistil of a new bloom. After the flower has wilted and dried, let students compare it (especially the ovary section) with a pollinated flower from the same plant. Through dissection, they can compare the differences in the ovules and the surrounding area.

Pollinating flowers. Obtain a plant, such as a tomato plant, that has been growing indoors and is now flowering. Wrap a small amount of absorbent cotton around one end of a thin piece of stiff wire. Make sure the cotton is firmly wound and not as thick as a commercial cotton swab. Select a full flower, and gently separate two petals so that the cotton tip can be inserted. Gently move it within the flower, touching first the stamens, then the tip of the pistil. Students will see the transfer of the pollen to the stigma. Tell them that when a plant is pollinated within its own flower, it is said to be *self-pollinated.* Using a fresh cotton-tipped wire for each flower, let students continue this process with other flowers, leaving some flowers unpollinated as controls for later comparisons. On a different plant, use the same cotton-tipped wire to touch all the flowers. Students will realize that this type of pollination is different from the previous type. Tell them that when pollen is moved from one flower to the stigma of another flower, it is said to be *cross-pollinated.* Experiment with each type of pollination. After four or five days (for most flowers), observe the flowers. Compare them with those that were not pollinated. **13 Bb Ec Ga**

Generalizing that pollination can take place naturally with the aid of wind. An analogy can be made to the traveling of pollen by wind if students lightly dust some flour inside the petals of a flower, then blow on it. They can see where the dust travels and note that some dust sticks to the stigma of the flower (self-pollination) and that some sticks to the stigma of other flowers that are close by (cross-pollination). **14 Fa Fc**

Collecting wind-borne pollen. Coat glass slides with petroleum jelly and suspend them in various locations outdoors. Each day, note the pollen and dust that accumulates on them. Different pollen that is collected can be compared under a low-power microscope. Slides can be fixed by adding a bit of crystal violet and aniline oil to the pollen and placing the slide over a flame until the pollen is deeply stained. Students can also sort the pollen by physical characteristics such as shape, size, or color. **15 Db Ge**

Comparing flowers pollinated by animals with those pollinated by wind. Have students look at the flowers of grasses, grains, and weeds to see if they are colorful or fragrant. Compare the pollen from these flowers with the pollen from more colorful flowers. Explain that flowers that depend upon wind to transfer pollen are not usually colorful or fragrant. Students can infer how differences in pollen grains and flower appearance are appropriate to the way they are pollinated. **16 Ca**

Contributing Idea G. Seeds are dispersed in various ways that contribute to a plant's survival.

Collecting seeds. Because of their wide dispersal, seeds can be collected from many locations. **17 Ge**

a. Dip a hand in a creek or stream to collect any seeds floating there.
b. Scoop up a spoonful of mud from around a drainpipe or a low place in a garden where seeds have been washed.

c. Stroll across a field or lot, dragging a woolen cloth. Examine the cloth for seeds that have collected on it. The cloth can be compared to animal fur or clothing.

Collected seeds can be stored in cellophane bags or envelopes labeled with data concerning the location, description of the seeds, and the date they were collected.

Examining seeds by the way they are dispersed. Collect various kinds of seed cases or capsules. Examine the capsules to determine methods of seed dispersal (e.g., columbines have a pitcher-like mouth at the top from which seeds are poured into the wind; poppies have tiny holes at the top with trap doors beneath them so that seeds are slowly scattered as the capsule shakes in the breeze; witch hazel, violet, touch-me-not, and jewelweed have capsules that swell and burst to pop their seeds in many directions). Depending upon locale, various seeds can be obtained and observed. **18 Aa**

a. Students can collect branches of Scotch broom that still carry pods. Allow the branches to dry. Listen and watch for signs of seed dispersal. Examine the pods before and after drying.
b. Place wild geraniums or witch hazel fruits in a covered jar with moist cotton. Place others of the same kinds in a dry, uncovered jar. Observe both for several days to see what happens as the fruits dry.
c. Obtain two mature blossoms of poinsettia, thistle, and dandelion. Watch them as they dry and go to seed. Have one student blow on a dry dandelion head, and observe how the seeds travel.
d. Collect pine cones. Heat them in an oven or set them in a warm place for several days. Students can observe the seeds as they pop out.

Generalization II. Animals are adapted to their environment for survival. Various structures of animals operate in special ways which keep the animal alive and growing.

Contributing Idea A. Animals are adapted to enable them to obtain and eat food in their environment.

Observing that the physical characteristics and actions of fish are related to their acquisition of food. Prepare an aquarium containing fish that feed in different ways (e.g., goldfish, catfish, and sunfish). Observe the feeding habits as commercial food is dropped into the container. Students will see that sunfish dart to the surface and snatch the food, goldfish take it more or less in the middle of the container, and catfish wait until it settles on the bottom. Let students describe the physical characteristics of the fish to see how they relate to each behavior (e.g., catfish have mouths on their undersides and long feelers for searching out food in crevices at the bottom of an aquarium). Research might be done on other fish to see how their characteristics enable them to obtain food. **19 Aa Bd**

Observing that the physical characteristics of birds are related to the food that they eat. Select pictures of birds on the basis of their bills and/or feet. Discuss how certain characteristics seem specialized for gathering and eating certain foods. The following illustrations will be helpful in identifying the relationships. **20 Aa Db**

a. Bill is like a long, pointed nail—useful for

digging into tough tree bark to pull out insects (woodpecker).

b. Bill is small, short, and pointed—useful for cracking open seeds and nuts (canary, meadowlark).

c. Bill is like a sharp hook—useful for tearing meat from the bones of animals (owl, hawk).

d. Bill is like a shovel—useful for scooping plants and small fish from water (duck, swan).

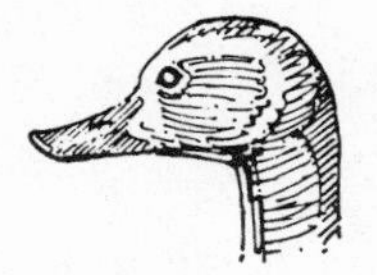

e. Feet have long toes, two forward and two backward—useful for holding onto vertical tree trunk (woodpecker).

f. Feet have short, curved toes, three forward and one backward—useful for perching on round tree limbs (robin, swallow).

g. Feet have sharp talons—useful for grasping and holding animals securely (eagle, hawk).

h. Feet are webbed—useful for paddling in water and walking on mud (duck, pelican).

On the basis of bills and feet, students can tentatively sort bird pictures into such categories as seed-eaters, insect-eaters, and meat-eaters. Let them check their classifications by researching various bird books.

Observing that the physical characteristics and actions of animals are related to their acquisition of food. Let students look at numerous pictures of animals and discuss the types of food each eats and whether or not each has a physical characteristic or ability that enables it to capture and eat the food. For example, meat-eaters and plant-eaters have teeth specially adapted for the type of food each eats; the long neck of the giraffe enables it to feed on high leaves in trees; the sacks on the legs of

21 Aa

bees enable them to carry pollen to their hives.

> *Contributing Idea B.* Animals are adapted to enable them to obtain oxygen from their environment.

Observing respiration in an earthworm. 22 Aa Ec Fa
Obtain four identical jars. Leave two empty and fill two with water. Select four similarly sized earthworms, placing one in each jar. Set one empty and one full jar on a shelf out of direct sunlight where they will not be disturbed. Set the others in a refrigerator (not the freezer). After one day, take the earthworms from all the jars, and put them on a sheet of waxed paper. Students can look for pulses of blood in the large blood vessel on top of each worm's body. Count the number of pulses per minute for each and compare the rates. Explain that an earthworm has no special organ for breathing. It obtains the oxygen it needs directly through its skin to its blood. Because of this, it can live under water. If the skin becomes dry, gases cannot pass through it, and the animal will die. Since the worms from the refrigerator can be seen to breathe more slowly than the others and are less active, students can realize that greater activity requires more oxygen.

Observing respiration in an insect. 23 Aa Fa
Leave a jar of water outside when mosquitoes are plentiful. When students see little wigglers hanging from the surface or moving through the water, put a piece of screen across the top of the jar. Count the number of times per minute a wiggler comes to the surface for air and the number of seconds a wiggler will stay down without coming up for more air. Explain that a mosquito wiggler hangs head down in the water with an air tube just above the surface. Use a hand lens to find the openings along the sides of a large insect's body. Explain that these openings lead to tubes that go into the insect's body. Oxygen from the air slowly enters the body through these tubes. This method of taking oxygen is a characteristic among insects.

Observing respiration in a fish. 24 Aa Fa
Have students observe a fish in an aquarium. Count how many times a minute it opens and closes its mouth, then count how many times the flap or gill covers just behind its head move per minute. Students will discover a relationship between the movements. (When the mouth is open, the gill covers are closed and vice versa.) Explain that a fish breathes by means of gills. As water enters the fish's mouth and passes over the gills, oxygen that is dissolved in water passes through thin tissues inside and enters the fish's blood. The water leaves the body by way of the gill cover. Now fill a beaker with water and boil it for five minutes to remove any dissolved air. When it cools to room temperature, place a fish into the water and observe. Return the fish quickly to the aquarium so that it does not die from lack of oxygen. Students will realize that fish depend upon oxygen in the water in order to live.

Observing respiration in an amphibian. 25 Aa Fa
Let students observe a frog in a terrarium. An indication that it breathes can be seen by the open nostril vents near its snout and the movement of its throat. Explain that a frog breathes by "swallowing" air. (The floor of its mouth raises and lowers, causing a pumping effect.) Frogs also absorb a good deal of oxygen through their skin. Count the number of times a minute its throat moves in and out, then hold the frog securely by its hind legs and submerge it completely in a jar of water for a minute. Students will see that the

nostril vents close and the throat movements stop. Do not hold the frog under water too long, for it will drown. Tell students that frogs also have the ability to breathe through their skins.

Observing respiration in people. Have students exercise and note the effect it has upon their rate of breathing. They can determine the capacity of their lungs by filling a large jug with water, holding a hand over the mouth, and inverting it in a pan of water. Insert a plastic or rubber tube into the jug. Use a clean piece of glass tubing as a mouthpiece for each student. Have them take turns exhaling fully into the jug and measure the amount of water forced out each time by refilling the jug. Compare capacities to the rates of breathing before and after exercising.

26
Aa
Fa

300-399
Energy

310 motion

311 CHARACTERISTICS

Generalization I. Motion has indentifiable characteristics.

> *Contributing Idea A.* Motion is the act of changing place or position.

Defining motion. Have students find evidences of motion they can see inside and outside the classroom. Have them then give examples that are not motion (e.g., thinking, objects that are not moving). From the examples, help students see that motion is always a coming or going from place to place. Let them use magazines to locate and collect pictures depicting motion. The pictures can be arranged on bulletin boards or in scrapbooks. **01 Bb Ga**

Identifying distance and time as two characteristics of motion. Have students list examples of motion that clearly involve moving from place to place. Ask them if motion is possible without distance being involved. (An object cannot go from one place to another if there is no distance between the two places.) Ask if an object can travel a distance without time being involved. (No matter what is moved, the journey takes some time.) Give examples of motion that clearly involve distance and time, then ask students to find other examples (e.g., students travel some distance from school to home, and it might take ten minutes, twenty minutes, or more to travel the distance). Identify in each example the parts that indicate that motion covers distance and also takes time. **02 Bc Ga**

Measuring motion in terms of distance and time. Let students work mathematical problems involving motion such as those given below. **03 Bb Cd Ch**

a. If you ride a bicycle ten miles in two hours, how far can you travel in one hour (or at what rate does the bicycle move per hour)?
b. If a man runs 100 yards in ten seconds, what is his average speed (or rate of distance covered per second)?
c. If an airplane travels 600 miles in one hour, what is its speed (or rate of distance covered per minute)?

Tell students that rates of motion are called *speed.* Explain that speed involves distance and time. To figure the average speed of a moving object, one must find the rate at which it travels a known distance in a known time. For each problem, the average speed is obtained by dividing the total distance by the total time:

$$\left(\text{Speed} = \frac{\text{Distance}}{\text{Time}}\right)$$

Contributing Idea B. The direction of motion is a straight line unless acted upon by other forces.

Observing straight-line motion. While students observe, place a checker on a table and flick it with a finger, roll a toy car in a straight line across the floor, and drop a ball from a table. Ask if each of the objects moved in a straight line. Repeat the experiences if necessary. Students will realize that each object moved in a straight line. Tell them that scientists call such motion *regular straight-line motion.* Recall other examples for the students: a bicycle continues moving in the same straight line if the front wheel is not turned, people lurch forward in a bus, train, or car when it stops quickly.

04
Aa
Bb

Observing influences on straight-line motion. Roll marbles across a smooth, level surface. Students will see that the marbles always go in straight lines. Now let them roll a single marble, and blow on it from the side as it travels. Discuss how the motion was changed. Next, hold a strip of thin cardboard on edge and curve it slightly. Roll a single marble into the curve of the strip, and discuss any change in its direction. Compare the effect of the cardboard to the blowing. Students will begin to realize that all moving objects travel in a straight line (e.g., bowling balls, hockey pucks, rain drops) unless influenced by other forces.

05
Aa
Ba
Ca

Observing periodic motion. Have students work in small groups. Let each group tie a string around a small weight such as a fishweight, metal washer, or nut. Suspend the string from a stand or the edge of a desk. Pull the weight back, and let it go. Explain that the apparatus is a simple *pendulum.* Now use stop watches or clocks with second hands to determine how long it takes for each pendulum to swing forward and back. After students have determined the time, tell them that one forward and back motion is called a *period* and that such repetitive motion is called *periodic motion.* (Grandfather clocks have pendulums with one-second periods.) Students might be asked to think of other examples of periodic motion (e.g., vibrating objects such as tuning forks, some clocks, swings on playgrounds, metronomes, some lawn sprinklers).

06
Bb
Ch
Ec

Measuring periodic motion. Suspend a large metal washer from a 1 foot (30 cm) length of thread and determine the time of its period with a stop watch or a clock with a second hand. One method for doing this is to count the number of periods (back and forth swings) in 30 seconds, then divide the number of periods by the number of seconds. For example, if one counts 60 periods in 30 seconds, then $60 \div 30 = 2$ periods per second or 1 period per half-second. Now provide students with rulers, scissors, more thread, and identical washers so that they can experiment to see what they must do to a pendulum to make its period twice as long as it was. Students might try varying the weight while keeping the length of the thread constant, or try varying the length of the thread while keeping the weight constant. Results can be recorded on a table. Students will find that the weight does not influence the period, but the length of thread does. If some students would like to see what effect the starting position of the weight has, let them rule a sheet of cardboard with horizontal lines 1 inch (1 cm) apart and stand the cardboard behind the pendulum. The weight can then be lifted to the 6 inch (12 cm) mark and released. (The height of the upward swing can be recorded as well as the period.) Repeat by starting the swing from different heights. (The height of

07
Bc
Ch
Ec

TABLE 311.07. Testing the Influence of Length and Weight upon a Pendulum

	Length	Weight	*Number of Swings*			
			Trial 1	Trial 2	Trial 3	Avg
TEST 1	6 in. (12 cm)	1 washer				
	6 in. (12 cm)	2 washers				
	6 in. (12 cm)	3 washers				
TEST 2	8 in. (16 cm)	1 washer				
	10 in. (18 cm)	1 washer				
	12 in. (24 cm)	1 washer				

the swing will have no effect on the period, however, the higher the starting point, the higher the swing.)

Observing rotary motion. Turn a bicycle upside down and spin the front wheel. Tell students that a continuous spinning motion traveling in the same direction is called *rotary motion.* You might explain that rotary motion is similar to periodic motion in repeating itself, but different because there is no change in the direction of rotary motion unless force is applied to change it. Ask students to think of other examples of rotary motion (e.g., rotary lawn sprinklers, Ferris wheels, merry-go-rounds, minute and second hands on clocks).

08
Aa
Bb

Measuring rotary motion. Turn a bicycle upside down, mark one tire with chalk or tie a white cloth around the wheel. Spin the wheel, and let students count the number of complete turns in one second. Now have them obtain a smaller wheel, such as one on a wagon, and determine its speed or rotation. For both wheels, students can find out roughly how far the wheel would roll along the ground in one minute by measuring the distance around the outside of the wheel and multiplying by the speed. *(Note: In actuality the wheel will not rotate at the same speed when rolled on the ground due to other factors such as friction.)*

09
Cd
Ch

Contributing Idea C. Motion travels at a uniform rate of speed unless acted upon by other forces.

Graphing uniform rates of motion. Give students the following problems and tables of data. Help them transfer the data to a graph form, labeling the horizontal and vertical dimensions as shown. The result will be a straight-line graph that depicts a uniform rate of speed. The questions at the end of each problem will help students in reading graphs and extrapolating information from them.

10
Bf
Fb

a. Mr. Adams measured how long it took to fill a cup from the water tap in his sink. He then measured how long it took to fill a pint bottle and then a quart bottle.

TABLE 311.10a Measuring the Flow of Water From Mr. Adams' Tap

Volume of Water	*Time in Minutes*	*Volume in Pints*
1 cup	¼	½
1 pint	½	1
1 quart	1	2
1 gallon	?	8

Make a graph of Mr. Adams' measurements and tell how long it would take water from the tap to fill a gallon-sized container.

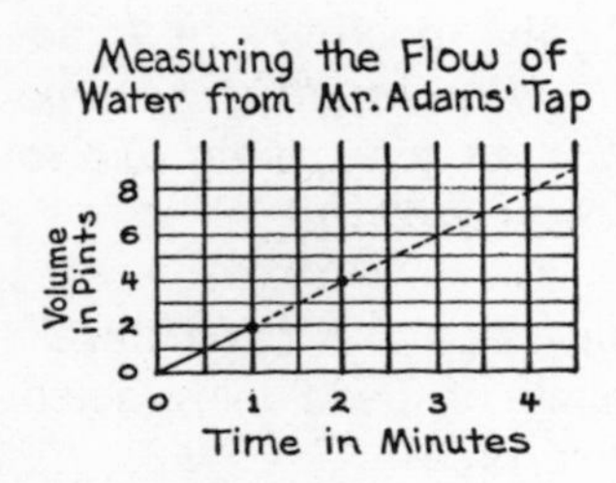

GRAPH 311.10b

b. Mr. Adams left home at 9 A.M. After he had been driving for an hour, he looked at his odometer and found that he had gone 50 miles. At noon he had traveled 150 miles, and at 2 P.M. he had gone 250 miles. Graph Mr. Adams' trip. How far had he gone by 11 A.M.? What time was it when he had traveled 200 miles? If he kept traveling at the same rate of speed,

TABLE 311.10c Mr. Adams' Trip

Time of Day	*Distance in Miles*
9 A.M.	0
10 A.M.	50
11 A.M.	?
12 M.	150
1 P.M.	?
2 P.M.	200

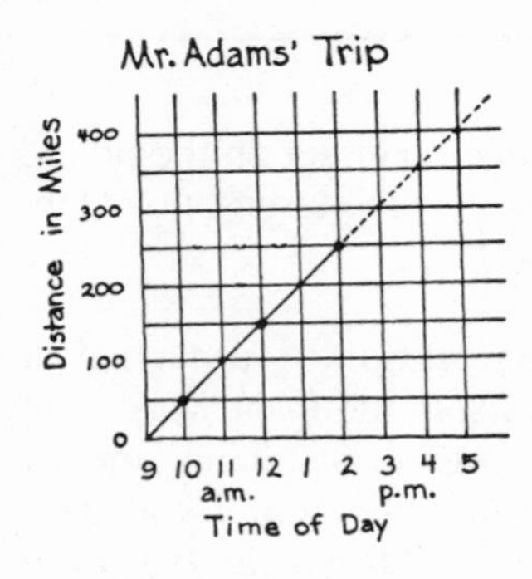

GRAPH 311.10d

what time would it be when he traveled 400 miles and how far would he have gone by 5 P.M.?

Defining uniform motion. Let students determine the average speed of vehicles in each of the mathematical problems given below. **11 Bb Cd Ch**

a. Two speedboats, *A* and *B,* pass a rock at the same time, then travel directly north for one hour to a bridge. Boat *A* travels 5 miles in the first 20 minutes, 5 in the

TABLE 311.11a Speedboats up a River

Boat A			*Boat B*		
Time in Min.	*Distance Miles*	*Cumulative Miles*	*Time in Min.*	*Distance Miles*	*Cumulative Miles*
20	5	5	10	5	5
30	5	10	30	10	15
40	15	25	40	5	20
60	5	30	60	10	30

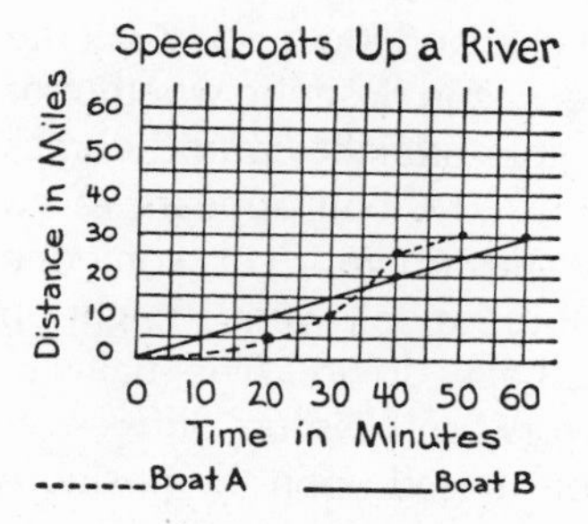

GRAPH 311.11b

second 10 minutes, 15 in the third 10 minutes and 5 in the final 10 minutes. Boat *B* travels 5 miles in the first 10 minutes, 10 in the next 20 minutes, 5 in the next 10 minutes, and 10 in the last 20 minutes. What is the average speed per hour of each boat and how would you describe the motion of Boat *B?*

b. Two cars travel a long, straight highway across a desert for two hours until they reach the outskirts of a city 60 miles away. Car *A* has car trouble and travels 20 miles per hour for the first hour, 60 miles per hour for the next 30 minutes, and 20 miles per hour for the last 10 minutes. Car *B* travels at exactly 30 miles per hour throughout the trip. What is the average speed per hour of each car and how would you describe the motion of Car *B?*

TABLE 311.11c. Trip Across a Desert

Car A			*Car B*		
Time in Min	*Miles*	*Distance Cumulative Miles*	*Time in Min*	*Miles*	*Distance Cumulative Miles*
60	20	20	60	30	30
90	30	50	120	30	60
120	10	60			

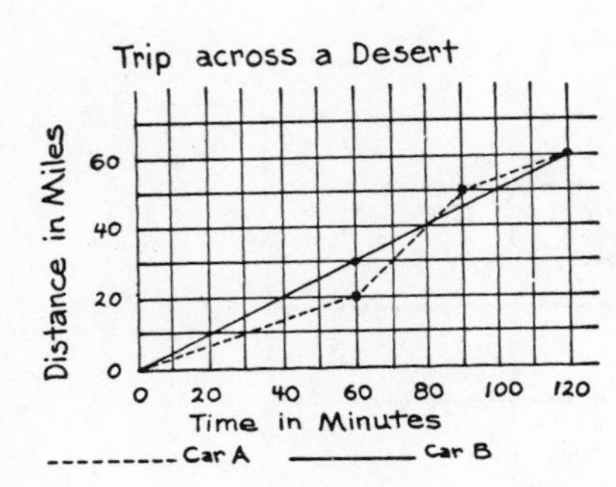

GRAPH 311.11d

In each of the above problems, the speed of vehicle *A* is not steady during the entire trip, while the speed of vehicle *B* is steady. Tell students that although each vehicle had the same average speed, only vehicle *B* had a motion in a straight line and at a steady speed. Such motion is so regular that it is called *uniform motion.*

Increasing and decreasing rates of speed. Make a trough by taping two rulers together. Support one end so that it is raised ½ inch (1 cm) above a flat surface. Roll a marble **12 Aa Ba Ec**

FIGURE 311.12

down the trough and allow it to coast to a stop. Ask students to tell when the marble accelerates (down the slope) and when it decelerates (throughout the coasting distance). Try tipping the trough to see if the marble will travel faster. Now set a strip of cloth or strips of sandpaper along the coasting route, and roll the marble again. Students will see that the marble will not travel as far. Let them try other materials to see how each affects the speed of the marble.

Contributing Idea D. Descriptions of motion depend upon the frame of reference used.

Observing motions from different positions. Several activities will help students realize that the position of the observer is important to know when describing motions because all motions are relative. **13 Ca**

a. On the playground, have a student sit in a wagon and throw a ball 8 to 10 feet (2 to 3 m) in the air and catch it. Have another student pull the wagon in a straight path and continue having the first student toss the ball. Be sure the velocity of the wagon is not changed too quickly. Let the person in the wagon describe whether or not the ball goes straight up and comes straight down. (The ball appears to go straight up and down.) Let someone watching tell if the ball went straight up and straight back down. (The ball appears to arc forward.) Discuss differences in descriptions based upon the positions of the observers.

b. Use chalk to draw a large circle on the playground. Have one student stand in the center of the circle while another sits in a wagon parked on the circle. As the wagon is pulled along the line of the circle, let the students in the wagon and center of the circle play catch. Students will find that the one in the wagon must aim slightly to the right or left of the student in the center, depending on which direction the wagon is moving, to hit the target. The student in the center must throw ahead of the wagon so that it gets there just as the student in the wagon does. Because of positions of observers, the ball appears to go in a curved path to the observers on the side, but in a straight path to the student in the moving wagon. Again, students can note that the interpretation of motion depends upon the position of the observers.

Generalization II. Motions can be organized.

Contributing Idea A. Motions can be seriated.

Seriating motions by rates of speed. Since all motions involve distance and time, students can time the movement of several moving objects along a given distance and order them from fastest to slowest. For example, runners, horses, or cars in a race can be timed and their times seriated. Similarly, the speed of the wind taken at the same hour, but on different days, can be ordered. The amount of water running from a faucet can be compared to the amount from other faucets.

14
Cd
Ch
Da

Seriating motions by distances traveled. Roll objects such as marbles, skates, or racing cars down an inclined plane. Start each object at the same position and measure the distance it travels. Order the objects by the distance each travels. Challenge students to find an object that will travel farther than any of the others.

15
Cd
Da

Contributing Idea B. Motions can be classified.

Classifying motions. Have students list examples of forces that cause motion. They will find that any force, living or nonliving, that starts something moving can be classified as a push or pull on something else.

16
Bc
Dc

312 INTERACTIONS

Generalization I. Motion is a form of energy produced by unbalanced forces.

Contributing Idea A. A force is any push or pull action upon an object.

Feeling forces that put objects into motion. Have students put the palms of their hands together and push. Next, have them hook two forefingers together and pull. In each case, let them describe what they felt. Using these activities, explain that a force always has a direction. (When hands are placed together and pushed, the direction of the right hand force was to the left and vice versa; when the hooked forefingers are pulled, forces are exerted in opposite directions.) Now have students move various objects by pushing or pulling. Tell them that words like push, pull, shove, and tug are often used to describe what makes objects move, and scientists use the word *force* to mean all of these actions.

01
Ab
Bb

Feeling forces that stop the motion of objects. Have every student stretch a rubber band tightly between two fingers of one hand and push against the rubber band with one finger of the other hand. Discuss what they feel. (The rubber band will seem to push back on the finger.) Now have them push harder, and ask what they feel. (It will seem to push back harder.) Explain that objects stop motion by pushing back, even though this is hard to see. Have one student push against a wall and compare what is felt to the pushing of the rubber band. Have the student place a postal scale with a square platform top on its side with the dial up and its base against the wall. Place a second scale so that its platform is flush against that of the first scale. Now push against the wall again by pushing on the second postal scale. If the student pushes with a force of 3 ounces (100 g) indicated on the scale, the scale against the wall will push back with an equal force. By testing, students will find that the two scales will always push back with

02
Ab

an identical force. (If one pushes against a wall or other surface, it will push back with equal force, thus stopping the forward motion.) This activity can be repeated by attaching a spring scale to a wall and pulling on it with another spring scale.

Applying forces to start, stop, and change the direction of a motion. Have a student place a ball on a table and push it with a finger. While the ball is rolling, push it from the side. Next, place a finger in front of the ball and apply only enough force to stop it. Point out to students that a force was used in each case. One force put the ball in motion, another changed the direction of the motion, and a third stopped the motion. Let students tell how the application of forces to change the direction of an object's motion takes place in various sports (e.g., kickball, baseball, football, tennis). **03 Af**

Contributing Idea B. Gravity is a force that can produce motion.

Feeling the force of gravity. Have a student hold an empty bucket out in front of her. Let another student pour sand or water into the bucket slowly. Have the student holding the bucket describe what is felt. (As the material is added, the bucket feels heavier and is pulled downward.) Explain that the more material there is, the greater the pull of the earth upon it. Tell students that the pull is called *gravitational attraction.* **04 Ab Bb**

Using gravitational attraction to produce motion. Tie a thread to a pencil and suspend the pencil from a table top by holding the thread in place with a book. When the pencil hangs motionless, two equal and opposite forces are being exerted upon it. Now ask students if they can put the pencil into motion without directly acting upon it themselves (the pencil might be held in the air then released; the book might be lifted from the thread; the thread might be cut). Students will find that in each case the gravitational attraction caused the pencil to move by pulling it toward the center of the earth. Explain that this is a natural force and not a human force. Repeat this activity using other objects such as wads of paper, paperclips, or fishweights. Students will find that the force exerted by gravity is a constant and affects all objects. Some children might be interested in looking up the English mathematician, astronomer, and philosopher, Sir Isaac Newton (1642–1727), in an encyclopedia to learn more about gravitational attraction. **05 Bd Ga**

Contributing Idea C. Magnetism is a force that can produce motion.

Using magnetism to produce motion. Slowly bring a magnet close to an iron nail and observe as the nail is pulled toward it. Now hold the magnet under a thin board or sheet of paper and place the nail above it. By moving the magnet, the nail is also moved. Next, bring two like poles of two magnets close together. The magnets will move apart. One magnet can be used to push the other across a table top. **06 Ga**

Contributing Idea D. Muscular activity can produce motion.

Using muscular activity to produce motion. Have students take a sheet of paper, crumple it into a ball, put the ball on top of a desk, and snap it gently with a finger. Discuss what made the ball move, whether or not it moved in a straight line, and what stopped its movement. Discuss how to make the ball move faster. Put the paper ball on **07 Ga**

a desk again; this time blow on it and repeat the previous discussions. Through these experiences, students will realize that motion of an object takes place only when some force is applied. They will realize that in these cases, muscular activity produced the motion.

Using muscular and mechanical activity to produce motion. Clamp a hook screw into a hand drill. Attach a 1 foot (30 cm) length of string to the hook, and tie a large nail to the free end. Steadily crank the drill and let students observe what happens to the nail. Discuss how the nail was put into motion, and describe the type and direction of the motion. Other mechanical apparatuses can be constructed and used to produce motion (e.g., cranks, levers, pulleys, gears, windmills, water wheels). **08 Ga**

Generalization II. The mass of an object affects the object's motion as long as the speed remains constant.

Contributing Idea A. Mass is a measure of the quantity of matter present in an object.

Observing that mass is independent of size. Obtain two balls of different sizes, such as a bowling ball and a beach ball. Let students push each with a finger and tell which is harder to move. They will find that the bowling ball is more difficult to move and that it has the greater mass or quantity of matter. Next, ask students to bring balls of different sizes and masses to school. A table can be used to record observations. *(Note: The only time there will be any correlation between size and mass is when the objects being compared are of the same material.)* **09 Af Bc**

Contributing Idea B. Mass affects motion as long as the speed remains constant.

Measuring the effect of mass upon motion. Obtain two small boxes of equal size and shape. Label one box *A* and the other *B*. Fill to equal levels, box *A* with sawdust and box *B* with sand. Let students take turns pushing each box with one finger and discussing which was easier to move. Have them decide whether it was the shape, size, or mass (quantity of matter) that made the difference. They will realize that the mass must be the factor because the shape and size of the boxes were similar. Now let them measure this difference by placing each box on a roller skate, tying the skate to a spring scale, and pulling each box by the scale. **10 Aa Cc**

TABLE 312.09 Comparing Masses

Objects	*Harder to Move*	*Greater Mass*	*Greater Size*
Beach ball Bowling ball	 x	 x	x
Shot put Basketball	x 	x 	 x

11 Aa Bb

Defining acceleration. Let a marble or racing car roll down a track from a height of 1 inch (2 cm), and have students observe to see when the marble moves the most slowly and most rapidly. Tell them that when the speed of a moving object is increasing, scientists say that it is *accelerating,* and when it is decreasing, it is *decelerating.* Have students tell when they accelerate and decelerate while riding a bicycle or other vehicle.

12 Aa Bb Cd

Applying the same accelerating force to objects of different masses. Obtain two identical lengths of brass curtain rods and two wooden dowels that equal the rods in length, shape, and diameter. Use heavy thread to tie a clothespin so that it is held

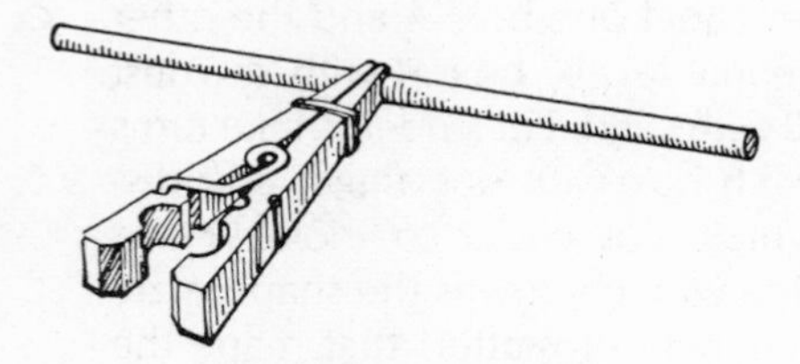

FIGURE 312.12

open. Set it on its side on the floor, and place the two wooden dowels against it. Let students observe as you cut the thread and the dowels are put into motion. Repeat, letting students mark the spots where the dowels come to rest. Tell them that when something is put into motion, it is said to be *accelerated.* Next, replace the dowels with the two rods and repeat. (The brass rods will move a shorter distance.) Students will realize, since the materials were of the same length, shape, and diameter and only different in mass, that when given the same force, the difference in mass must have caused the difference in acceleration. Now have them prepare the clothespin again but this time place a wooden dowel against one side and a brass rod against the other. Let them predict what will happen. (The wooden dowel with the lesser mass will have the greater acceleration.)

Generalization III. Factors that influence motion can be controlled.

> *Contributing Idea A.* Friction is a resistance to motion.

13 Af Bb Cd

Observing that friction impedes motion. Saw a 1 inch by 1 inch (2 cm by 2 cm) piece of lumber into cubes. Cut 1 inch (2 cm) squares of different materials (e.g., rubber, foam plastic, sandpaper, cloth, aluminum foil, leather) and glue them carefully on the faces of the cubes. Be sure the edges are trimmed so that each face rests on only one kind of material at a time. Set a cube on a smooth board resting on a flat surface and gradually raise one end of the board until the cube starts to slide. Students can measure

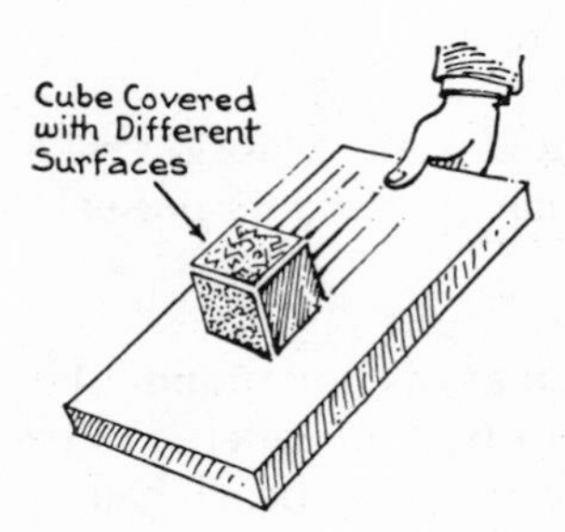

FIGURE 312.13

and record the height of the raised end when the cube begins to slide. Repeat several times to obtain an average. Then repeat with the different faces of the cube resting on the board surface. Discuss why there were differences among the different materials. Tell students that there is a resistance to motion when surfaces are rubbed together, and the resistance is a force called *friction.* Let them rub each material with a finger to feel the frictional differences in the surfaces.

Testing the effect of different surfaces on motion. Glue or tape light tagboard to make a pair of identical boxes like the one shown *(a).* Place one box on top of a flat table, and have a student blow on it. Let

14
Aa
Bc
Ec

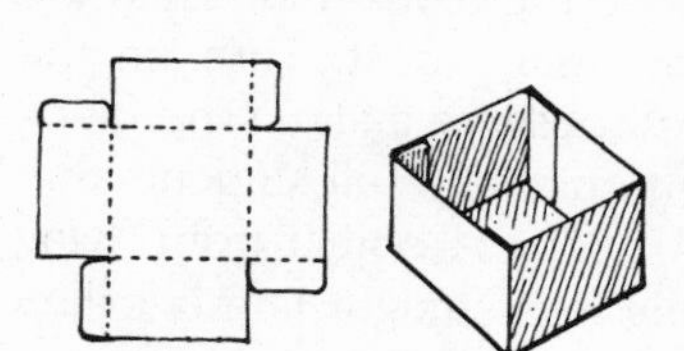

FIGURE 312.14a

another student push it with one finger. Now put ten identical pieces of chalk, crayons, or marbles into the box and push again. Students will feel that they have to apply more force to make the box move when it is heavier. Have them work in pairs to connect two paper boxes with 1 yard (1 m) length of string *(b).* Let one box hang over the edge of a table and place three crayons in the box on the table. Students can see how many crayons must be placed in the hanging box before the box on the table starts to move. Repeat this several times, then increase the number of objects in the box on the table

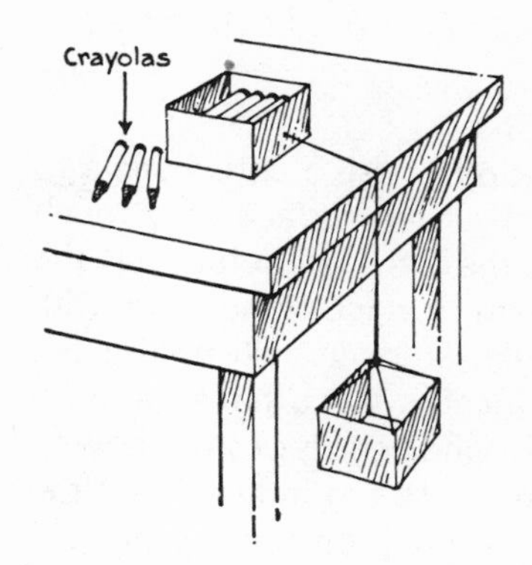

FIGURE 312.14b

and repeat again. *(Note: Be sure the students are careful that they do not give an extra push to the hanging box when they put the crayons in it.)* Now test different surfaces to see what effect they have on the motion of the box. For example, students can set a large sheet of sandpaper with the rough side up on top of the table, place the box on the sandpaper and find out how many crayons in the hanging box are needed to move it. Repeat the procedure for other surfaces, then table and discuss the findings.

TABLE 312.14. Testing Different Surfaces

Surface	*Number of Crayons in the Box*								
	3	*4*	*5*	*6*	*7*	*8*	*9*	*10*	*11*
table top									
sandpaper									
waxed paper									
construction paper									
other surfaces									

Contributing Idea B. Friction can be reduced.

Reducing friction. Use a spring scale to pull a brick across the surface of a rough board. Record the force needed to pull the brick. Now rub the surface of the board with soap and repeat. Students will note a decrease in the friction. Next, have them arrange a row of round pencils or small dowels (about 6 inches or 15 cm in length) on the board. Place the brick on the pencils and pull again. Some students may be interested in investigating early efforts in moving large stones to build pyramids, castles, cathedrals, and temples. **15 Cd**

Comparing sliding and rolling motion. Place some blocks or fishweights in a shoebox, and pull it along a table top by a rubber band or spring scale. Now set the box on twenty-four parallel drinking straws and pull again. Students will realize that the difference in effort needed to pull the box is caused by substituting rolling motion for sliding motion. Similarly, they can compare sliding and rolling motion by: **16 Ca Cd**

a. filling a tall, slender can with sand, setting it on a jar lid, and trying to spin it. Repeat after putting a circle of marbles in the lid.
b. rolling a skate down a playground slide, then fastening the wheels with tape so that they cannot turn and sending the skate down the slide again.

If possible, obtain samples of ball bearings to show how they are used to reduce friction.

Contributing Idea C. Circular motions tend to impel objects outward from a center of rotation.

Observing the effects of circular motion. Place a block of wood on a phonograph turntable. Turn on the phonograph at 78 RPM, and let students observe what happens to the block. (It flies off the turntable.) Now tie a string around the block and fasten the free end to the center post. Turn on the phonograph and observe again. (The block turns with the turntable.) Tell students that the force that keeps the object from flying off the turntable in a straight line is called *centripetal force.* Explain that in this experience the pull of the string on the block exemplifies the centripetal force. Try other objects of different masses and surfaces on the turntable. Students will find that different masses make no difference, although surface friction does in some cases. *(Note: The tendency for objects to break away from centripetal force is sometimes called* centrifugal force; *however, the latter is really the inertial tendency of a body in motion to travel in a straight line and is not a force.)* **17 Aa Bb**

Observing the tendency of objects to break away from circular motions. Clamp a small hook screw into the bit of a hand drill. Attach two nails to the end of a ½ yard (½ meter) length of string hanging **18 Aa**

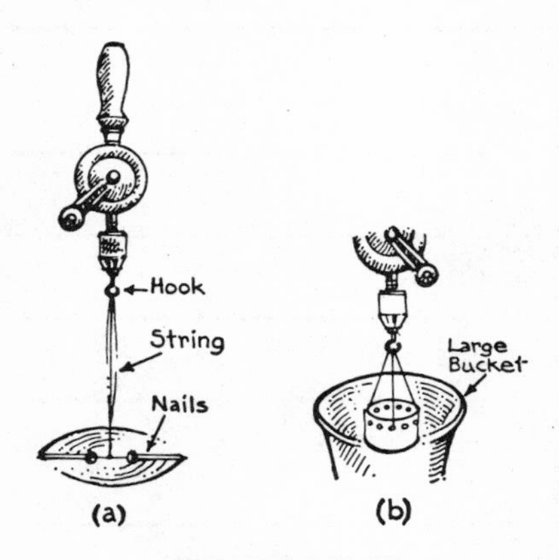

FIGURE 312.18

from the hook, then rotate the drill steadily *(a).* Students will see that the nails have a tendency to fly straight outward from the center as the speed of rotation increases. Point out that the nails are held in place by the inward pull or centripetal force of the string. To show that liquids are similarly affected, punch several equidistant holes around the top of a can. Suspend the can from the hook screw with three strings *(b).* Hold the can inside a larger container such as a bucket, filling the can one-fourth full with water. As the rotation of the drill steadily increases, students will see the water rise in the can and fly out through the holes.

Contributing Idea D. Some effects of circular motion can be controlled.

Observing how the inertial tendency of an object to travel in a straight line can be controlled. Obtain three containers of similar diameter but of different depth (e.g., saucer, soup plate, deep bowl). Place a marble in the shallowest container and have a student hold the container in the palm of his hand and rotate it until the marble moves in a circle. Students will see that as the rotation increases, the marble moves faster and eventually leaves the container in a straight line. Repeat the activity with the other containers. Students will find that the deeper or more vertical the sides of the container, the better the confinement of the marble, even at increased speed. Relate the effect on the marble to how people lean when they ride around curves (the tendency to continue traveling in a straight line). Tell how road engineers control this problem by banking the road around curves. If possible, find pictures of banked freeway curves designed to keep automobiles on the road at high speeds.

19
Aa

321 CHARACTERISTICS

Generalization I. Sounds have identifiable characteristics.

Contributing Idea A. There are many kinds of sound.

Describing outdoor sounds. Take your class to different areas of the school grounds at different times of the day. Have students sit or stand quietly for thirty seconds with their eyes closed to listen to and remember each sound they hear. When the time is up, have various students report on the sounds they heard by describing them, identifying them, and telling where each came from. **01 Ac Ba**

Describing household sounds. Prepare a tape recording of various household sounds (e.g, a whistling tea kettle, a typewriter, a vacuum cleaner, an alarm clock, an egg beater, the winding of a clock, a washing machine and dryer). Play the recording and have students describe the sounds; then have them identify the objects that made the sounds. **02 Ac Ba**

Contributing Idea B. Sounds vary in volume—some are loud and some are soft.

Hearing variations in rhythm and volume. Play a tapping game. Students can listen with their heads lowered or their eyes closed while you tap evenly several times on the chalkboard or desk. Ask how many taps were made or have students repeat the number of taps on their own desks. With practice, they will be able to reproduce different rhythms and changes in volume. You might tell them that the loudness or softness of a sound is called *volume.* **03 Ac Bb**

Making sounds louder or softer. Let students produce a sound by tapping a drinking glass or bottle with a pencil, by plucking a one-string instrument, or by blowing a whistle. Have them make the sound again, putting more effort into the tap, pluck, or blow. Next, repeat putting in less effort. After they describe differences in each sound, they will realize that because the object produces only one note, the difference must be primarily caused by the effort put into its production—loud sounds require more effort than soft sounds. Let them discuss the efforts they put into their voices when they whisper and when they call someone across a street. **04 Ec**

Making sounds louder. Students can hold a comb in one hand and stroke its teeth with a thumb. Have them do this again by placing one end of the comb against differ- **05 Bb Ec**

ent surfaces such as a cleared desk, doors, chalkboards, windows, bottoms of waste baskets, and so forth. They will find that certain materials and surfaces make the sound louder. You might tell them that a surface that makes a sound louder is called a *sounding board.* Ask them to find the sounding board portions of some musical instruments such as guitars and pianos.

Contributing Idea C. Sounds vary in pitch—some are high and some are low.

Hearing variations in pitch. There are several experiences in which students can hear differences in high and low sounds. After these experiences, you might tell them that the highness or lowness of a sound is called *pitch.* 06 Ac Bb

a. Have them run a fingernail over the fibers of a cloth-covered book or stroke a pencil or stick across the teeth of a comb, wood file, or corrugated cardboard. Students can note that the faster the movement, the higher the sound.

b. Have them listen as you use a hand saw to cut through a board. They can note that the faster the teeth cut across the board, the higher the sound.

c. Have them hold a playing card against a turning bicycle wheel and note the change in the sound as the wheel is speeded up or slowed down.

Making sounds higher or lower. Working in small groups, students can place a plastic ruler on their desks so that it extends about 10 inches (5 cm) beyond the edge. Holding one end of the ruler firmly on the desk, press down on the free end, then let go. Repeat with the ruler sticking out 8 inches, 6 inches, 4 inches, then 2 inches (5 cm intervals) from the edge. Ask them what differences they heard, and how they relate to the extended length of the ruler. 07 Ac Ec

Contributing Idea D. Sounds vary in quality.

Hearing variations in quality. To help students realize that the quality of a sound is neither its loudness nor pitch, prepare a table top barrier large enough so that different soundmakers can be used behind it without being seen. Tell students to listen carefully as you make a sound behind the barrier. You can play a record loudly while sounds are being made; thus, students will need to tune out the record sound to concentrate on other specific sounds. Follow this activity with a discussion of the fact that sometimes we hear things and sometimes we do not (e.g., a spoon dropped on the floor sometimes makes us jump, while at other times we hardly notice it). 08 Ac Bb

Contributing Idea E. Sounds can be reflected and absorbed.

Hearing reflected sounds. Locate echo-producing areas in or around your school (e.g., auditorium, large assembly room, hallways). Place students where they can best hear reflected sounds, then make sounds from different positions around them. Ask if they can hear better in some places than in others. Explain that sound sometimes bounces off surfaces and that if the sound returns to the sender it is called an *echo.* You might discuss the advantages and disadvantages of reflected sounds (e.g., people addressing audiences in large rooms can be hindered by reflected sounds that interfere with the direct sounds of their voices; ships rely upon reflected sounds to indicate the 09 Ac Bb depth of water by bouncing sounds from the ocean floor).

Seriating materials by the way they absorb sounds. Strike a tuning fork with a rubber mallet and place its base against a wooden box. Repeat, but each time place a different material between the base of the fork and the box. Students can compare the sound produced each time. They can test a variety of materials (e.g., squares of cloth, felt, paper, metal, an empty and a full can of water, styrofoam) and seriate them by their ability to absorb sound.

10
Ac
Da
Ec

Generalization II. Sounds can be classified on the basis of their characteristics.

Contributing Idea A. Some characteristics of sounds can be seriated.

Seriating sounds by volume. Have students collect pictures of machines. Let them order the pictures by the loudness of sound each produces. You might tell them that the intensity of sound is measured in units called *decibels* that are ordered along a logarithmic scale. The following scale may help students seriate their pictures.

11
Bb
Da

TABLE 321.11. Volume Scale

Decibels	*Examples*
0	Threshold of hearing
10	Ordinary breathing
11–20	Whispers
21–30	Ordinary household sounds; Natural country settings
31–40	Ordinary classroom activities; Turning pages of newspapers
41–50	Automobile engines; Vacuum cleaners
51–60	Noisy offices or stores
61–70	Ordinary television sounds or conversations
71–80	Heavy street traffic
81–90	Trains; Subway cars
91–100	Boiler factories; Air drills; Riveters
101–110	Claps of thunder; Jet engines
120	Threshold of pain

12
Bb
Da

Seriating sounds by pitch. Have students tap different objects and listen to the pitch each produces. Seriate the objects from the lowest to highest pitch. Students can also make simple musical instruments that require the seriation of sounds in order to produce a musical scale, or you might cover the frequency numerals (vibrations per second) on a set of tuning forks with tape and have them seriate the forks by pitch, then remove the tape so that the order can be checked. The numerals indicate the number of vibrations the fork makes per second (VPS); and the higher the sound, the greater the number

TABLE 321.12. Pitch Scale

Frequency: Vibrations Per Second	*Examples*	
About 16	Lower limit of human hearing	
20– 200	Deep bass tones (27 VPS = lowest note on the piano)	
256	"do" Middle C	"Middle" Musical Scale
278	C-sharp	
294	"re" D	
312	D-sharp	
330	"mi" E	
349	"fa" F	
370	F-sharp	
392	"sol" G	
416	G-sharp	
440	"la" A	
466	A-sharp	
494	"ti" B	
512	"do" C	
525 – 3,000	Normal adult conversation	
4,000	About the highest tone used in music	
8,000	High pitched, shrill tones	
About 20,000	Upper limit of human hearing	
About 30,000	Upper limit of hearing for dogs and cats	
About 100,000	Upper limit of hearing for bats	

of vibrations. Tell them that *frequency* is the term used when describing how many vibrations there are for a particular tone. The illustrated scale can help students order various sounds on the basis of pitch.

Contributing Idea B. Some characteristics of sounds can be classified.

Selecting categories for grouping sounds. Have students list some categories for sounds. Some headings might be: pleasant sounds—unpleasant sounds; loud sounds—soft sounds; school sounds—home sounds; outdoor sounds—indoor sounds; sounds people make—sounds animals make. Let different groups of students use different headings as they listen for sounds around the school grounds. Compare listings for different locations and for different times of the day. Let them decide if some headings are more useful than others.

13
Db

Grouping sounds. Various activities allow students to group sounds by similarities.

14
Db

a. Ask what the last sound was that they remember hearing before going to sleep last night. Ask what kind it was (e.g., machine made or natural), then ask what the first sound was that they heard this morning. Have them group the sounds by listing them under appropriate headings on a chalkboard.

b. If you can obtain a collection of pictures of typical daytime sound sources (e.g., auto traffic, people working, people moving, dogs barking, wind blowing), students can group the pictures according to headings of their own choosing.

c. Have them group pictures of various musical instruments according to the method used to produce sounds on them (e.g., striking—percussion instruments; plucking or bowing—string instruments; blowing—wind, reed, or flute instruments).

Classifying sounds. Have students prepare a table to keep a record of soundmakers, the kinds of sound they produce, and the duration of the sound. When finished, let students devise sets of categories and subcategories for the soundmakers. Figure 321.15 suggests soundmakers that can be used, as well as possible sets of categories.

15
Bc
Dc

Classifying Sounds

Soundmaker	Description of Sound	Duration of Sound
Gong	Ringing, High	Continues after Effort Stops
Flute	Shrill, High	Stops when Effort Stops

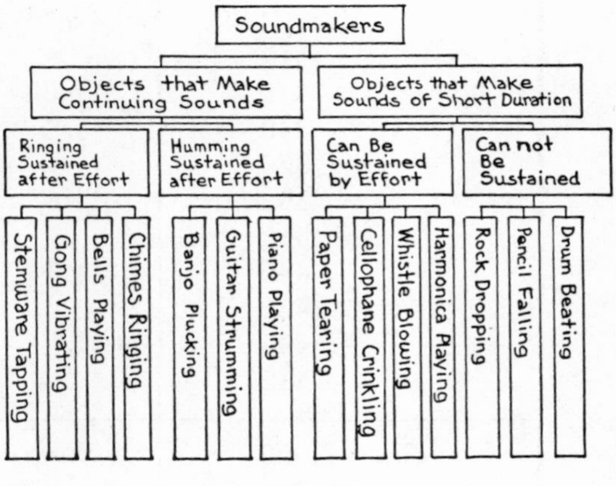

Soundmakers

- Objects that Make Continuing Sounds
  - Ringing Sustained after Effort: Stemware Tapping; Gong Vibrating; Bells Playing; Chimes Ringing
  - Humming Sustained after Effort: Banjo Plucking; Guitar Strumming; Piano Playing
- Objects that Make Sounds of Short Duration
  - Can Be Sustained by Effort: Paper Tearing; Cellophane Crinkling; Whistle Blowing; Harmonica Playing
  - Can not Be Sustained: Rock Dropping; Pencil Falling; Drum Beating

FIGURE 321.15

322 INTERACTIONS

Generalization I. Sound is a form of energy produced by vibrating matter.

Contributing Idea A. Vibrations are the only source of sounds.

Sensing sound vibrations. There are many ways by which students can sense vibrations as sound is produced. For each of the following suggested experiences, have different groups of students describe what is felt, heard, and seen. **01 Af Ba**

a. Fasten a rubber band to a door knob, pull it taut, then pluck it.
b. Touch the strings of a piano or stringed instrument as it is played.
c. Half-fill a drinking glass with water, wet a finger, then rub it around the edge of the glass. For more consistent results, vinegar can be used in place of water. If the glass is steadied with the other hand, hold it near the base. *(Note: Not all drinking glasses will produce a good sound.)*
d. Place the tip of a spatula blade (or hacksaw blade) on a desk with the handle extending over the side. Pull the handle down, then let go.
e. Fold a 2 inch by 6 inch (5 cm by 15 cm) strip of paper in half, cut a small semicircle in the fold, hold between two fingers, and blow hard through the semicircle.
f. Place some fine sand or salt on a drumhead, then tap the drum gently and with a steady beat.

After these experiences you might explain that whenever something moves back and forth or up and down, it is said to *vibrate.* From these and similar experiences the students will realize that sound is produced by vibrations that are usually too rapid to see except as a blur.

Feeling vocal vibrations. Have students feel their lips and throats during each of the following experiences. **02 Ab**

a. Hum with mouths and noses open (air passes through them) while lightly touching throats and lips.
b. Hum with mouths open and noses closed.
c. Hum with closed mouths.
d. Try to hum with both mouth and nose closed.

In each case, let them describe and discuss what they felt. They will realize that all vocal sounds are vibrations involving air passing out of the lungs.

Contributing Idea B. Sound vibrations occur in wavelike patterns.

Making sound vibrations visible. Sound vibrations are usually too rapid to be seen as more than a blur. The following experiences suggest some ways students can more easily observe sound vibrations. From the experiences they will also learn that when an object stops vibrating, it no longer produces a sound. **03 Gc**

a. Stretch a rubber band tightly around three nails in a board. Hang several small strips of paper, creased in half, over one section of the band. Pluck the section and observe how the paper strips move. Similarly, strips can be placed over the different strings of string instruments. Students can note how long a string vibrates after it can no longer be heard.

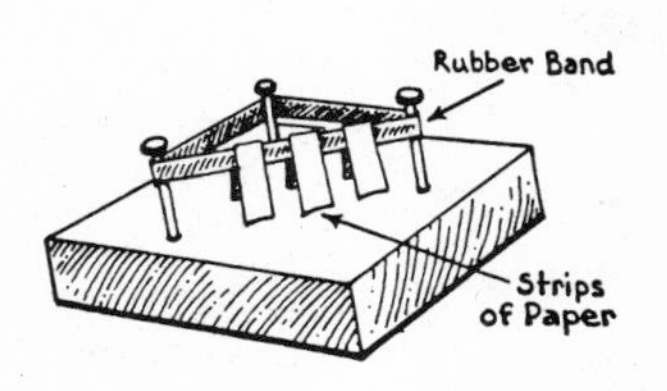

FIGURE 322.03

b. Strike a tuning fork and place one of its tines into some water or against a suspended ping-pong ball.

c. Cut an opening in one end of a round cardboard carton and paste a thin sheet of onion skin paper over it. Glue a small mirror to the center of the paper. Have one student speak into the other, open end of the container while shining a light against the mirror so that the reflection is cast against a screen or wall. Let other students observe the reflected light as it dances on the wall as the sound of a voice causes the onion skin to vibrate. *(Note: Different cartons perform in different ways—some require a strong voice or loud sound before the onion skin will vibrate.)*

Changing sound vibrations into wave pictures. Smoke a piece of glass pane by holding it in a candle flame. Using a drop of wax or piece of clay, fasten a very fine wire to one of the tines of a tuning fork *(a)*. Attach

04
Bb
Ca

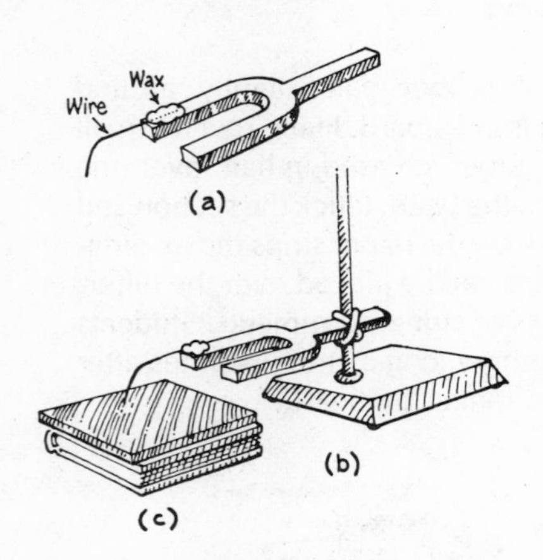

FIGURE 322.04a,b,c

the tuning fork to a ring stand or other upright to hold it in a constant position *(b)*. Have a student set the smoked glass under the point of the fine wire and gently bend the wire so that it just touches the pane *(c)*. Strike the tuning fork with a rubber mallet and observe how the fine wire moves back and forth as the tine vibrates. Students will see a streak appear on the glass as the wire scratches away the carbon. A similar apparatus can be prepared by firmly gluing sharpened, soft pencil leads to the tuning fork and suspending the fork over a sheet of paper *(d)*. Several tests can now be per-

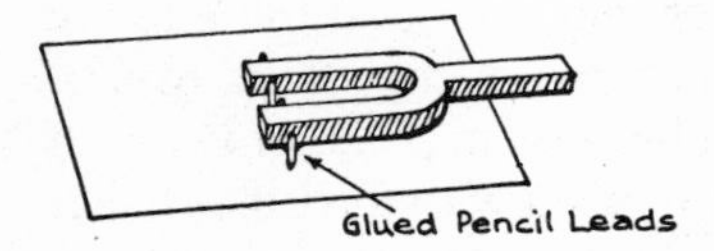

FIGURE 322.04d

formed using the materials. *(Note: The wave pictures should not be confused with transverse waves—sound waves are longitudinal waves.)* You might explain that each sound produces a regular pattern of vibrations and that the number of vibrations that occur in one second is called *frequency*.

a. Have a student slowly pull the glass away from the fork when it is not vibrating. (The fine wire should make a straight line on the glass.) Replace the glass under the wire at the starting position. Now strike the tuning fork and have the student pull the pane away again *at the same rate*. (A sequence of waves will be produced along the first line.) Discuss why this took place, then repeat the procedure, this time striking the tuning fork with more force. As different intensities are tested, students will see that the number of waves remains the same over the same distance but that they rise higher on each side of the straight line.

b. Repeat the above test using different tuning forks. For each new test, different smoked pieces of glass should be used and labeled so that the wave pictures can be compared. With practice, students will be able to pull the pane of glass at a steady rate over one second of time. If the pulling time and the distance the glass travels are kept constant, comparisons will be more fruitful.

c. Each time the fork is struck, have a student pull a freshly smoked pane of glass away at a different rate. Be sure to strike the tuning fork with equal effort each time. Compare the wave pictures that are produced. (The heights of waves will remain the same on each side of the straight line, but some will appear to be more spread out.)

> *Contributing Idea C.* Sound vibrations can be produced by striking, plucking, or blowing.

Producing different sounds. Have students tap on a variety of objects such as large and small pans, pans made of different materials (aluminum, cast iron, stainless steel), blocks of wood of various sizes, metal cans, glass jars or cereal boxes—filled, partially filled, and empty. Challenge them to find something that makes no sound when tapped. By testing many objects and keeping track of the sounds made, they will form some understanding of how size, shape, and the materials that make up the objects relate to the sounds produced. **05 Gc**

Creating sound effects. Let students use a tape recorder to present a radio script or play. Have them select stories from their readers and library books, or let them make up stories of their own. Be sure that they select stories that allow many sound effects to be added. You might also challenge them to produce the sound of an egg frying, a freight train starting up, or a horse trotting. Here are some hints for producing some sound effects: birdseed falling on a ping-pong ball—rain; beating coconut shells—horse trotting; cellophane being slowly crumpled—egg frying; cellophane being quickly crumpled—forest fire; sandpaper pieces being rubbed together in an accented rhythm—freight train; two thin flat boards being slapped together—gunshot; BB's being shaken in an inflated balloon—thunder. **06 Gb**

> *Contributing Idea D.* The volume of sound vibrations depends upon the amount of energy input (intensity) and the distance the vibrations travel.

Listening to the effect distance has upon sounds. Place about ten students at 10 yard (10 m) intervals across the school playground. Have them turn their backs toward you so that vision will not influence their judgments. Now make various loud and soft sounds and have students raise their hands if they hear the sound. They will discover the distances that sounds of different intensities travel by the number of hands that are raised. You might tell them that the amount of energy a sound has when it reaches the ear is called *loudness.* Let each of the ten students describe the same sound in terms of its loudness. (The loudness of a sound diminishes over distances.) From these experiences, they can theorize that loud sounds are produced by strong vibrations that travel for long distances, while soft sounds are produced by weak vibrations that travel shorter distances. **07 Ac Bb Fa**

Testing factors that influence the volume of a sound. Wire a plastic or metal spoon to a mouse trap. Use a protractor to make a semicircle on a heavy piece of cardboard **08 Ec**

and mark it every ten degrees. Paste the semicircle to the side of the mouse trap. Hold the apparatus so that the back of the spoon just touches the side of an empty glass. Taps against the glass can be controlled by pulling back the spoon to a mark on the semicircle and letting go. Now have

FIGURE 322.08

students listen for differences in volume and in the duration of the sound as they test weak to strong taps. Keep records for the following suggested tests.

a. Some students can produce taps of equal strength while others move measured distances from the glass. The sound will become fainter as the distance is increased. *(Note: Some students will find that the pitch changes from low to high as the distance increases—scientists have discovered that the closer, stronger vibrations seem to affect the signal our ears send to the brain and lower sounds are realized.)*

b. Experiment with the apparatus by tapping other objects or by placing the glass on different surfaces such as tin pans, porcelain plates, wood, or corrugated cardboard.

c. Systematically tap the glass with measured amounts of water added to it. *(Note: Tapping makes the glass and the water vibrate, while blowing makes the air column in the glass vibrate.)*

d. Study the ripples in the water for taps of different intensities. Heavier liquids (syrup) or lighter (alcohol) can be tested for visual and auditory comparisons. Students will discover that the weight of the material influences the sound.

e. Systematically dissolve measured amounts of salt or baking soda in measured amounts of water in the glass and note differences in the ripples and sounds that are produced.

Contributing Idea E. The pitch of sound vibrations depends upon the number of sound vibrations per second (frequency).

09 Bb Bf

Comparing differences in the pitch of sounds. Have a student hold the tip of a spatula blade tightly against the edge of a table with one hand, pull down the handle with the other, then let go. (A low raspy tone should be heard.) Repeat the action several times, but each time slide more of the blade onto the table. (The sound will become higher.) You might tell students that the change in sound from high to low or from low to high is called *pitch.* If they observe carefully, they will note that as the vibrating portion of the spatula is shortened, it vibrates faster. They might try counting the number of vibrations in a one-minute period for different lengths of the spatula. In place of the spatula, test plastic spoons, hacksaw blades, straightened-out bobby pins, and rulers. If a ruler is used, it can be clamped

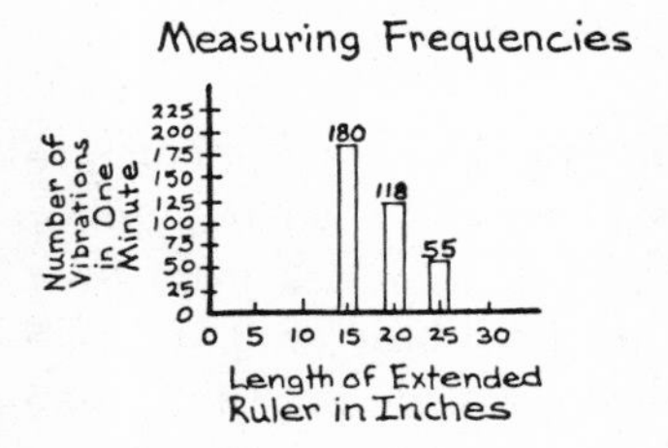

GRAPH 322.09

to the edge of the table, then systematically shortened for each trial. Data can be recorded on a graph. Explain that the term *frequency* is used to describe the relative number of vibrations among sounds of different pitch.

Testing factors that influence the pitch of a sound. Pitch depends upon several factors—usually the composition of the material, the tension it is under, its length, and its diameter or thickness. These factors can be tested for their influence upon the production of sounds by having students stretch three rubber bands of the same length around a small wooden box. The bands should range in width. If rubber bands of the same length and different width cannot be found, bands can be cut from an inner tube with widths of ¼ inch, ½ inch, and ¾ inch (.5 cm, 1 cm, 1.5 cm). Students might explore the factors freely, or you might systematically guide them from one factor to another. Some suggestions follow:

10
Bc
Ec

a. Students can pluck each band, one at a time, while listening. Be sure they pluck each time with the same amount of effort. Have them describe in what ways the sounds differed and record their findings on a table. They will realize that since the length, effort, material, and other factors were identical, the primary factor influencing any difference was probably the width (the wider or thicker the material, the lower the sound; the narrower or thinner the material, the higher the sound).

b. Twist a piece of wood near one end of one band to make the band tighter. Pluck the band, twist the wood, pluck, and twist again while noting changes in the pitch. Repeat for the other bands. Students will realize that the tension was the primary factor influencing the different sounds produced by the band.

c. Place a wedge under the rubber bands and have students pluck the longer portion of each. Move the wedge to make the length of each portion shorter or longer and continue plucking. Students will realize that the length was the primary factor in the production of the sound differences they heard.

d. Substitute wires, fishline, or other materials and test each of the above suggestions again.

e. Experiment by combining each of the factors in different ways to note what happens to the sound production (e.g., tautness and length, thickness and length, thickness and tautness).

TABLE 322.10. Comparing Pitch and Width (Thickness)

Factor	*High Pitch*	*Medium Pitch*	*Low Pitch*
Thin Band			
Medium Band			
Wide Band			

Generalization II. Sound travels through solids, liquids, and gases at different speeds.

Contributing Idea A. Sounds travel through solid materials.

Hearing sounds through solids. Have students listen to the sound made as they scratch or tap a leg of their desks. Have them repeat the action with an ear to the desk top. Compare the sound coming through the solid top to the sound coming

11
Ac
Ca

through the air. You might have them discuss why Indians put their ears to the ground to listen for hoofbeats and why railroad maintenance workers place an ear to a rail to listen for an approaching train. Several other experiences can help students realize that sounds travel through solids—in each case they can judge that sounds can be heard better through solids than through air.

a. Place a watch in the center of a table and try to hear its ticking from a yard (meter) away. Next, rest one end of a yardstick on the watch and let students take turns placing an ear against the other end. (They should hear the ticking more clearly.) You might also let them experiment by holding the watch (or a tuning fork) against various parts of the head (e.g., chin, teeth, jawbone) to discover that sounds may reach the ear through solid parts of the body (e.g., bones). This can also be demonstrated by holding a pencil between the teeth, tapping it lightly with another pencil while holding the ears shut, and comparing the effect when the pencil is held only by the lips.

b. Tie a spoon or coathanger to the center of a 3 foot (1 m) length of string. Suspend the object from the ends of the string, strike it against a chair or desk, and listen to the sound. Now have students wrap each free end of the string around one finger on each hand, and place the two fingers into their ears. Swing the object so that it taps the chair or desk again. Discuss differences between the sound transmission through the string and through the air. Test the transmission using one ear at a time. Also, experiment with spoons of other sizes and shapes and materials (e.g., plastic, wood, metal) and strings of various lengths and materials (e.g., short and long; fishline and wire).

FIGURE 322.11

c. Lightly scratch one end of a broomstick with a fingernail while students take turns listening at the other end. To be sure the sound is being transmitted through the wood, the students can listen while raising their ears from the handle, but keeping the same distance from the sound source so that a comparison can be made.

d. Send International or Morse code messages through pipes in the school building, stair railings, or metal fences by tapping them with a piece of hardwood.

Contributing Idea B. Sounds travel through liquids.

Hearing sounds through water. Fill a large bowl, bucket, or aquarium with water. Make various sounds under the water (e.g., tap spoons together, strike two rocks together, snap fingers, ring a bell) or float a large block of wood on the surface and touch it with a vibrating tuning fork. Students can listen to the sound first through the air, then with their ears against a side of the container. They will find that sounds can travel through a liquid and judge that they travel better through water than through air.

12
Ac
Ca

They might discuss sounds heard while swimming under water.

Contributing Idea C. Sounds travel through gases.

Hearing sounds through air. Students tend to take for granted the idea that sounds travel through the air because they hear people talk. A few experiences, however, will stimulate them into considering this assumption more carefully. 13 Ac

a. Stretch a piece of balloon rubber over the open end of a tin can, fasten it with a rubber band, and sprinkle some salt on it. Let one student strike a tuning fork with a rubber mallet and hold it about 1 inch (3 cm) over the salt. (Students will see the salt bounce and infer that the sound vibrations are transmitted from the fork through the air to the rubber.) Students can note that the effect does not take place if the fork is moved away from the rubber—indicating that the sound energy diminishes with distance.

b. Suspend a piece of breakfast cereal from a string, bring a vibrating tuning fork close to, but not touching, the piece of cereal. (Students will see the cereal being moved and infer that the sound vibrations are being transmitted through the air.)

c. Students can lay an empty wastebasket on its side, set a lit candle in front of the open end, tap the bottom of the basket to produce a sound, and watch the candle flame. (The flame will move as the sound vibrations are transmitted through the air.)

d. Students can feel the sides of long paper tubes or empty cereal boxes as they speak through them. (The air inside vibrates and pushes with some force.) Discuss how some sound vibrations traveling through the air can cause damage (e.g., windows breaking from sonic booms or claps of thunder; glasses breaking from high-pitched musical notes).

Communicating through a tube of air. Have students add funnels to each end of an empty garden hose, and speak to one another. They will find that sounds coming through the hose, even when bent around a corner, are louder than identical sounds traveling through free air. They should especially note that the hose contains a column of trapped air. Let them experiment with different lengths and diameters of tubes. You might tell them that ships sometimes use a similar device for communicating between sections and that some apartment houses use it between entryways and the apartments. 14 Ba Gb

Contributing Idea D. Sounds travel through some materials faster than others.

Hearing that sounds take time to travel. Lay out an empty 100 foot (30 m) or more length of garden hose in a large loop so that the two ends are near each other. Students can take turns placing an ear to one end while tapping the metal portion of the other end with a pencil. They will hear two taps—one through the air and the other through the hose. Explain that the second tap is due to the distance the sound had to travel through the hose. Discuss similar experiences indicating that it takes sound time to travel across a distance (e.g., distant loudspeakers compared to close ones). 15 Ac Ch

Hearing that the speed of sound varies through different materials. Pound a wooden stake into the ground at one end of 16 Ch Da a playground while students place their ears against the ground some distance away from the stake. They will hear the sound twice—first through the ground, then through the air. They should realize that it takes the sound longer to travel through the air than through the ground. Repeat this activity at various distances to determine if sound always travels faster through the ground.

Contributing Idea E. The speed of sound through air can be measured.

Determining the speed of sound in air. 17 Ch Prepare a pendulum timer (see *432.09*). On a day when there is no wind, set the timer in motion and have a student strike the bottom of a large can with a stick each time the pendulum is at one end of its swing. (This may take some practice.) When synchronized, have other students move away from the pendulum. They will notice that the sound does not remain synchronized with the swing. As they move farther and farther away, they will find that the sound again becomes synchronized, but a half-swing late. To determine the speed of the sound, measure the distance between them and the sound source. For example, if the distance is 550 feet (168 m), then 550 feet (168 m) in ½ second (a half-swing) equals 1100 feet (336 m) in one second or 750 miles (1210 k) per hour. *(Note: Sound travels at 1,087 feet or 332 m per second at 32°F or 0°C. The speed increases 1.1 feet per second for each degree of increase in Fahrenheit temperature, or 3 mps for each 5 degrees C.)* If a very large open area is available, students can determine the speed of sound in another way by using a tape measure to carefully mark off a straight-line distance of 2,500 feet (800 m). Let one student with a stick and large can stand at one end of the distance, while a group of students stand at the other end with a stop watch. Have the first student strike the can with the stick. When the group sees the can being struck, start the watch. Stop it as soon as the sound is heard. To determine the speed of the sound, divide the distance by the time. This activity might be repeated several times to find the average time it took for the sound to travel the distance.

Contributing Idea F. The speed of sound through air is slower than the speed of light.

Comparing the speed of sound to the speed of light. 18 Ca Divide the class into two groups. Give one group a whistle and the other a flag, then station them about 500 feet (150 m) apart. Have the whistle group blow the whistle loudly and have the flag group signal as soon as it hears the whistle. The whistle group will observe that it takes some time for the sound of the whistle to reach the second group. Interchange the flag and whistle and repeat. Similarly, one group can strike the bottom of a large can with a stick. The other group will see the action completed before it hears the sound.

Estimating the distance of a flash of lightning. 19 Cb Lightning and thunder are illustrations of the fact that light travels faster than sound. Even though they occur simultaneously, we usually see the lightning bolt, then hear the clap of thunder. The distance between the lightning and the observer can be computed by timing the interval between seeing the bolt and hearing the thunder. For example, since each second of delay between the flash and the sound indicates that the lightning is approximately ⅕th mile (320 m) away, students can simply count the seconds that elapse after seeing a flash until the thunder is heard (e.g., 15 seconds between flash and thunder—15 seconds ÷ 5 seconds per mile or meter = 3 miles or 4800 m).

Generalization III. Sounds can cause objects to vibrate.

Contributing Idea A. A vibrating object can force another object to vibrate.

Observing forced vibrations. Have a student strike a tuning fork and press its base against a table top or other flat, clear surface. Students will hear that the sound becomes louder and infer that the vibrations produced by the sound forced the larger surface to also vibrate at the same frequency. (If a larger surface is made to vibrate, the sound is louder.) Similarly, students can hear that the sound of their voices becomes louder when they talk into a waste basket or megaphone as the sides of the object are forced to vibrate at the same frequency. You might explain that when this happens, the transferred sound is called a *forced vibration.* Examine the sound boxes of violins, guitars, and similar instruments and note that they are shaped to increase the loudness of musical sounds.

20
Ac
Bb
Fc

Using forced vibrations to amplify sounds. Have students use a hand lens to examine the groove on a phonograph record. Place a fingernail into the groove as it turns to sense the vibrations and to hear a faint sound. Now place the point of a toothpick, long cactus needle, or phonograph needle into the groove. (Students will hear the sound a bit louder as the object vibrates in the groove.) Repeat several times after placing the toothpick through the corner of an index card, then through the corner of an empty matchbox, and finally through the folded end of a paper megaphone. In each subsequent test, the sound will be amplified as the vibration of the toothpick forces the larger surfaces to vibrate at the same frequency. You might follow this experience by having some students find pictures of early phonographs that used horns to amplify sounds.

21
Ac
Ec
Ga

Contributing Idea B. Sympathetic vibrations occur when a vibrating object is in tune with another object.

Observing sympathetic vibrations. Mount two tuning forks of the same frequency on two identical wooden boxes about 6 inches (15 cm) apart with the open ends facing each other (or have two students hold them by their bases about 3 inches or 7 cm apart). Strike one fork with a rubber mallet, listen, then stop the vibration with your hand. If students listen carefully to the other fork, they will hear a faint sound and realize that the second fork was put into motion by the vibrations of the first. You might explain that when this happens, the transferred sound is called a *sympathetic vibration* (another term is *resonance*). If this activity is repeated with tuning forks of different frequencies, students can discover that sympathetic vibrations can only be set up using identically pitched forks. Similarly, they can try blowing across the top of a jar or bottle near an identically pitched jar or

22
Ac
Bb
Fc

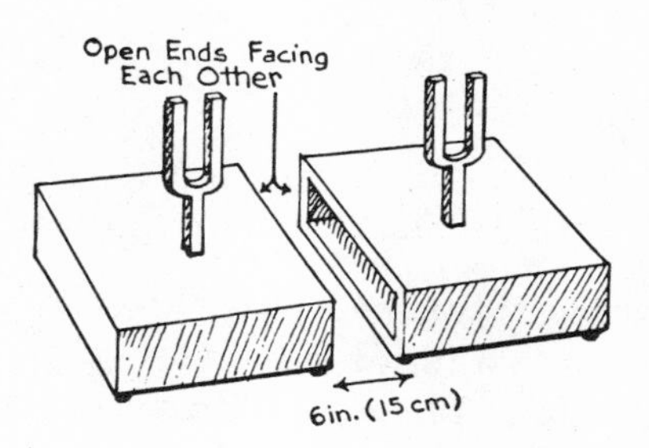

FIGURE 322.22

bottle and listen for the sympathetic vibration.

Generalization IV. Sounds can be controlled.

Contributing Idea A. Sounds can be directed to a particular place.

Controlling the direction of sounds. 23 Ac Gb
Have students put two funnels in a 1 yard (1 m) length of rubber hose. Using the device, let them listen to their heartbeats, to

FIGURE 322.23

themselves talking, to watches ticking, to engines running, or to a friend tapping his or her head. **(Caution: Be sure children do not shout or blow into funnel when someone is listening at the other end.)** Discuss how the sounds are funneled into the hose and through it in a controlled direction. Students can compare the same sounds being made without the hose and funnels or can make a two-ear stethoscope by connecting three funnels and rubber tubing to a Y-shaped glass or metal tube. This device can be compared to a one-ear stethoscope for effectiveness.

Contributing Idea B. Sounds can be reflected.

Observing that sounds can be reflected. 24 Ac Bb
Have students hold a curved index card behind a ticking watch while listening. They can progressively move the card back and forth or change the curvature to hear how the sound is reflected. Similarly, a student can turn her back to the class as she speaks loudly (or sings one note without interruption for about ten seconds) into the open end of a wastebasket that is alternately raised to her face and then lowered. The class will realize that the sound produced was reflected from the basket's base and sides. You might tell them that reflected sounds are called *echoes.*

Contributing Idea C. Sounds can be absorbed.

Determining that sounds can be absorbed. 25 Ac Cd
Place a loudly ticking clock in a box with the open side toward the back of a student who is listening. Have the student walk away from the box in a straight line until the ticking becomes inaudible. Record the distance, then fill the box with some sound-insulating material such as cotton, and have the student walk away again. Mark the distance where the ticking becomes inaudible again. Compare the two distances and discuss reasons for differences.

Testing materials that absorb sounds. 26 Da Ec
Pull a rubber band around a wooden box and pluck it. Next, place some cotton in the box and pluck again using the same amount of effort. (The cotton should deaden the sound.) Let students test other materials in the box (e.g., paper, sawdust, different cloths, pieces of rubber, sponges) to see whether some absorb sound better than others. Compare the sounds produced and rank the materials from the best sound absorbers to the poorest. These tests might be followed with a discussion of sound controls used in homes and buildings (e.g., carpeting, drapes, building materials). If possible, have the students look for examples of soundproofing or the need for it in your school building.

330 light

331 CHARACTERISTICS

Generalization I. Light has identifiable characteristics.

Contributing Idea A. Light travels in all directions from a source.

Observing that light travels in all directions from a source. Have students pour several inches or centimeters of sand into an oatmeal box. Turn on a small pen flashlight and insert it into the sand so that it stands facing upward. Cover the box, darken the room, and punch a hole in the top of the box with a large turkey needle or medium-sized nail. Students will see that the light from the flashlight travels straight up through the hole to the ceiling. Now punch other holes in the box along the sides as well as the top. Students will find that a beam of light travels from each hole no matter where it is punched. Have them note that the spots of light appear in every direction, indicating that the light travels in all directions from its source. **01 Aa**

Contributing Idea B. Light travels in a straight line.

Observing a straight line of light through the air. Cut a small hole in a large piece of construction paper, and place the paper entirely over a sunlit window. Darken the room as much as possible, and have students look at the light coming through the hole. If the light cannot be seen clearly, clap two chalkboard erasers together near the beam. Ask if the light is traveling in a straight line. How would it look if it was not traveling in a straight line? **02 Aa**

Observing that light travels in a straight line. Use a paper punch to make holes in the exact centers of three old playing cards or index cards. Find the center by drawing diagonal lines between opposite corners. The center will be at the point where the lines intersect. Now place the cards into chalkboard erasers, lumps of clay, or tack them to blocks of wood so they will stand upright. Next, light a candle that is just the height of the holes. Place it in the center of a table, and line up the cards so students can look through all three holes and see the flame. Ask them to slightly move any one of the cards. (The flame will no longer be seen.) Students will find that the flame can only be seen when the cards are in a straight line, no matter where they place the cards around the table. **03 Aa**

Observing light passing through a narrow opening. Cut pieces of cardboard as shown *(a)*. Glue a square of aluminum foil to each piece, cutting a narrow slit in it about **04 Aa**

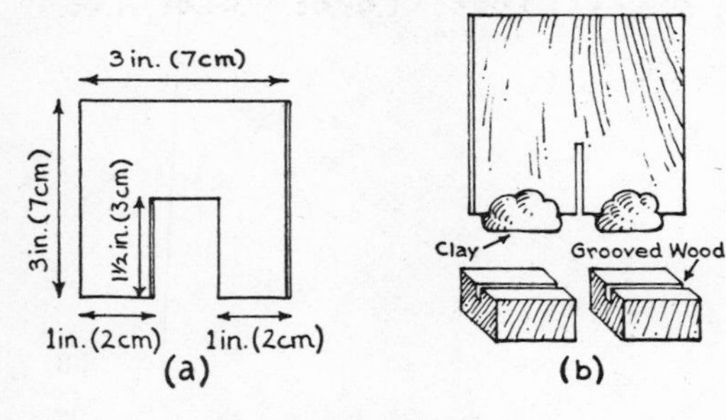

FIGURE 331.04a,b

1/16 inch (.2 cm) wide and 1 inch (2 cm) long *(b)*. Set the arrangement into lumps of clay or pieces of grooved wood to hold it upright. Now write the numerals from 1 through 5 about 1 foot (30 cm) apart on a chalkboard or large sheet of paper. Let a few students walk back and forth in front of the numerals and note that they can be seen from many places, as light reflects from the numerals in every direction. Students can set the cardboard pieces on desks about 12 feet (4 m) away from the numerals, and look through the narrow slits at the numeral 3. Place sheets of paper behind the slits, and have students mark a 3 on their sheets directly in front of their eyes *(c)*. Without mov-

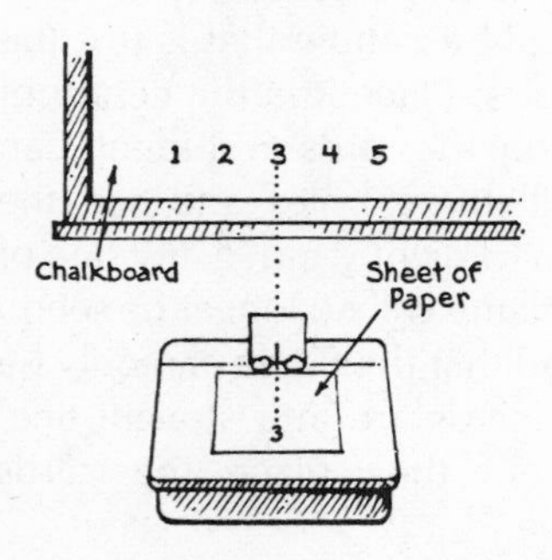

FIGURE 331.04c

ing the cardboard, let students move their heads until the numeral 2 can be seen with the same eye; and then mark a 2 on the sheet of paper in front of their eyes. Repeat this for each of the remaining numerals. *(Note: Be sure the aluminum foil is held steady.)* When finished, compare what has been written. Ask students what they notice about the numerals they wrote. (The numerals will be the reverse of the ones on the chalkboard.) Next, have each student hold the cardboard between an eye and the numeral 1 on the chalkboard. Students will find that only the beam of light from the numeral that is in a straight line with the eye can be seen and that if they move an eye to the right or left, they will see more to the left or right side of the numeral. They should realize that because light travels in a straight line, objects on the right side of an opening are seen on the left, and objects on the left side of an opening are seen on the right.

Contributing Idea C. Light can be diffracted or refracted. White light is composed of different colors.

Observing that white light can be separated into colors. In a darkened room, lean a small mirror in a full, clear glass of water. Shine a bright beam of light through the water at the mirror. Adjust the mirror so that the light reflects a number of colors on the ceiling or wall. Ask students what colors they see and whether or not the colors are spread out. From this experience, they will realize that the colors were made from ordinary light. You might tell them that the range of colors they observed is called the *visible spectrum*. **05 Aa Bb**

Observing diffraction. Place a strip of cellophane tape across the lower halves of two identical needles about ½ inch (1 cm) apart. Fasten the tape to a magnifying lens so that the tops of the needles are slightly higher than the middle of the lens. Shine a **06 Aa Bb**

narrow beam of light from a pen flashlight from one end of the darkened room. At the other end, have students hold the lens with the needle side toward the light and take turns looking at the beam with one eye. *(Note: They may have to move the lens back and forth until it is filled with light.)* They will observe that there are colored shadow bands around the sharp edges of the needles. You might explain that the bands appear because light bends around the sharp edges of obstacles. You might also tell them that the bending is called *diffraction.*

Making a spectroscope. A spectroscope is an instrument used to separate light into its component parts. To build such an instrument, obtain a cardboard tube from a roll of paper toweling or wrapping paper. Cut two identical circles from a piece of cardboard so that they will just seal the ends of the tube. Punch a ¼ inch (.5 cm) hole in one circle, and cut the other circle in half. Tape a piece of diffraction grating over the small hole and fasten it to one end of the tube so that the grating faces inward. *(Note: Inexpensive diffraction grating replicas, which are made from a special film imprinted with about 15,000 to 30,000 parallel lines to the inch, are generally available from scientific supply houses. If they cannot be obtained, a small prism can be substituted; however, it will refract the light rather than defract it.)* Next, tape the two half circles to the other end of the tube with a very narrow vertical slit left between them. Now use masking tape to seal both ends of the tube so that no other light enters. Aim the slit end at a light bulb. Looking through the slit, students will see the spectrum along the inside of the tube. (Light is broken up into its component parts as it passes through the narrow spaces on the diffraction grating.) Let them aim the spectroscope at other light sources. **(Caution: Do not aim it at the sun.)** They will soon realize that different light sources produce different patterns of light. You might tell them that a *continuous spectrum* shows all the colors of the rainbow, while a *line spectrum* shows only some of the colors.

07
Bb
Gc

Observing refraction. In a darkened room, shine the light from a slide projector through a prism. Have students locate the spectrum that is produced and hold a sheet of white paper over it so that the colors can be clearly seen. You might explain that when light passes into and out of a prism, the different colors (frequencies) are bent by different amounts-–the blues the most and reds the least—thus appearing separated as in a rainbow. Have students look for examples of the spectrum produced in more natural ways (e.g., rainbows, soap bubbles, oil slicks).

08
Aa
Ga

Making a prism. Tape the edges of three microscope slides together to form a triangle. Press the slides into some clay or plaster.

09
Gc

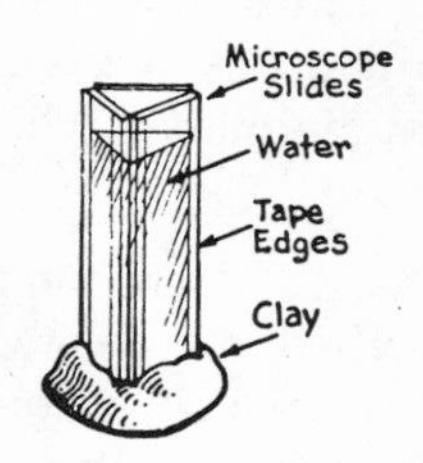

FIGURE 331.09

When the triangle is filled with water, the light transmitted through it will act like the light through a prism.

Observing that a spectrum of colors can be combined to make white light. You can show in several ways that white light is composed of the colors of the spectrum.

10
Aa

a. Place a shallow pan of water in direct sunlight or in front of a projector beam in a darkened room. Lean a mirror against the rim of the pan so that the light strikes part of the mirror beneath the water's surface. When the water is still, a spectrum will be found on a nearby wall or ceiling. Stir the water and note that the colors combine and white light appears on the wall.

b. In a darkened room, hold a prism in direct sunlight or in front of a projector beam and slowly rotate the prism until a spectrum appears on a wall or ceiling. Now place a second prism in the path of the light leaving the first prism so that all of the colored light strikes one facet of the second prism. Vary the angle of the second prism until the colored light passing through it is focused in a small spot. (Students can also try a hand lens in place of the second prism.) The spot of light will appear white. *(Note: It may take some practice to set the second prism properly.)* If three prisms and a very powerful light beam are available, challenge students to arrange them so that the light through two prisms is colored and the light from the third is white.

Contributing Idea D. Light can be reflected and absorbed; the color of objects depends on which colors are transmitted, reflected, and absorbed.

Observing the transmission and reflection of colors. In a darkened room, shine a projector beam through a prism set on a pile of books so that the spectrum appears on a white screen. Have students note the exact position of the red color in the spectrum, then place a double thickness of red cellophane between the prism and the screen. Students will see that all but the red part of the spectrum disappears and will realize that the cellophane prevented all but the red light from getting through. Repeat this activity using green and then blue cellophane. *(Note: It is difficult to obtain perfect filtering action with ordinary cellophanes, so some visual discrepancies may occur. Colored glass can be tested in place of the cellophane.)*

11
Aa

Observing the transmission, reflection, and absorption of colors. In a very dark room, challenge students to detect the actual colors of squares of colored construction papers as they are held in various colored lights one at a time. The students can keep their eyes closed, while the squares of paper and the cellophanes (blue, green, red) over the lens of a projector are changed. Stay with one color of cellophane before switching to another so that all the squares can be observed. After each square is observed, have students write down their guesses for the actual color, then remove the cellophane and let them see the actual color in white light to check. From this experience, they will understand that color is a function of the light striking an object as well as the color it reflects. For an additional experience, they can use red and green crayons to lightly print their names on pieces of white paper. Have them make some letters green and some red, then look at their names and at other red and green objects through the red and green cellophane pieces.

12
Aa
Ec

Generalization II. Characteristics of light can be classified.

Contributing Idea A. Some characteristics of light can be seriated.

Seriating colors of light. Have students list the sequence of colors produced in a spectrum of light from a prism. Let them use paints or crayons to reproduce the sequence on paper, then compare their representations to other rainbows of light that they can find. They will note that the sequence of colors always remains the same. 13 Da

Seriating light by intensity. In a darkened room, hold a light meter 1 foot (30 cm) away from various light sources (e.g., burning candle, light bulb, fluorescent tube) to measure the intensity of each. Let students order the sources from least to most intense. They might compare the intensity of one candle to the intensities of other sources of light. 14 Da

Contributing Idea B. Some characteristics of light can be classified.

Categorizing luminous and nonluminous light sources. Have students make two lists of as many light sources as they can: (1) objects that produce their own light (e.g., sun, burning fuel); (2) objects that reflect light (e.g., moon, mirrors). Discuss each item on the lists. They will realize that reflective sources tend to be cooler. You might explain that the objects that produce their own light are called *luminous* objects while those that reflect light are called *nonluminous* objects. 15 Bb Db

Classifying materials by the way they transmit light. Have students hold different transparent, translucent, and opaque objects up to a light source and divide them into three categories based upon the way light does or does not pass through them. 16 Dc

332 INTERACTIONS

Generalization I. Light is a form of energy produced by different sources.

Contributing Idea A. The sun is a primary source of light (nuclear propagation).

Observing that the sun is a source of light. While sitting in the classroom without the lights on, let students discuss from where they think the light comes. Ask how things look on dark cloudy days compared with clear sunny days. Have them close their eyes and tell why they cannot see anything (their eyelids block the light from entering their eyes). Make the room as dark as possible by covering the windows with shades and drapes. Ask what can be seen now as compared with when much light entered the room. If possible, take groups of students into a closet, ask from where the small sources of light come, and ask if anything could be seen if all the cracks were sealed up (without a source of light, nothing could be seen). From these experiences and discussions, students will realize that during the day, the primary source of light is the sun. 01 Aa Ba

Contributing Idea B. Heating can be a source of light.

Observing light produced by heating. Hold a thin nail or wire with a pair of pliers and heat it in a flame until it glows. Let students describe how the appearance changes as the object is heated. Ask them to guess what would happen if the object could be made hotter (it would glow more brightly—from "red hot" to "white hot"). Explain that many rocks and minerals glow when heated, and materials can become white-hot in the production of steel and other metals in factories. You might tell them that when a metal is glowing hot, it is said to be *incandescent.* Have students find pictures of glowing metals in magazines or reference books. 02 Aa Bb

Observing light produced by glowing wires. Place a thin piece of copper wire across two battery terminals and observe what happens. **(Caution: The wire will become too hot to touch.)** Discuss what caused the light to be emitted. Explain that electricity passing through the wire made it hot enough to glow. After students test other wires, tell them that light bulbs (sometimes called incandescent bulbs) work in a similar way. Let them examine the filament in a bulb to see that it performs similarly to their copper wire. *(Note: Bulb filaments are usually made of tungsten rather than copper; however, platinum, carbon, and tantalum were used in early bulbs.)*

03
Aa
Ec

Contributing Idea C. Burning can be a source of light (chemical propagation).

Observing light produced by burning. Light a match and a candle. Discuss the appearance of the light that is emitted by each. Explain that when something burns, a chemical action (combustion) takes place and light is given off. Have students list other combustible materials that can be used as light sources (e.g., kerosene, gasoline, and oils in lamps; wax and wicks; wood). Some students might be interested in researching the history of human-made lighting from early campfires and torches, to oil and candles, to gas and electricity.

04
Aa
Bd

Comparing candlelight to sunlight. Light a 1 inch (2 cm) thick candle in a closet or darkened room and note the appearance of some objects that are 1 foot (30 cm) away. Explain that the intensity of the light upon the object is a standard measurement for light called one *foot-candle.* Now light two, then three candles, and note the increase in brightness. Next, light one candle outdoors in the bright sunlight to see how effective one candle is. Tell students that on a clear day at noon, the intensity of sunlight is about the equivalent of 10,000 candles 1 foot (30 cm) away. *(Note: The intensity of sunlight 1 foot or 30 cm from the sun is about 2.5 billion billion billion candles.)*

05
Bb
Ca

Contributing Idea D. Electricity can be a source of light (electrical propagation).

Observing light produced by electricity. Have students examine neon lighting tubes. Have them note that compared to incandescent bulbs, the neon tubes have small separate filaments at each end. Light one tube, and let students describe how it lights up compared with an incandescent bulb. Discuss differences in the lighting (e.g., the incandescent light is yellower). Explain that neon tubes are filled with a gas (neon), and when electricity passes through the gas from one end to the other, it glows as light. The color of the light depends upon the kind of gas in the tube (e.g., neon gives off a red light; argon gives off blue). An interested committee of students can research to find out what other gases are put to use in lighting. Some may wish to study fluorescent and mercury-vapor lights. **(Caution: Fluorescent tubes made prior to 1963 are dangerous if broken.)**

06
Aa
Bd

Contributing Idea E. The intensity of a light depends upon its energy source and the distance the light travels.

Observing that light intensity diminishes over distance. In a darkened room, light a lamp with its shade removed. Hold a book close to the lamp and observe the brightness of a page. Have one student walk from the light while holding the book, and let others

07
Ca

note that the page becomes dimmer. Similarly, four identical pieces of white paper can be placed 1, 2, 3, and 4 feet (20, 40, 60, and 80 cm) from the lighted lamp and observed for differences in brightness.

08 Ca Ec

Determining factors that influence the intensity of light. Insert a piece of aluminum foil that is shiny on both sides between two identical blocks of paraffin wax. Be sure all three pieces are the same size. Hold them together with rubber bands. Set the blocks in the middle of a table and place a lighted birthday candle at one end and four identical candles the same distance away at the other end. Darken the room, and let students observe the blocks from the side to see which glows the most. Next, let them move the blocks back and forth between the two light sources until they glow equally brightly. They will find that the intensity of each glow depends on the energy source (four candles versus one candle) and on the distance the source is from the receiving object.

Generalization II. Adding or removing light affects many materials and organisms.

> *Contributing Idea A.* Adding or removing light causes changes in some materials.

09 Db Ec

Testing the effect of light upon some colored materials. Place one piece of dark blue construction paper or cloth on a window sill in direct sunlight. After two days, compare the colors on each side. Next, test different colors and other materials to see if they fade differently in sunlight. Each of the colors of the spectrum can be ordered by the amount it fades. Students can also test other sources of light to see how they affect the colors and materials (e.g., fluorescent lights; incandescent lights).

10 Ga

Using blueprint paper. Have students take a piece of blueprint paper, place an object on it, cover the object and paper with clear glass, and expose it to direct sunlight for two to five minutes. (The time needed will vary with the brightness of the day.) After the paper has been washed in clear water, observe how the light affected the portion not covered by the object.

> *Contributing Idea B.* Adding or removing light influences plants and animals.

11 Ec

Testing the effect of light upon plants. Place a plant in a dark closet for an extended period of time and note the gradual changes in it. For comparative purposes, be sure an identical plant is kept in sunlight and that all other conditions (e.g., amount of water, temperature) remain identical for both plants.

12 Ec

Testing the effect of light upon animals. Obtain two glass jars, and paint the outside of one completely black so that no light can enter it. Fill the unpainted jar with house flies or fruit flies, and cover it with a stiff piece of cardboard so they cannot escape. Set the painted jar upside down on top of the cardboard on the unpainted one, remove the cardboard, keep the jars on top of each other, and observe where the flies go. After about five minutes, turn the jars over so that the painted jar is below the unpainted one. After another five minutes, students will be able to suggest why flies are usually found near windows. (Flies respond to light—that is, they move toward light.) If possible, test other animals such as mosquitos, spiders, or worms in the same way to see how light affects them.

Generalization III. Light travels through some solids, liquids, and gases at different speeds.

Contributing Idea A. Light travels through some materials but not others.

Observing that some materials are transparent. Place a coin on a desk and cover it with a piece of clear plastic. Cover the plastic with clear cellophane, then an upright, clear drinking glass. Add water to the glass. Ask students if they could see the coin through each material. Tell them that materials that allow the light from an object to pass through them so that the object is clearly seen are called *transparent* materials. Next, have students observe the coin and other objects through colored glass or cellophane. They will realize that the colored materials are transparent, but allow only certain colors to pass through. Have them find examples of transparent materials in the classroom.

13
Aa
Bb
Ga

Observing that some materials are translucent. Shine a flashlight beam through various translucent materials such as oiled paper, frosted glass, or a glass of milk. Have students note that although light is transmitted, one cannot clearly see through the materials. Tell them that materials that allow some light to pass through, but transmit no clear image, are called *translucent* materials. Let them find examples of translucent materials around the school.

14
Aa
Bb
Ga

Observing that some materials are opaque. Stand a wooden ruler upright between two books on a desk in a darkened room. Lay a small narrow-beamed flashlight behind the ruler to make a shadow on a wall or screen. Have students describe what happens to the light beam. (It is blocked by the ruler and produces a shadow.) Tell them that objects or materials that block light so that none gets through are called *opaque* materials. Now have them move the flashlight closer to the ruler, then farther away. Let them try to explain why the shadow becomes taller and wider when they move the flashlight closer, and why it becomes shorter and thinner when they move the light farther away. Next, have them test various objects between the light source and the wall. They will find that most objects are opaque.

15
Aa
Bb

Contributing Idea B. Slanted light rays bend when traveling from one medium to another of a different density.

Observing refraction. From the following experiences, students will find that light can be "bent" under certain conditions. After the experiences, you might tell them that such bending of light rays is called *refraction.*

16
Aa
Bb
Ca

a. Set a lighted candle on a table about 2 feet (60 cm) away from a wall in a darkened room. Have one student stand across the room and shine a flashlight beam on the wall above the candle flame. (The flashlight beam will seem to waver.) This can also be done across the top of a heated hot plate. Let students describe similarities with what they have seen near the surface of roads on hot days.

b. Place a straight stick, pencil, or ruler in an empty clear thin-sided glass. Look at the object from all sides. (It will appear to be straight.) Now half-fill the glass with water, and look again (From certain positions slightly above the water level, the object will appear to be bent.)

c. Stand a glass of water at the edge of a table. Look upward through the side of the glass at the top surface of the water. While looking, have a student stick the pointed end of a pencil down into the water. (The point will look larger than it really is.)

Observing that light rays can be refracted as the rays move from a more dense to a less dense medium. Place a lighted narrow-beamed flashlight into a plastic bag, seal it with a rubber band, and submerge it (beam shining upward) in a large transparent container filled with milky water. In a darkened room, students can see what happens to the beam as it passes from the water to the air. Several additional activities can be experienced by students.

17 Aa Ca Fc

a. Have students work in pairs. Let one in each pair place a coin in the bottom of an empty, clear cup. Have the other stand up, look at the coin, and back up until it just disappears from view. While she keeps her head in this position, have the partner, without moving the cup or the coin, slowly pour water into the cup. (The viewer will see the coin reappear.) Let them switch places and repeat. Explain that the direct view from the eye would not detect the coin, but because the light rays from the coin leave the water and enter the less dense air at an angle, the rays that strike the air first travel faster than the rays still within the denser water—the result is the light rays are refracted (bent) into the viewing path.

b. Students can observe from various positions spoons or pencils placed in empty glasses of different thicknesses or glasses filled with different liquids. In each case, they can describe how the light travels from one medium to another. Let them decide from their observations which medium is more or less dense.

c. Some students might enjoy filling an aquarium with water and trying to touch objects at the bottom with a pencil or stick. Discuss why they are visually fooled. (The light is refracted as it leaves the water and enters the air; thus, the object in the aquarium is not in the position it appears to be.)

Generalization IV. Light can be controlled.

Contributing Idea A. Light can be refracted. Lenses in optical instruments refract light.

Observing the shapes of lenses. Obtain several convex and concave lenses. Let students feel the surfaces of each and note that some are hollowed and others are bulged outward on one or both sides. Let them look through the lenses at various objects and describe what they see. (Convex lenses magnify—double convex lenses provide more magnification than plano (single)-convex lenses. Concave lenses make things look smaller.) Tell students that a lens that curves outward on one or both sides is called a *convex lens* and that a lens that curves inward on one or both sides is called a *concave lens.*

18 Aa Bb Ca

Finding examples of how lenses are used. Have students research the topic of lenses in encyclopedias and other reference books. They may find that several different kinds of lenses are found in the eyes of animals and in manufactured devices such as telescopes, microscopes, projectors, and cam-

19 Ga

eras. Have them find examples of how different lenses are put to use in optical equipment. Double-convex lenses are used in hand magnifiers, slide projectors, and telescopes because they provide more magnification than plano-convex lenses. Plano-convex lenses are used in eyeglasses to correct farsightedness. Double-concave lenses are used in eyeglasses for nearsighted people, in some binocular eyepieces, and artists use them as "reducing glasses" to make images smaller. Automobile headlights have many small lenses distributed over their surface in a complex pattern to control the light to obtain superior illumination. The lenses in some simple cameras are combinations of convex and concave lenses—the side away from the film is convex and the side nearest the film is concave—and expensive cameras may contain eight or more lens combinations either next to each other to obtain superior optical qualities, or separated by space as in the objective/eyepiece combination used in refracting telescopes, binoculars, and microscopes.

20 Aa Ca

Observing that an outward curving surface refracts light to magnify the appearance of objects. Fill a tall, thin bottle, such as an olive bottle, three-fourths full with water. Let one student stick an index finger into the water and compare its appearance to his other index finger outside the bottle. Ask if the inserted finger appears larger near the back or front of the bottle. Point out that the outside curvature of the bottle is similar to that of a convex lens: as light travels from the water in the bottle to the less dense air, it is refracted by the curved surface between them. Next, cut a strip of graph paper or two strips of notebook paper—one with lines running lengthwise and one with lines running across the strip. Place the strip into the bottle of water and observe. Have students compare the spacing between the lines both vertically and horizontally, and relate any observable differences to the curvature of the bottle. Repeat these examinations using bottles of different curvatures (diameters). You might discuss which bottle students would use if they wanted a product to look larger (e.g., olives in a jar).

21 Aa Ca

Observing that a convex lens refracts and magnifies light. Have students print a small letter "F" on a piece of paper, measure its height and width, and look at the letter through a convex lens. (The letter will look larger—at times it will be clearly upside down.) Now have them remove the shade from a lamp, find the trademark on the bulb, face the trademark toward the ceiling, notice the arrangement of the numbers and letters in the trademark, then turn the lamp on. Have one student move a convex lens up and down over the top of the bulb until a clear image of the trademark is projected on the ceiling. (A magnified, but reversed image will be seen.) Let students find examples of instruments in which convex lenses are used to magnify (e.g., binoculars, microscopes, telescopes), and discuss whether the viewed image is upside down or not.

22 Cc

Measuring the magnifying power of a convex lens. Focus a convex lens over some narrowly lined paper. Compare a single space seen through the lens with the number of spaces seen outside the lens. For example, if there are four spaces outside the lens to one through the lens, the power is four times (4*x*).

23 Aa

Observing that a convex lens refracts and focuses light. In a darkened room, shine a narrow beam of light through a convex (magnifying) lens. Clap two chalkboard erasers together so that the light beam can be seen clearly. Students will see how the beam of light comes to a point after it passes through the lens.

Locating the focal point of a convex lens. Hold a convex (magnifying) lens about 1 foot (30 cm) above a large sheet of paper at such an angle that the direct rays of the sun (or other bright light source that is at least 25 feet away) pass through the lens. Tell students that the point where the light under the lens comes together at its greatest brightness is called the *focus* or *focal point.* Have them measure the distance between the paper and the magnifying lens when it is focused, and tell them that this distance is called the *focal length* of the lens.

24
Aa
Bb
Cd

Observing that a concave lens refracts and spreads out light. Hold a concave lens about 1 foot (30 cm) above a large sheet of paper at such an angle that the direct rays of sunlight pass through the lens. Slowly move the lens back and forth to try to bring the light to a focus. Students will find that this cannot be done. (Convex lenses bend light inward while concave lenses bend light outward.) In a darkened room, shine a narrow beam of light through the lens, clap chalkboard erasers together to put some dust in the air, and observe that the beam is spread out after it passes through the lens.

25
Aa

Testing to see how different pieces of optical equipment affect light rays. Obtain a thick sheet of corrugated cardboard that is about 1½ feet by 1 foot (1½ m by 1 m) in size. Mount the cardboard to a stand to hold it upright (a). Cut holes in the cardboard for the placement of different pieces of optical equipment such as lenses, prisms, and mirrors. Paint the face of the cardboard white so that it will serve as a viewing screen. Now set a slide projector about 6 feet (2 m) to one side of the cardboard so its beam just strikes the center of the screen at a very slight, glancing angle. A projector slide that will produce rays of light against the screen can be made in two ways: (1) Using a pen and India ink, rule a ¼ inch by 1 inch (1 cm by 3 cm) area of frosted acetate with fine, closely spaced lines; (2) Using a white background, tape strips of ½ inch (1 cm) black tape at ½ inch (1 cm) intervals, then photograph the strips with a 35mm camera from a distance of about eight feet (2.5 m). By either method, the slide that is made will produce parallel rays of light when placed horizontally in the projector's slide mount. Be sure any unlined area on the slide is blacked out with tape. Darken the room, project the rays so that they strike the screen, then place pieces of optical equipment, singly or in combinations, in the screen openings and in the path of the rays. Observe the results. Students can test combinations of differently angled prisms, plano- and double-convex lenses, plano- and double-concave lenses, convex and concave mirrors. Many studies can be carried out with this simple equipment (e.g., colored glass or cellophane can be used to produce colored light rays; quantitative data can be obtained by attaching paper to the screen,

26
Ec

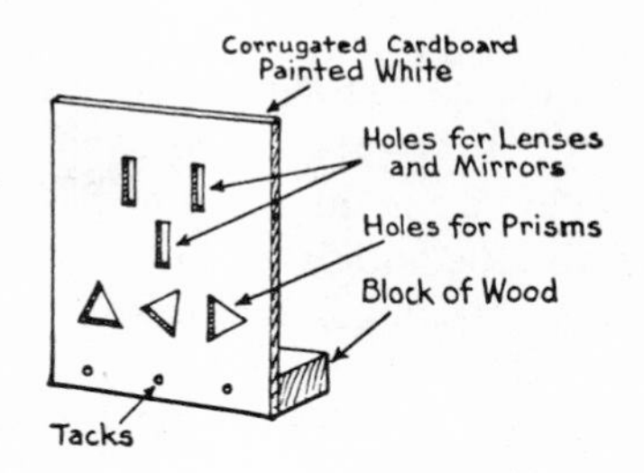

FIGURE 332.26

tracing the rays, then measuring the angles and distances after the paper is removed; cylindrical liquid lenses can be tested by filling bottles with various liquids and placing them in the beams of parallel light rays).

Contributing Idea B. Light can be reflected; mirrors in optical instruments reflect light.

Observing reflections. From the following experiences, students will find that light can be "bounced" under certain conditions. After the experiences, you might tell them that the bouncing of light rays is called *reflection.*

27
Aa
Bb

a. Fill a pan with water, set it on a table in the sun, and look for a bright spot on the wall or ceiling. Students will realize that water can serve to reflect light.

b. Clean a piece of window glass, lay it on white paper, and let students try to see their reflections in it. They can repeat their observations after laying it on black paper. From this they will be able to explain why a window makes fairly good mirror reflections at night when it is bright inside and dark outside. Let them try other colored papers under the glass.

Reflecting light. Have students take turns holding a mirror in the sunlight and flashing the light on a wall or ceiling. Let other students take turns holding another mirror in the line of reflection. Using two mirrors, challenge them to direct a spot of light to particular locations in the room. You might tell them that signals are sometimes sent at sea from one ship to another in a similar way and that the United States Cavalry once used reflectors to send signals from one hill to another.

28
Aa

Making a seismograph using reflected light. Place a pan of water on a table in such a way that a beam of sunlight strikes the surface of the water and reflects a spot of light on the ceiling or a wall. Now have a student stamp a foot on the floor, and observe the spot of light. Explain that the device is similar to a seismograph that detects vibrations—every time there is a small quake, vibrations travel through the floor and air to the water and set up ripples on its surface, thus causing the spot on the wall to shimmer. If the quake is weak, the shimmering is slight, if it is strong, the shimmering is great. Students can set up vibrations in many ways (e.g., clapping their hands, closing a door or window, dropping a book). Have them see if their seismograph is sensitive enough to detect the hum of a motor in a school appliance or the rumble of vehicles passing by the school.

29
Ga
Gc

Testing to see how different surfaces reflect light. Use poster paint to blacken the inside of a shoe box. Cut a slit 1/8 inch by 3 inches (.2 cm by 8 cm) in one end as shown. Stand the box on its long side in the

30
Da
Ec

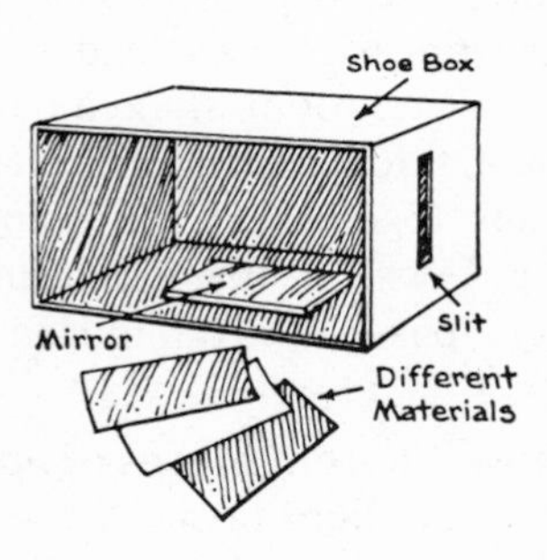

FIGURE 332.30

sunlight, and observe the beam of light that enters through the slit. Now place a mirror in the bottom of the box and observe again. (The light will reflect and strike the opposite end.) Next, place the mirror with other materials such as white paper, black paper, colored paper, aluminum foil, and so forth to see how each reflects light. Students will see that some of the materials, like white paper, scatter the light in all directions (irregular reflections), while more polished surfaces reflect it with little scattering (regular reflections). Have students order the materials by the way they reflected the light.

Observing that particles in gases (air) and liquids (water) reflect light. Prepare a flashlight so that it projects a thin beam. To do this, cover the bulb with masking tape, leaving a pinhole in the center. Next, tape a paper cup over the end of the flashlight, leaving a pinhole in the bottom. Now darken the inside of a shoe box with black paint and cut viewing holes in each end and the top. Place a white index card opposite one end and shine the thin beam from the flashlight through the box so that it can be seen on the card. Have students take turns looking down into the box through the top hole at the passing beam of light. (The beam cannot be seen.) Explain that light is generally invisible and that we see it only at its source or when it is reflected. For example, when one looks into the sky at night, only the stars (sources of light) and the moon and planets (reflectors of light) can be seen, while the light that fills the sky from the sun and stars remains invisible. Next, blow some smoke into the box or clap two chalkboard erasers together to fill it with dust. Students can observe the interior of the box again. (The beam will be clearly visible.) They will realize that the beam is now visible because it is reflected from countless tiny particles in the air. Now shine the beam through a clear glass of water, then add a spoonful of milk to the water. Students will see that the beam was invisible (except for the tiny dust particles) until the particles of milk were added.

31
Aa

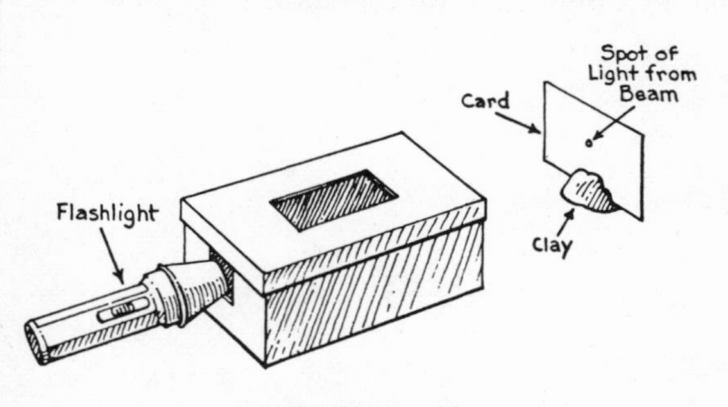

FIGURE 332.31

Measuring angles of reflection. Prepare a flashlight so that it projects a thin beam (see *332.31*). Tape mirrors inside three identical cardboard boxes at the center of one of the long sides. In one box, cut a hole in the opposite side so that the flashlight can be inserted. Paint the inside of the boxes black, and cover the open top of each with transparent plastic (e.g., sandwich wrap or a dry-cleaning bag). Before closing the first box, set a burning string, piece of punk, or incense on a metal jar lid so that the box will fill with smoke (the smoke will make the light beam easier to see). If the flashlight holes are placed in different positions in the remaining two boxes, students can vary the angle at which incoming beams strike the mirror. By

32
Cd
Ec

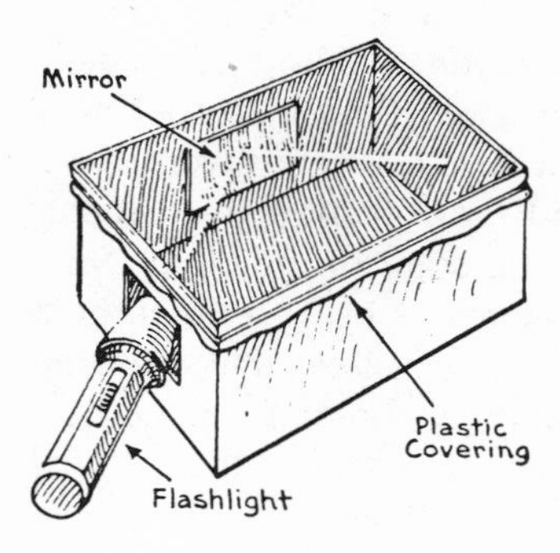

FIGURE 332.32

looking straight down through the plastic, they will see how the light beam strikes and reflects from the mirror. A string can be stretched across the top of the plastic to match the lines that form the angles of incidence and reflection. A protractor can be used to measure the angles. Students will discover that the beam always travels in straight lines, and the angles of incidence and reflections are always equal no matter how the beam strikes the mirror. Test other materials in place of the mirror to see how they reflect. From the tests students will find that rougher surfaces tend to scatter or diffuse the light beams because of their varying angles of incidence.

Contributing Idea C. Light can be absorbed.

Observing that colored materials absorb most of the colors of the spectrum. Place a piece of green cloth in a completely dark room. Put a red bulb in a socket or cover a white bulb with several layers of red cellophane, and turn on the light. Let students describe the color of the cloth. Now repeat the observations using a green bulb in the socket. Students will see that when the red light falls on the green material, no light is reflected (it is absorbed), and the material appears black; but when the green light strikes the material, the color is reflected. Test other colored light sources and colored materials. *(Note: Objects have different colors because they contain different chemicals that have the ability to absorb [change into heat energy] certain colors [wavelengths of light] and reflect others.)*

33
Aa
Ec

Observing what happens to light that is absorbed. Place a piece of black cloth about a foot beneath an electric light bulb. Use books to support a pane of glass between the bulb and the cloth to prevent radiant heat from reaching the cloth. Turn the light on, then after fifteen minutes let students feel the cloth. Repeat this using a piece of white cloth, and compare the findings. Students will discover that when light is absorbed by a material, the light energy is changed into heat energy. As a supplemental activity, they can lay a thermometer under the black cloth, check its reading every three minutes for a half hour, then plot the change in temperature on a graph.

34
Ab
Bf
Ch

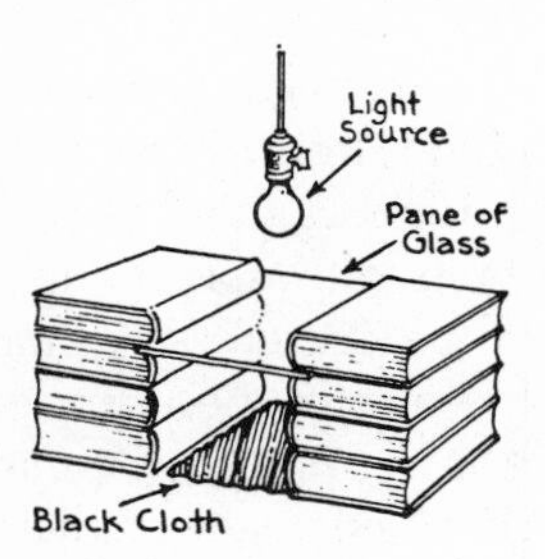

FIGURE 332.34

341 CHARACTERISTICS

Generalization I. Heat has identifiable characteristics.

Contributing Idea A. Temperature is a characteristic that can be measured with a thermometer.

Feeling and describing temperatures. 01 Ab Ba Prepare three bowls of water: one three-fourths filled with water that is about as hot as a student's hand can stand; one three-fourths filled with water that is about room temperature; one three-fourths filled with very cold water. Have students place one hand in the very hot water, one hand in the very cold water, and describe the sensation felt by each hand. Remove hands from the two bowls and place them simultaneously into the bowl of lukewarm water. Let them describe the sensations. (The hand from the hot water will feel cold while the hand from the cold water will feel hot.) Students will realize that telling temperatures by the sense of touch is relative and not accurate. For example, it may seem hot in a schoolroom, but after being out in the sun for a period of time the same schoolroom will seem cool.

Observing how temperatures affect thermometers. 02 Aa Obtain various kinds of thermometers (e.g., mercury, alcohol, gas, metal). Have students notice that if they touch a thermometer, the heat of their hands changes the temperature reading. Let them observe what happens to each thermometer as it is touched with an ice cube or when it is inserted into bowls of water of different temperatures.

Standardizing temperature measurements 03 Ce Prepare two or more containers of water that are very nearly the same temperature. Without using thermometers, let students ascertain which water is warmer. Try to obtain a general agreement, then measure the temperature of each with a thermometer so that universal agreement can be obtained. Point out the need for universal agreements in measuring temperatures.

Measuring temperatures. 04 Ce Have students use graph paper to prepare a map of the school grounds showing the arrangement of buildings, lawns, shrubs, trees, etc. Let them use thermometers to measure the temperatures of the air, water, and soil in different locations around the grounds. These measurements can be placed in the appropriate squares on the map. *(Note: Air temperatures are more accurate if the thermometer is allowed to swing freely in the air from the end of a string. Ground temperatures can be taken by carefully inserting the bulb end of a thermometer into the soft earth. Sur-*

face temperatures are more accurate if the bulb is shaded from the sunlight.) Compare the temperatures. Temperatures can also be compared for different seasons of the year.

Making a liquid thermometer. Completely fill a milk carton with water colored with red dye. (The dye is used to make the water more visible.) Make a hole in the center of the carton's top, and insert a clear plastic straw. Use an eye-dropper to add more colored water into the straw until it rises half way up the tube. Seal the straw into the carton with clay, gum, or wax. Add a few drops of light machine oil to the top of the colored water in the tube to prevent evaporation. Fasten a card behind the tube, and mark the level of the liquid on the card each day. Test the instrument in various areas around the school grounds.

05
Gc

Calibrating a liquid thermometer. Have students fill a flask or softdrink bottle with colored water. Fit the container with a one-hole stopper and a plastic straw or glass tube. Press the stopper into the container so that the colored water rises in the tube about 3 inches (7 cm) above the stopper. Now place the container in a pan of water and heat it on a hot plate. When the colored water stops rising, mark the water level on the straw. Let the container cool to room temperature. Next, place it into a deep bowl filled with ice and pack the ice around the container. When the colored water stops dropping, mark the level. Students will realize that the two marks provide the high and low points for a temperature scale (the fixed points at which water expands when heated and contracts when cooled). Explain that the most common thermometer is the Fahrenheit thermometer. Gabriel Fahrenheit (1686–1736), a German physicist, established his scale by marking 0°F for a mixture of ice and salt. His high point was what he thought was the normal temperature of the human body (96°F). This range was extended for measuring other temperatures. On his scale, the freezing point of water is 32°F, the boiling point 212°F. Anders Celsius (1701–1744), a Swedish astronomer, established his scale by marking the freezing and boiling points of water as 0°C and 100°C. Scientists all over the world use the Celsius scale. *(Note: This scale is also called the Centigrade Scale.)* Students can divide the range of their scales into tenths and hundredths and compare temperature readings with those from a commercial scale. Some students might be interested in researching the work of Fahrenheit and Celsius.

06
Bd
Gc

Making an air thermometer. Obtain a bottle with a screw cap. Punch a hole in the cap so that a plastic straw or glass tube will fit into it snugly. Screw the cap on the bottle and drip candle wax around all connections to seal them. Fasten the bottle upside down on a support so that the straw sticks into colored water in a second bottle. Have a student warm the first bottle with her hands

07
Gc

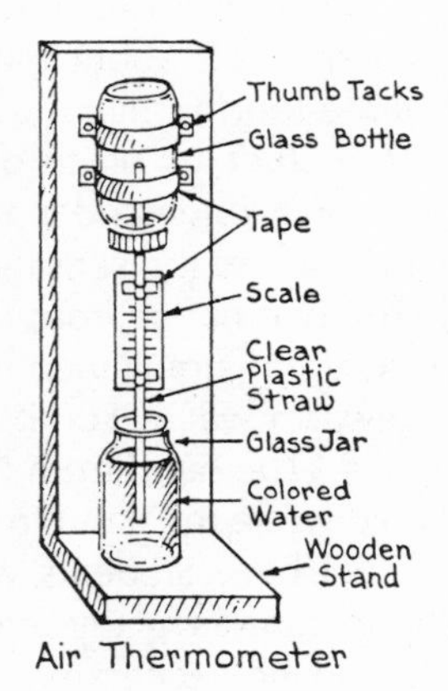

FIGURE 341.07

to force some air from it. (Air bubbles will be seen in the colored water.) When the bottle cools, a column of water will rise inside the straw to replace the air that was forced out. A scale can be attached behind the straw and calibrated.

Making a copper-wire thermometer. Build a wooden stand as shown. Suspend a length of copper wire (#12 to #16 in size) from a nail that is bent slightly downward.

08
Gc

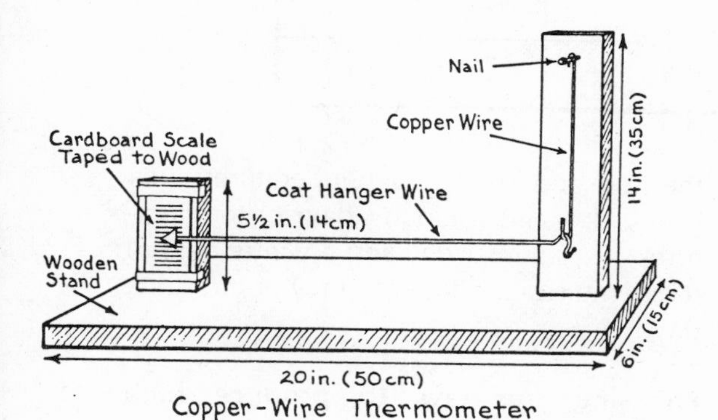

FIGURE 341.08

Drill a hole about 1 inch (2.5 cm) deep in the base of the upright, just below and to the right of the lower end of the copper wire. Cut and bend a coat hanger wire and slip the bent end into the hole. Be sure it turns freely in the hole. Tape a scale to a small block of wood and set it at the far end of the coat hanger. As temperatures change, the copper wire will expand and contract causing the hanger to rise and fall.

> *Contributing Idea B.* Temperature is a measure of the average amount of energy in a material; heat is the total amount of energy in a material.

Comparing differences between temperature and heat. Obtain an aluminum nail and an aquarium nearly filled with warm water. Have students compare the nail and the aquarium (which has the greatest amount of material?). Measure the temperature of the water in the aquarium with a thermometer. Ask how the temperature of the nail might be taken. (If the nail has been sitting untouched in the room for a period of time, it is probably the temperature of the air in the room.) Record the two temperatures. Use a clothespin or a pair of pliers to hold the nail in a flame until it glows red hot. Ask which now has the highest temperature, the nail or the aquarium. Plunge the hot nail into the aquarium and allow it to remain for a few minutes. Take the temperature of the aquarium again. Students can realize that even though the nail was extremely hot, it hardly affected the water because the water originally contained a greater amount of heat due to its quantity.

09
Ce

Determining which metals hold more heat. On one side of a balance arm, place five or six pennies. *(Note: Although pennies are not made of pure copper, they can be used as a basis of comparison for other metals.)* Next, fold several large sheets of aluminum foil into small lumps and place enough aluminum on the other balance arm to balance the pennies. Now place the pennies in a tea bag or small plastic bag with a string attached. Lower the bag into a pot of boiling water. While the pennies are warming, take the temperature of a one-fourth cup (100 ml) of cold water and record it on a table. Remove the heated pennies and quickly lower them into the cold water. Stir, and record the highest temperature reached by the water. Calculate the number of degrees the water temperature rose because of the hot pennies. Next, let students repeat the activity using the aluminum foil, then equal weights of iron or other metals. They might also see how "sandwich" type coins com-

10
Bc
Ce

TABLE 341.10. Measuring the Heat Held by Metals

	Water Temperature *Cold*	*Heated*	*Rise In Temperature*
Copper			
Aluminum			
Silver			
Nickel			
Iron			

pare to "older" types in terms of heat. Challenge them to find a way to test which liquids, such as cooking oil and water, hold the most heat. They can try placing equal weights of metals in a freezer, then adding them to equal amounts of warm water and determining the cooling effect.

342 INTERACTIONS

Generalization I. Heat is a form of energy produced by different sources.

Contributing Idea A. The sun is a primary source of heat (nuclear propagation).

Feeling heat produced by the sun. Have students sit in the sun for some time, then move to the shade to feel the difference in heat. Discuss how burned one can get if one stayed in the sun too long. *(Note: Burning can also take place on a cloudy day and has nothing to do with apparent sunlight.)* **01 Ab**

Comparing differences in temperatures from the sun. Have students use thermometers to compare: a) temperatures in the sun and shade; b) temperatures of a sunny day and a cloudy day; c) temperatures of a summer day and a winter day; d) temperatures at different times of the same day. **02 Ce**

Focusing sun rays to produce high temperatures. Obtain magnifying hand lenses, small aluminum pans, and small sheets of tissue paper. Have students hold the hand lenses in the sunlight to focus sun rays on a ball of tissue paper in the aluminum pan until the paper bursts into flame. Hand lenses can also be used to focus sun rays on a thermometer to observe the temperature rise. **(Caution: Be sure the rise in temperature does not exceed the limits of the thermometer.)** **03 Aa**

Contributing Idea B. Motion can be a source of heat (mechanical propagation).

Feeling heat produced by motion. Heat produced by motion can be observed in many common experiences. Have students feel the blade of a saw after sawing a piece of wood; feel the break hub of the rear wheel after applying the brakes while riding **04 Ab**

down a hill on a bicycle; feel the warmth after sliding down a rope or when a rope or string is pulled through the hands. **(Caution: The heat generated in this manner can get hot enough to burn.)** Have them feel the sides of an air pump after it has been pumped briskly for a short time; feel how warm a nail gets after hammering it quickly several times or after it is pulled from a block of wood; feel a wire after bending it back and forth several times.

Rubbing objects to produce heat. Have students rub various objects together: rub hands together briskly until warmth is sensed; rub a brass button (or similar object) energetically against a piece of wool cloth, then touch the button; rub a small piece of metal on a piece of paper and sense how warm they get; rub a pencil eraser briskly against a piece of paper, then quickly touch the eraser to the upper lip or the tip of the nose (a sensitive part of the body) to feel the warmth produced. If possible, have the class observe a scout (or someone with similar experience) start a fire by rubbing pieces of wood together.

05
Ab

Measuring the heat produced by motion in a solid (sand). Fill jars or tin cans about one-fourth full of dry sand. Let students measure and record the temperature of the sand in each container. Cover the containers, and shake them vigorously about 100 times, then measure the temperature again. Compare the two temperatures. Let students experiment to see if shaking the containers more or less times changes the temperature. The data can be graphed.

06
Bf
Cd
Ec

TABLE 342.06. Measuring Heat Produced by Motion
(jar filled with sand)

Number of Shakes	*Temperature in °F (or °C)*

GRAPH 342.06. Measuring Heat Produced by Motion

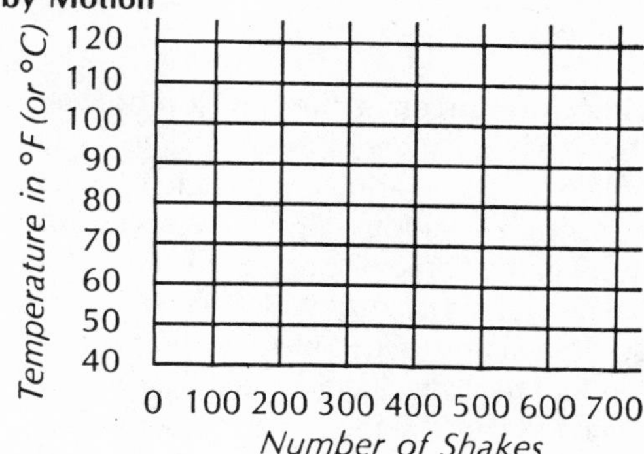

Measuring the heat produced by motion in a liquid (water). Fill a mixing bowl three-fourths full of cold water, measure and record the temperature, then run an electric beater in the bowl for about ten minutes. Measure the temperature again. Compare the temperatures. Students will realize that the motion of the beater produced the amount of heat that was measured. Experiment to see if a time difference in using the beater changes the measurements. The data can be graphed.

07
Bf
Ce
Ec

Contributing Idea C. Burning materials can be a source of heat (chemical propagation); some burning materials can be used as fuels.

Feeling heat produced by burning. Light a match, then a candle. Students can feel the heat given off by the flames. Use pliers, forceps, or a long fork with a wooden handle to hold various flammable objects in the

08
Ab

candle flame. Students will sense that heat is given off as the objects burn.

Comparing differences in the ignition of materials. Light a candle or alcohol lamp, and use pliers to hold various objects (e.g., celluloid, twists of various papers, wood, coal, cotton) in the flame one at a time. You might explain that the point at which a material bursts into flame is called its *kindling point,* and that heat-producing materials do not all burst into flame at the same temperature. Have students order the materials by the amount of time it takes to ignite each.

09 **Bb** **Ca** **Da**

Comparing kindling points. Kindling points can be compared by placing various objects (e.g., match head, coal, sulfur, wood, paper) on a sheet of copper, iron, or aluminum. Bridge the metal sheet across a flame source, make sure the objects are about the same size and equidistant from the flame, then have students note the order in which the materials ignite. *(Note: Some materials may not ignite because their kindling points are higher than can be reached with this method.)*

10 **Ca**

Contributing Idea D. Electricity can be a source of heat (electrical propagation).

Feeling heat produced by electricity. Have students observe the filaments in various appliances (e.g., electric toaster, stove, heater, iron) and feel that heat is radiated from them when they are turned on. Students can also observe the filament in a nonfrosted 100-watt electric lightbulb. Place a sheet of paper against the bulb, turn it on, and observe as it scorches.

11 **Ab**

Comparing electrically produced temperatures in different materials. Obtain two or more dry cell batteries and connect them in series (positive pole of one dry cell to the negative pole of another) by attaching, one at a time, pieces of copper wire that vary in thickness. Let students feel the differences in temperatures (the thicker the wire, the lower the temperature). **(Caution: Some thin wires can get hot enough to melt.)** Students might begin with two dry cells, test the wires, then add more dry cells, one at a time, to compare what happens as the current is increased.

12 **Ca**

Generalization II. Adding or removing heat affects materials and organisms.

Contributing Idea A. Adding heat causes nearly all solids to expand; removing heat causes them to contract.

Observing the expansion and contraction of some solids. Expansion and contraction can be observed through a variety of classroom activities. Following the suggested activities, you may wish to introduce the terms *expansion* and *contraction.*

13 **Aa** **Bb**

a. Hammer a nail into the base of an empty can. Gentle remove the nail, then replace it to make sure the hole is just large enough for it. Hold the nail with a clothespin or pair of pliers, heat it in a flame, then try to put the nail back into the hole.

b. Obtain two 1 foot (30 cm) dowels with 1 inch (2.5 cm) diameters, one large eye-screw, and a flat-head screw that barely passes through the eye-screw. Screw the pieces into the ends of the two dowels, then using a candle flame or alcohol lamp, heat the flat-head screw and try to pass it through the eye-screw. (Expansion will keep it from passing through.) Now heat the eye-screw and try again.

(Expansion will allow the flat-head screw to pass through.) Cool the eye-screw and try again. (The solid eye-screw contracts and keeps the expanded screw from passing through.)

Measuring the expansion and contraction of some solids. Build a stand. On Block 1, place a wooden pencil so that it will roll easily. Put a pin in the eraser end of the pencil and hang a threaded fish-weight from

14
Cd

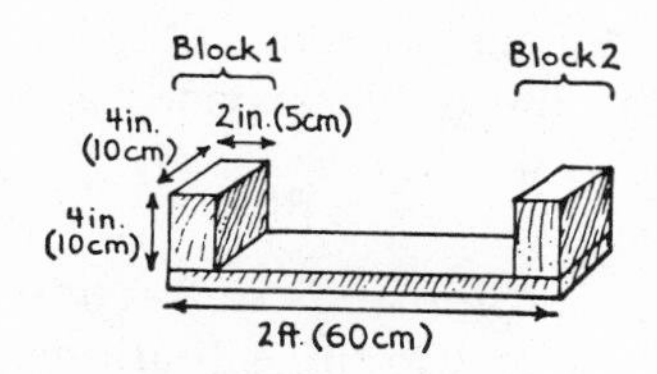

FIGURE 342.14a

the pin. Let the weight hang freely over the edge. Mark a card in ⅛ inch (3 mm) intervals for a scale and mount it to the block behind the weighted thread. Now as the pencil is rolled back and forth, the thread will move along the scale, and the distance it moves can be measured. To Block 2, attach one end of a rod or strip of metal, such as a brass curtain rod, with U-nails or straight nails that are bent over the rod. Set

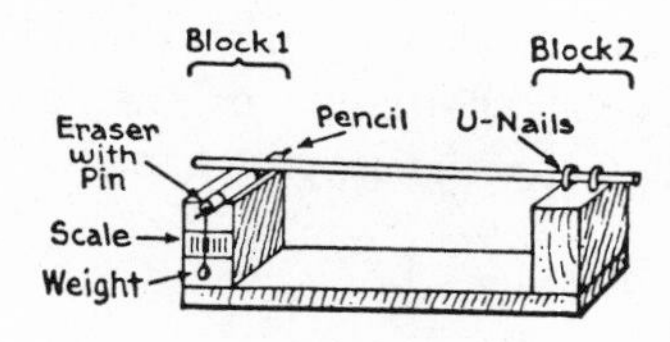

FIGURE 342.14b

the other end of the metal rod so that it rests on the pencil near the inside edge of Block 1. Heat the metal with a flame. As the metal expands, students will see it move across the pencil and cause it to roll. The thread will move along with it, and the distance can be measured and recorded. Now place several different flame sources to the metal at the same time to see how far the metal can be made to expand. Other metals such as iron, copper, or aluminum can also be tested and compared.

Measuring the expansion and contraction of some wires. Obtain a board about ½ inch by 2 inches by 12 inches (1 cm by 5 cm by 30 cm), one large nail and two small nails, one drinking straw, a rubber band, and a white card. Assemble them as shown. A

15
Cd

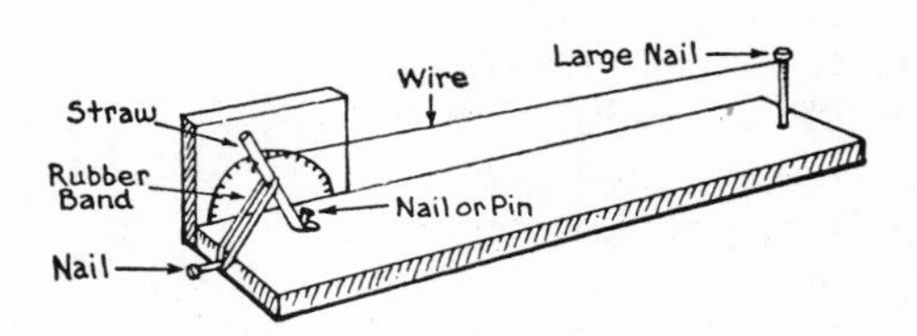

FIGURE 342.15

scale can be drawn on the card with the aid of a protractor. With this device, students can compare the expansions and contractions of different wires quite accurately.

Observing how people adjust materials to allow for the expansion and contraction of solids. Take a field trip to notice the spaces between the concrete slabs of sidewalks, the spaces between the steel rails of railroad tracks, or the spaces between the steel girders in bridges. Discuss the need for this allowance for expansion and contraction of solid objects.

16
Aa

Observing that expansion and contraction can break some solids. Thoroughly heat some small marbles or rocks in a pan over a heating unit. Using pliers, pick them from the pan one at a time, and plunge each

17
Aa

into a glass or beaker of cold water. The expanded objects will crack or break as the cold water quickly contracts their exterior. Similarly students can observe in a demonstration that a hot liquid poured into a cold glass container will sometimes cause it to crack or break because the quickly heated inside surface suddenly expands, pushing out the outside unheated surface.

Contributing Idea B. Adding heat causes nearly all liquids to expand; removing heat causes them to contract.

Measuring the expansion and contraction of some liquids. Fill two or more flasks or bottles to the brim with various liquids such as colored water, alcohol, gasoline, and oil. Press into the flasks one-hole stoppers fitted with about 12 inch (30 cm) lengths of glass tubing. **(Caution: Glass tubing is fragile—a paper towel wrapped around the tubing while it is being inserted will protect hands should the glass accidently break.)** As each stopper is pressed into a flask, the liquid will rise part way up the tube. Let the flasks sit in a shady part of the room for several hours so that the liquids will attain the same temperature, then mark the levels of the liquids with a string or rubber band (or a card can be placed behind the tubes for marking). Now move the flasks into a sunny spot in the room or take them outside into the sunlight. After the liquids have stopped rising, mark the height of each liquid as before, and compare the expansions of the different liquids. Students can order them by degrees of expansion. *(Note: A greater range between expansion and contraction can be found for each liquid by placing each flask first in a bowl of ice, to mark the near freezing point of water, then into warm water that is brought to a boil for several minutes, to mark the near boiling point of water.)*

18 Cd Da

Contributing Idea C. Adding heat causes all gases to expand; removing heat causes them to contract.

Observing the expansion and contraction of a gas (air). Place the open end of a bottle under water. Heat the bottle with a flame or pour fairly hot water over it. Students will see air bubbles coming out of the bottle. (The air in the bottle expands and forces some of the air out.)

19 Aa

Observing the expansion and contraction of a gas (air) in different situations. Stretch a sheet of rubber cut from a balloon over the mouth of a can or jar. Attach the rubber tightly to the can with a rubber band. Students can realize that the gas (air) in the can will now be in a closed system, isolated, and generally independent of the outside air. Influences upon the isolated gas (air) can create observable changes in the flexible rubber lid. Heat the container with a flame and observe the lid. (The air inside will expand and force the lid upward.) Next, cool the container with a damp cloth or ice cube and observe the lid. (The air inside will contract and the outside air will force the lid downward.) Several alternative or supplementary activities can be provided for the students.

20 Aa

a. Fasten a rubber balloon over the opening of a flask, test tube, or pop bottle. Set the container in hot water or heat it gently with a flame. Have students observe the balloon. (It will become partially inflated.) Now cool the container, and observe again. (It will deflate.) *(Note: If the balloon is stretched before attaching, it* *will expand and contract more easily during the activity.)*

b. Dip the mouth of a medicine bottle into soapy water to produce a film across the opening. The film will expand if warm hands are placed on the bottle.

c. Students can discuss experiences with the popping of bottle tops. Have them recall the sequence of events prior to the popping. (Usually a cold bottle is taken from the refrigerator, used, and stoppered—as the bottle sits in the warm room, the air inside the bottle becomes warm, and as it expands it presses harder against the sides and top of the bottle, causing the top to pop open.) If a refrigerator is available at school, have students see if they can cause the event to happen.

Contributing Idea D. Adding heat causes some solids to melt into liquids; removing heat causes some liquids to freeze into solids.

Observing phase changes. Basically, matter exists in states—solid, liquid, and gas. Adding or reducing heat causes matter to change from one state to another. You might explain to students that a change from one state to another is called a *phase transformation.* Observation of water best depicts the various phase transformations. Students can cite examples for and discuss each of the following: (liquid to solid) ponds, lakes, and rivers solidify in cold weather; (solid to liquid) ice forms melt in warm weather; (liquid to gas) puddles of water evaporate on warm days; (gas to liquid) unseen water vapor condenses against a cold mirror in the bathroom; (solid to gas) wet wash on a clothesline in freezing weather will freeze, then sublime dry; (gas to solid) water vapor in the air will frost on plants on cold mornings.

21
Aa
Ba
Bb

Observing freezing and melting (liquid ⟷ solid). Completely cut the covers from several small milk cartons. Prepare each carton with the same amount of water. Add different concentrations of salt, measured in level spoonsful, to the containers. Stir and label the containers, then place them in a freezer and check every hour to determine the order in which the solutions freeze. *(Note: Some may not freeze.)* Students will realize that liquids do not all freeze at the same temperature. In this activity the salt lowers the freezing point of water. You might explain that in the winter saltwater harbors are often unfrozen, while nearby freshwater lakes are frozen—it takes a much lower temperature to freeze salt water than fresh water. For further comparisons, students can repeat this activity using other liquids and solutions (e.g., oil, alcohol, sugar water, syrup). After the liquids have been frozen, see how long it takes for each to melt under the same temperature conditions.

22
Aa
Ca

Contributing Idea E. Adding heat causes liquids to evaporate into gases; removing heat causes gases to condense into liquids.

Observing evaporation (liquid → gas). The phenomenon of evaporation can easily be observed in several ways.

23
Aa

a. Students can observe the effect of hot weather upon wet sidewalks, puddles, or spilled water. They can also observe wet clothes hung on a clothesline (or wet cloths on the school fence) and feel them periodically as they dry in the sunlight.

b. Students can set liquids such as alcohol, water, or milk in open dishes in the sunlight and watch them on a daily basis.

Measuring the cooling effect of evaporation. Wrap the bulbs of three identical thermometers with small pieces of cotton—one saturated with water, one with alcohol, and one with ether. Let students fan them rapidly with a sheet of cardboard. Observe the temperature changes. You might explain that people with high temperatures are sometimes given alcohol rubdowns to lower their temperatures.

24
Ce

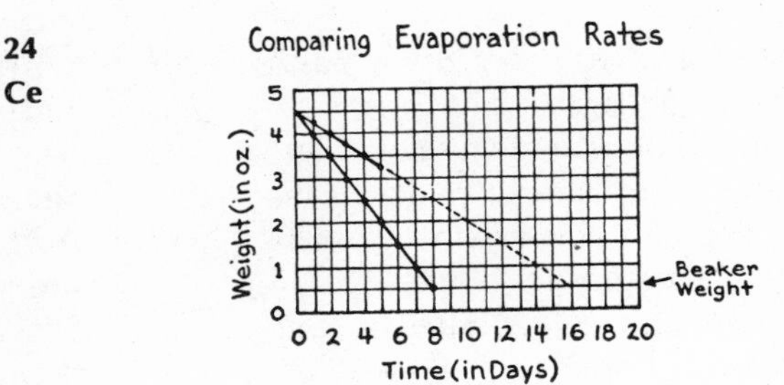

GRAPH 342.25

Graphing evaporation rates of different liquids. Have students weigh two beakers or paper cups and record the weights on a table such as the one shown. Fill one beaker with 4 ounces of water and the other with 4 ounces of alcohol. The cups and their contents should then be measured at equal intervals—preferably daily at the same time. *(Note: If a measurement is taken one day at noon and the next measurement is taken the following day at 3 P.M., the first point on the graph could be plotted at TIME= 0, and the second point on the graph could be plotted at TIME= 1 1/8.)* An alternative method of measurement for this activity would be to put the liquids in graduated cylinders or measuring cups. (The depth of each liquid could be measured with a ruler.) The volume could be recorded at successive intervals rather than the weight. *(Note: The tabled data for this activity will produce a straight line graph—if all conditions remain constant, predictions concerning total evaporation time for each liquid can be*

25
Bf
Cf
Ch

TABLE 342.25. Comparing Evaporation Rates
(Beaker weight = ½ oz.)

Alcohol		*Water*		*Remarks*
Days	Wt. (oz.)	Days	Wt. (oz.)	
0	4	0	4	Started Observations
1	3½	1	3¾	
2	3	2	3½	
3	2½	3	3¼	
4	2	4	3	
5	1½	5	2¾	Made a prediction

made after five points have been graphed.) Many thought-provoking questions can aid students in interpreting graphs of this type: which liquid takes the longer to evaporate; after four days, how much water and alcohol was left; when there was 1½ ounces of alcohol remaining, how many days had passed; when the alcohol had evaporated, how much water was left?

Observing condensation (gas→liquid). 26 Aa
Ask students to exhale against a pane of glass or a mirror. (Since the warm air in one's breath contains much moisture, the cool or cold glass will cause the moisture to condense, making it visible on the glass.) Some students can recall the fogging of car windows on cold nights or the sight of their own breath in the air on cold days. Students who wear glasses can relate some of their experiences.

Observing the effect of removing heat on condensation. 27 Aa
Obtain two milk bottles or large coffee cans. Fill one with warm water and one with very cold water. (Water vapor in the air around the cool bottle will condense from the air as droplets of water on the surface of the bottle.) If students think the droplets came from the inside of the can, some dye can be added to the can's water. The colorless droplets on the outside of the can should convince students that the water came from the outside air. You may wish to discuss condensation on pitchers of ice water, bottles of cold softdrinks, and glasses of cold drinks.

Contributing Idea F. Adding heat causes some solids to sublime into gases; removing heat causes some gases to frost into solids.

Observing subliming (solid→gas). 28 Aa
Put a small piece of dry ice into a small flask. Because the dry-ice gas (carbon dioxide) is odorless and colorless, put a balloon over the neck of the flask and fasten it securely with a rubber band. As the dry ice sublimes, the balloon will inflate and perhaps pop.

Observing subliming and frosting (solid ⟷ gas). 29 Aa
Place iodine crystals in a flask, cover the flask with a saucer or water glass, and set the materials in the sunlight. Students will see that the heat from the sun causes the iodine to sublime (seen as a violet-colored gas). As the gas contacts the cooler saucer and sides of the flask, crystals of iodine will form. **(Caution: Care should be taken in handling iodine—do not inhale the fumes or handle the crystals with bare hands—rinse all equipment thoroughly with cool water after use.)** The formation of morning frost can be compared to this activity.

Contributing Idea G. Adding or removing heat influences plants and animals.

Testing the effect of heat upon seed germination. 30 Bf Ec
Have students plant identical seeds in three labeled flower pots. Place the pots respectively in cold, normal, and very hot locations. As an alternative, the temperatures can be created artificially using dry ice or crushed ice, a heater, or a light bulb. Thermometers can be placed on each pot to take the temperatures in each location. Students should see that each condition is given equal measures of water. They can compare the germination pattern and growth in each pot and judge what effect heat or the lack of it has upon the plants. This activity can also be done without soil (see *441.01*) so that students can observe the temperature effects on root growth. If measurements are taken, the data can be graphed.

Testing the effect of heat upon insect metamorphosis. Have students gather caterpillars that are feeding on plants. Place equal numbers of caterpillars, together with parts from the plants, into separate jars. Cover the jars with nylon mesh to keep the caterpillars from escaping. Now place one jar in a warm, but not hot, location and another in a cool location. Be sure all other conditions, such as the amount of sunlight and food, remain the same for both. Thermometers can be used to keep daily temperature records. Record any differences in the time it takes the caterpillars to go into pupation and for the pupas to change into adult butterflies or moths. *(Note: Be sure to keep a wad of moist cotton in each jar during the pupa stage to prevent the insects from drying out.)* From this experience students will be able to make some judgment concerning the relationship of heat to stages of metamorphosis. **31 Ec**

Generalization III. Heat travels through solids, liquids, and gases in different ways and at different speeds.

Contributing Idea A. Heat sometimes travels through materials by conduction; some materials conduct heat better than others.

Feeling the conduction of heat. Have students recall that the metal handles of pans become hot as the pan is heated, that spoons become hot when placed into hot liquids, that heat can travel gradually through an object's entire length. Several similar experiences can be carried out in the classroom. **32 Ab Bb**

a. Place spoons of different metals into cups of hot water. After several minutes let students feel the handle of each spoon to see which conducted the heat best. (Silver spoons are excellent conductors of heat.)

b. Fill an aluminum cup and a china cup with very hot water. Let students touch the handles to compare the heat flow through aluminum and china.

c. Have a student hold a nail or brass curtain rod in the flame of a candle to see how rapidly the heat is conducted throughout the length. **(Caution: Let the student be ready to drop the item if it gets too hot.)** Different kinds of nails (aluminum, galvanized, etc.) or rods can be compared.

Following these experiences, you may wish to tell students that when heat travels in this manner, the traveling is called *conduction.*

Observing differences in heat conduction through various materials. Several activities can help students realize that some materials conduct heat better than others. **33 Aa**

a. Obtain a pencil with a metal cap on one end (or with a metal band around an eraser). Tightly wrap a piece of white paper around the end of the pencil so that it covers the wood and the metal. Heat the pencil in a flame at the metal tip end. Observe what happens, then remove the pencil from the flame when the paper starts to char. Open the paper and have students note that there is less char area where the paper was wrapped around the metal. (The heat was conducted away from the paper by the metal, but not by the wood.)

b. Partially fill a nonwaxed paper cup with water. Hold or suspend it over a flame, and have students observe as it is brought to a boil. (The water will reach the boiling point before the cup is burned because the heat is conducted away from the paper and dispersed through the water.)

c. Lower a small piece of aluminum screen over a candle flame. Have students observe how the flame is suddenly cut off even though the screen has many holes in it. (The metal wires conduct the heat away from the flame.) An interesting reversal of the phenomena can be observed if a Bunsen burner is available. Set the screen across a tripod or on a ring stand, and place the unlit burner under it. Turn on the gas and light the gas *above* the screen. The gas will burn only above the screen because the metal wires conduct the heat away from the gas below.

d. Place a wooden match between two tines of a fork, and light the match. Students will observe that the match burns only until the flame reaches the metal. (The metal conducts the heat away from the match so that part of the wood never reaches its burning point.)

Observing that liquids are generally poor conductors of heat. Drop a piece of ice into a test tube, and wedge it in securely with a piece of steel wool. Fill the tube with water and tilt the test tube diagonally in order to heat the water in the top half of the tube. Students can observe the water boiling at the top while the ice remains undisturbed at the bottom. Other liquids can be tested in a similar way to show that they are also poor conductors. *(Note: Liquid metals are not poor conductors of heat.)* A dramatic variation of this activity can be done by filling a test tube with water, holding it at the base with a bare hand, and heating the top of the tube. Students will observe the boiling water while the tube is held in a bare hand.

34
Aa

Determining that gases (air) are poor conductors of heat. Light a candle, and have students take turns holding their hands about 10 inches (23 cm) above the flame. Let them describe what they feel. Have them move their hands to 10 inches (23 cm) from the side of the flame, and describe what they feel. (If air were a good conductor, there would be no difference at the same distance above and to the side of the flame.) As a variation, hold the head of a wooden match about ½ inch (1 cm) from the side of a candle flame for about ten seconds. (The match will not burst into flame.) Next, hold the match about ½ inch (1 cm) above the flame. (The match will burst into flame immediately.) *(Note: Because gases are poor conductors of heat, porous and fibrous materials, such as asbestos, fur, and feathers, are usually good insulators because they have air spaces throughout their structures.)*

35
Ab

Contributing Idea B. Heat sometimes travels through materials by convection; the exchange of heated and cooler gases and liquids creates currents.

Determining the locations of heated and cooler air in the classroom. To help students realize that warmed air is lighter (thus near the ceiling) and cold air is heavier (thus near the floor), have them stand on ladders, hold their hands high up, and describe what they feel. Next, have them kneel on the floor and feel the temperature of the air for comparison. As an alternative experience, they can use individual thermometers to measure the temperature in different parts of the room. For example, a thermometer can be put near the floor of the room away from the radiator, while another can be put on a window pole and held near the ceiling for several minutes. A chart of the room's various temperatures can be made. Students will realize that the highest temperatures are

36
Ab
Be
Ce

generally in the highest places in the room because heated air is lighter and rises, while the coolest temperatures are generally the lowest in the room because cool air is heavier and settles downward.

Observing that heated air rises and cooler air falls. Several experiences can help students realize that heated air rises and that cooler air falls to take its place. 37 Aa

a. Clap two chalkboard erasers over a hot radiator, then do the same close to an open window on a cool or cold day.
b. Light a rope and blow out the flame so that it will smoke. Hold the rope a short distance above a candle flame and away from other air currents. The smoke will travel upward. This can be repeated over a hot radiator, electric hot plate, near an open refrigerator door, or at the top and bottom of an open window.
c. Hold strips of aluminum foil above a lit lamp chimney and observe what happens to them when they are dropped. Repeat near the base of the chimney.

Locating air currents. Various individualized devices can aid students in making observations of air movements. 38 Aa Be

a. Cut a 6 inch (15 cm) square of paper as shown, and fold the corners toward the center. Put pinholes through the folded over corners and the center of the pinwheel. Suspend the pinwheel from a knotted thread, and let students explore the room and schoolgrounds for rising and falling air. They can chart the air currents on a school map. Several maps can be used to show the currents at different times of the day.

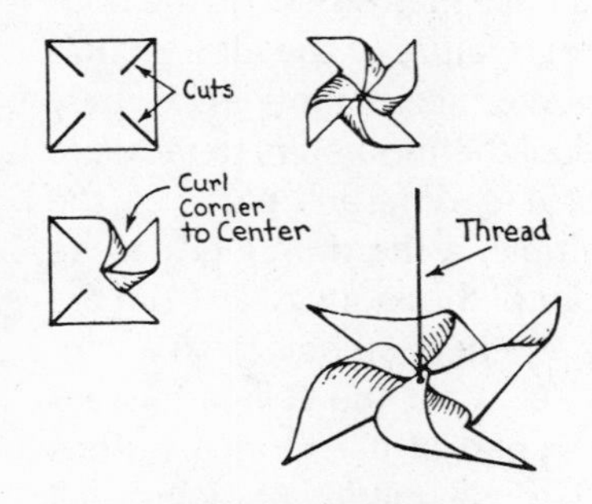

FIGURE 342.38a

b. Cut a circle of paper about 6 inches (15 cm) in diameter and make a pinhole in the center. Cut the paper in a spiral as shown, and hang the spiral from a knotted thread. The spiral can be held over a radiator or in other parts of the room or schoolyard to find the movements of air currents.

FIGURE 342.38b

c. Tissue paper can be cut into narrow 6 inch (15 cm) long strips and attached to a stick or ruler with tacks, glue, or tape. The stick can then be held in various locations to aid seeing air movements.

Observing convection currents in gases. Several experiences can help students realize that warmer and cooler air tend to exchange places. After several experiences you may wish to tell them that when heat travels in this manner, the traveling is called *convection.* 39 Aa Bb Ec

a. Wrap a few turns of wire around a candle, allowing enough wire for a handle so that the candle can be lowered into a large bottle. Light the candle and lower it into the bottle. (The candle will go out quickly because the air in the bottle is heated, rises, blocks the mouth of the bottle, and prevents fresh air from entering.) Now cut out a T-shaped piece of cardboard with a stem that just fits into the neck of the bottle. Remove the candle from the bottle, turn the bottle upside-down so that the "candle air" (carbon dioxide) can escape, and swing the bottle in the air a few times to fill it with fresh air. Light the candle again, and lower it into the bottle. Quickly put the cardboard T in place and observe. Now light one end of a piece of rope, blow out the flame, and while the rope is smoldering, hold it first on one side of the cardboard T and then on the other side. (On the heated side the smoke will rise; on the cooler side the smoke will go down into the bottle.) From observing the drifts of smoke, students will begin to realize that nearly all upward and downward drafts of air are caused in similar ways and that such differences in temperatures cause winds.

b. Put three holes in a large wooden (or sturdy cardboard) box as shown. Plug

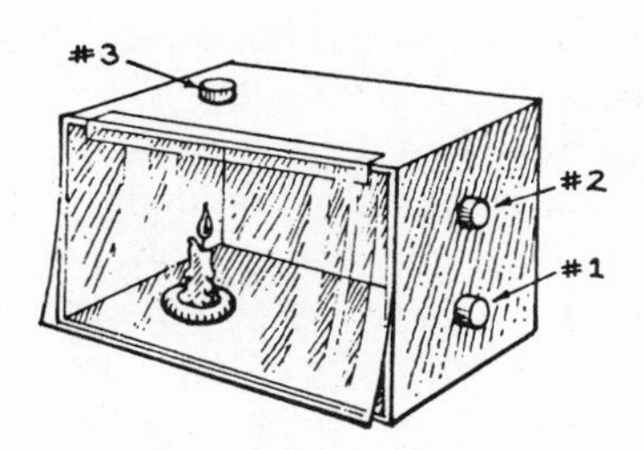

FIGURE 342.39

the holes with corks or rubber stoppers. To the face of the box, attach a sheet of glass or transparent sandwich wrap so that it can be lifted as a door, and observations can be made of the interior when it is closed. Place a candle in the box under the top hole. Light the candle, remove any two plugs, and hold a smoking rope near the lowest opening. (The smoke will travel from the lowest opening through the box and out a higher opening.) The openings simulate some of the possible openings in a room for proper ventilation (e.g., #1 and #3 simulate an open window or an open chimney; #1 and #2 simulate a window opened at top and bottom). Students can check their own room for comparisons. Let them experiment with the box to create other air movements.

Observing convection currents in liquids. To help students understand that convection currents take place in liquids as well as gases, the following activities can be done using water.

40 Aa Be

a. Obtain a clear heat-resistant pot, fill it with cold water, and place it on a heating unit. To make the convection currents visible, shred some blotter paper into the water by rubbing it against the fine side of a food grater. Let the shreds settle to the bottom, turn on the heating unit, and have students observe the movements of the pieces of paper. Next, turn off the unit, and have students draw the movements they saw. After the water has cooled, repeat the activity, but this time place the pot only partly on the heating unit. Have students observe and draw what they see. (In both instances the warmed water rises to the surface and is replaced by the cooler water.) As an alternative to blotter paper, sawdust or wood shavings from a pencil sharpener can be used. Similarly, cold milk can be gently poured down the inside of the clear pot filled with warm water. If the water is quiet, the heavy cold milk will settle to the bottom. When the heat is turned on low, students will see the movements of the two liquids.

b. Obtain two identical milk bottles, one filled with hot, deeply colored water, the other with uncolored, cool tap water. Put a card over the hot-water bottle opening. Invert the bottle and place it gently over the second bottle. Remove the card. (There should be almost no mixture, because the cold water is heavier.) Now hold the bottles carefully at their mouths and reverse their positions. (The heavier, cold water will quickly displace the hot water.)

Contributing Idea C. Heat sometimes travels by radiation.

Feeling radiant heat energy. Obtain enough electric lamps so that students can work in small groups. Gooseneck lamps with shades removed work best, because they can be turned in various directions. Have students turn on the light bulb, and put their hands near it but not touching it. If they hold their hands in different positions around the bulb, they will find that heat travels from the bulb in all directions and realize that the heat reaches their hands by some means other than conduction or convection. (If the heat traveled by convections, it would be felt only above the bulb, and gases, such as air, are poor conductors.) By turning the bulb off and on, they will find that the heat they feel is immediate. (Heat traveling by conduction or convections takes time.) For alternative radiant heat sources, electric irons, toasters, or a warm radiator can be used. Have students recall and discuss the warmth of a fire after the fire has gone out or the warmth radiated from sidewalks and concrete buildings shortly after the sun has set. You might explain that when heat travels in this manner, the traveling is called *radiation.*

41 Ab Bb

Feeling that radiant heat energy will pass through transparent materials without heating them. Place a pane of glass between a lit bulb and bare hands. The hands will sense the heat immediately, but the pane of glass will remain cool. *(Note: This experience also emphasizes that, because of the glass barrier, the heat must have traveled by a means other than conduction or convection.)* Students might recall having felt heat through a closed window while the pane of glass remained cool. Additional tests can be tried using various transparent materials (e.g., sandwich wraps, plastics) or variously colored cellophanes.

42 Ab Ec

Comparing the effect of dark and light materials upon radiant heat energy. Several activities will help students realize that dark objects absorb radiant heat more readily than light ones.

43 Ca

a. Place a sheet of white and a sheet of black paper on the ground in the sunlight. After several minutes, let students feel the sheets of paper. Ask them to describe what they feel. (The black paper will feel warmer than the white paper.) Similarly, pieces of light and dark cloth of identical size and material can be set on the ground or on snow in winter to note the differences in temperature. (The snow under the dark cloth will melt more rapidly.) These activities may bring about discussions on the effects of light and dark clothing on a hot summer day as compared with a cold winter day.

b. Obtain two identical softdrink bottles. Snap a balloon over each one, and paint one bottle with black paint. Place both bottles in the sunlight and have students observe and compare them. (The black bottle will absorb the sun's rays more quickly, and its balloon will inflate first.)

c. Paint two tin cans, one black and the other white. Invert them over ice cubes in the sunlight. Uncover them every two minutes to let students observe and compare the effects on the ice cubes.

Observing that heat builds up within some materials. Obtain five identical sheets of black paper. Place the sheets, one at a time, in the sunlight at one-minute intervals. At the end of four minutes, have students feel the papers and judge which is the warmest. (The temperatures will range from warmest to least warm in relation to the time in the sunlight.) This activity can also be done with tin cans painted black and covered with cardboard lids that have thermometers inserted through them. **44 Ce**

Determining whether dark or light materials reradiate heat faster. Paint two identical tin cans, one black inside and out, the other white inside and out (or leave it shiny). Pour hot water into the two cans and cover each with a cardboard lid that has a hole cut in it just large enough to hold a thermometer without slipping through. Have students record the temperature changes within both cans at regular intervals, then compare the results. (The darker color helps to increase the heat radiation—the water temperature drop is greatest in the dark colored container.) Students can also observe that the most efficient radiators (automobile and room radiators) are usually painted black so as to give up as much heat as possible (silvered or light colored ones are not as efficient). **45 Ce**

Contributing Idea D. Heat travels through some materials faster than through others.

Comparing materials by conduction rates. Obtain rods of the same diameter and length, but of different materials (e.g., brass, iron, copper, glass, wood, rubber). Attach thumbtacks to them at 1 inch (2.5 cm) intervals using candle wax. Insert the rods vertically through holes in a coffee can lid. Add 1 inch (2.5 cm) of water to the can and place the lid on it so that the rods rest in the water at the same depth. Place the can on a heating unit. Students can note the speed of conduction by the dropping of the tacks. **46 Ca**

Generalization IV. Heat can be controlled.

Contributing Idea A. Poor conductors of heat make good heat insulators.

Comparing the insulating differences of some materials. Have students fill four jars to the top with hot water, record their temperatures, and cover them tightly. Wrap three of the jars in different materials (e.g., wool, cotton, nylon). Leave the fourth jar unwrapped for later comparisons. Set the jars aside for about thirty minutes, then take the temperatures again and compare the changes. From their findings, students can discuss the best types of clothing to wear at certain times of the year. A similar activity can be done by comparing various vacuum bottles that students bring to school. Compare the better insulators for their common characteristics. Students can also examine and test hot pads, insulated bags to carry ice cream, and handles on pots and pans. You might challenge them to use their understanding to see who can preserve an ice cube the longest. **47 Ce**

Generalizing about the insulating quality of some materials. Obtain three or more large tin cans and three or more small cans that fit in the larger ones leaving about a 1 inch (2.5 cm) space between the sides. (Beakers or jars can be used instead of cans.)

48
Ce
Ec
Fa

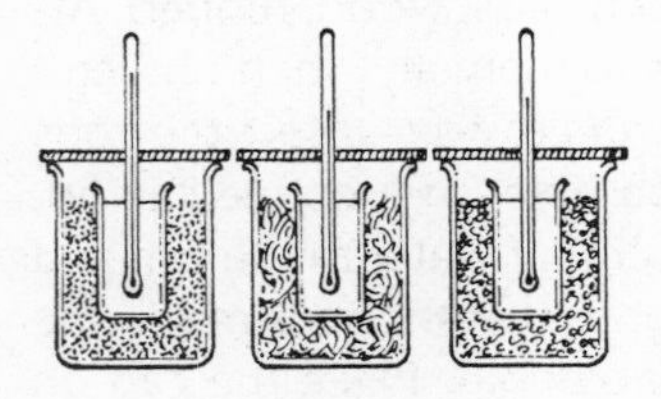

FIGURE 342.48

Obtain some materials that might be used for insulation (e.g., sawdust, rock wool, ground cork, styrofoam pieces, packing materials). Place about 1 inch (2.5 cm) of each material in the bottom of each of the three large cans. Set the smaller cans on top of the material and fill in around them with about a 1 inch (2.5 cm) thickness of the material. Use pieces of heavy cardboard to cover the large cans. Punch a small hole in the center of each piece and insert a thermometer so that it will reach about half way into the small can, making sure the hole is small enough so that the thermometer does not fall through. Remove the cardboard pieces and pour equal amounts of boiling water into the small cans until they are almost full. Cover the cans with the cardboard pieces. Let students observe the thermometers and record the temperature of the water in each can at one-minute intervals. Graph the results, and elicit some inferences about the insulating power of the tested materials. (The material that retained the heat in the water for the longest period of time is the poorest heat conductor and the best insulator.) *(Note: The best insulators are generally a combination of a poor conducting material and trapped air spaces.)* Using the same cans, many other tests can be made on other materials such as fur, feathers, and water. Air can also be tested if the small can is kept from directly touching the larger can by placing three or four 1 inch (2.5 cm) high rubber stoppers or corks under the small can. The can will then be suspended from the larger can and surrounded by air. From these experiences, students will be able to state some generalizations concerning insulating materials.

Contributing Idea B. Heat can be refracted.

Observing that radiant heat energy can be focused and reflected. Have students hold magnifying glasses in the sunlight and focus the sun's rays to a point. Now have them tilt a mirror about halfway between the point of focus and the magnifying glass. One student can adjust his hand until the new point of focus is found. A sheet of paper can be placed at this new point and burned. Students will realize that the sun's heat was refracted by the lens and reflected by the mirror.

49
Aa

Contributing Idea C. Heat can be reflected.

Observing that reflective materials can be used as insulators. Obtain a sturdy square wooden or cardboard box that will fit over an upright lightbulb with a few inches (5 to 6 cm) to spare on all sides. Obtain several rectangles of glass such as 2 inches by 2 inches (5 cm by 5 cm) or 3 inches by 5 inches (8 cm by 13 cm) mounting slides. Cut two square holes exactly opposite each other on the box. Be sure they are smaller than the pieces of glass. Cut a piece of aluminum foil to cover one glass

50
Ec

pane. Glue or tape it as smoothly as possible to the pane. Mount the glass against one hole in the side of the box with tape with the aluminum on the outside. Attach the other glass, without aluminum, in the same way to the other hole. Use a drop of wax to fasten a tack to the center of each pane.

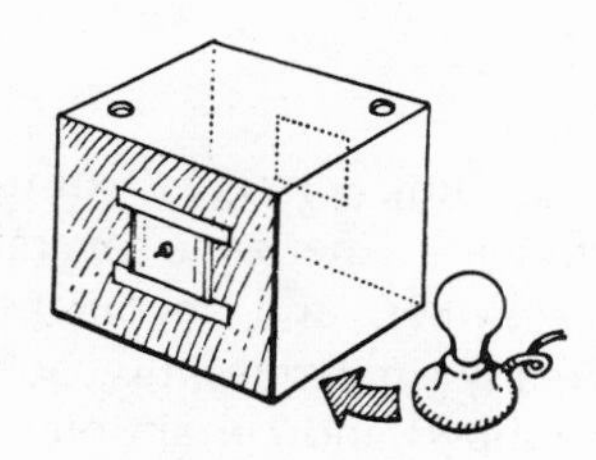

FIGURE 342.50

Put two ventilation holes in the top of the box at opposite ends and place the box over the lightbulb. Turn on the bulb and have students observe or time how long it takes the tacks to drop. (The tack on the reflective aluminum will drop last.) Similarly, other materials can be tested for their reflective power, and different bulb wattages can be tried.

Contributing Idea D. Heat can be absorbed.

Observing the insulating quality of dark and light surfaces. Blacken one-half of a large tin can with paint, inside and out. Do not paint the other half (it should be shiny). Use small drops of wax to fasten tacks on the can at opposite sides. Place a lighted candle or light bulb inside the can in the center, and have students observe which tack falls off first. The surfaces of the can can also be tested by touch. Students will realize that the dark surface absorbed the heat while the shiny surface reflected it. They can discuss the advantages and disadvantages of light and dark surfaces in building homes and clothing.

51
Aa
Ga

FIGURE 342.51

350 electricity

351 CHARACTERISTICS

Generalization. Electricity has identifiable characteristics.

Contributing Idea A. Electrical energy can be stationary (static electricity).

Seeing, feeling, and hearing evidence of electrical charges. In a darkened room on a dry, cool day, have a student pull a piece of tape quickly away from a roll of black, sticky, bicycle tape. (A spark will appear at the place where the tape pulls away from the roll.) Similarly, in a dark room, long hair can be combed with long strokes to produce visible sparks, or students can rub the leather soles of their shoes against a wool rug and touch a metallic object. When hair is stroked, they can often hear a crackling sound. (Stroking a cat or removing clothes recently tumbled in a dryer also produces the crackling sound.) If hair is stroked near a radio that is tuned between stations with the volume turned high, they will hear the crackling over the radio. The spark that is seen is similar to lightning (both are discharges of static electricity), and the sound is similar to thunder (but on a much smaller scale). You might ask what must happen to produce the phenomenon (certain materials must be rubbed or stroked).

01
Aa
Ab
Ac

Observing that rubbing charges some materials. Let students experience several of the following activities, and encourage them to determine any similarities. (In each case, objects are rubbed and certain other objects are attracted to them.) Tell students that when this phenomenon takes place, the materials are said to be *electrically charged* and that this form of electricity is called *static electricity.* For each of the following, students can test various other materials to see which ones can produce charges and which ones cannot. *(Note: The activities work best on dry, cool days.)*

02
Aa
Bb
Fb

a. Hold a newspaper flat against a table, a wall, or a chalkboard. Rub the entire surface with a wool cloth. (The paper will remain attached to the surface for quite some time.) Let students listen as they pull one corner loose, then quickly let it go. (They may hear a crackling sound, and the paper will be attracted back to the wall.)

b. Turn on a water faucet to produce a slender stream, rub a comb against a wool cloth, and hold one edge close to the stream. (The static electricity in the comb will attract the water.)

c. Set two books on a table about 4 inches (10 cm) apart. Place a sheet of glass across them. Cut confetti pieces out of tissue paper and put them on the surface between the books under the glass. When students rub the glass vigorously with a silk cloth, they will see the pieces move. Similarly, place confetti pieces in small glasses or plastic dishes, such as petri dishes, cover them with plastic wrap or glass, and rub the covering surface. (The pieces will jump about and cling to the surface.)

d. Hold a hard rubber comb near confetti pieces made from tissue paper or grains of puffed rice (nonsugarcoated), then compare what happens when the movement is repeated after rubbing the comb vigorously with a piece of wool cloth. Challenge students to see if rubbing the comb harder or longer makes any difference in what they observe. Similarly, move a plastic ruler or glass rod near the pieces, then compare what happens when done again after rubbing the object with a piece of plastic wrap or silk cloth. Students can also test metal rods by rubbing them with a wool cloth, plastic wrap, or a silk cloth.

e. Blow up a balloon and close it with a rubber band or string. Rub the balloon with a piece of wool cloth and bring the balloon near the pile of confetti. Challenge students to see what happens when the rubbed balloon is touched to a wall, then released, or what happens when it is brought next to their hair.

Determining what materials can produce electrical charges. Place several sheets of newspaper on a flat, wooden table. Rub each sheet vigorously with a different material (e.g., wool, silk, fur, paper, plastic wrap, pencil, foil). After rubbing, try to lift each sheet from the table. (Some will have acquired an electrical charge and will tend to be attracted to the table.) Tell students that when certain materials are attracted to others after rubbing, they are said to be *electrically charged* and this form of electricity is called *static electricity.* Students can sort the materials according to whether or not static electricity was produced.

03 Bb Ec Fa

Determining that not all electrical charges are alike. Place two 6 inch (9 cm) lengths of wool yarn on a flat surface and rub them lengthwise several times with an inflated rubber balloon. When the lengths are held up by one end, about 1 inch (2.5 cm) apart, they will repel each other. Now prepare two 6 inch (9 cm), narrow lengths of rubber from a balloon and rub them with wool. When held about 1 inch (2.5 cm) apart, the lengths will repel each other. Students will find it reasonable to deduce that each of the paired lengths have the same kind of charge since each are of the same material and receive the same treatment. Next, hold up a charged length of wool near a charged length of rubber. Because the lengths behave differently, students will realize that the charges must be different. Tell them that after the wool has been rubbed with rubber, the charge is called a *positive charge;* the charge on the rubber is called a *negative charge.* Knowing this, students can now identify charges produced when different materials are rubbed together.

04 Bb Ec Fa

Contributing Idea B. Electrical energy can move (current electricity).

Generating a direct current. Obtain or build an electric cell and a galvanometer (see *352.15*). When the cell and galvanometer are connected, the galvanometer will indicate a current moving in the wire. Tell students that when a current moves continuously in one direction it is called a *direct current.* Now have them reverse the wires on the cell and observe any changes

05 Aa Bb

in the galvanometer. (The galvanometer will indicate that the current reverses its direction.)

Generating an alternating current. Obtain or build a galvanometer. Wind fifty turns of insulated copper bell wire to make a coil. Tie the coil with string so it will not come apart, then connect it to the galvanometer. Move one end of a bar magnet through the coil and notice that the galvanometer indicates a current moving in the wire. When the magnet is removed, students will observe that the galvanometer indicates the current moves in the opposite direction. They will also note that no current is produced when the magnet and the coil are motionless. Tell them that a current moving in first one direction and then the other is called an *alternating current.* Now let them test to see if the current is greater when the movement of the magnet is vigorous, whether it matters if the magnet or the coil is moved, what happens when the other end of the magnet is used, and how different magnets of different strengths and coils with different numbers of loops affect the current. Interested students can research information about how electricity is generated to run machinery and light cities. **06 Aa Bb**

352 INTERACTIONS

Generalization I. Electricity is a form of energy produced by different sources.

> *Contributing Idea A.* Chemical reactions can be a source of electricity (electro-chemical propagation).

Making an electric cell. Have students bare about 1 inch (2.5 cm) at the ends of two 2 foot (60 cm) lengths of insulated copper bell wire. Attach one end of one length to a silver knife. Attach one end of the other length to a sheet of aluminum foil. Now place a paper towel that has been soaked in salt water between the two metals. An electric current can be detected using a current detector (see *352.15*). Tell students that what they have made is called an *electric cell.* Explain that in an electric cell there are always two different metals called *electrodes* and a fluid called an *electrolyte.* Next, let them try making other cells by substituting magnesium or zinc for the aluminum and copper or gold for the silver. They can also substitute vinegar or boric acid for the salt solution. **01 Bb Gc**

Producing electricity from a "wet" cell. Wrap one bared end of a 2 foot (60 cm) length of insulated copper bell wire around a ½ inch by 6 inches (1 cm by 15 cm) strip of clean copper; then do the same with another wire and a piece of clean zinc (a strip of zinc can be obtained by cutting the outer casing of an old dry cell). Connect the free ends of the wires to a galvanometer, then let one student touch the two metal strips to her tongue. A slight deflection of the galvanometer will indicate a slight current (the saliva and the two dissimilar metals make an electric cell). Now use the galvanometer **02 Gc**

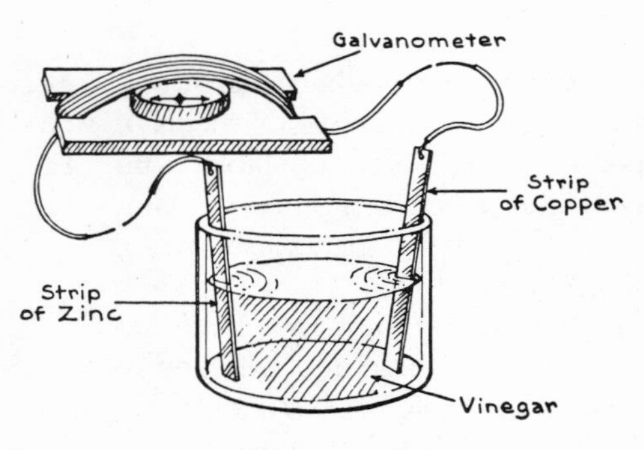

FIGURE 352.02

after suspending the two different metals in a glass of vinegar. Try other metals and other liquids such as cooking oil, sugar water, mouthwashes, detergents, and various concentrations of salt solution. Test to see if there is a stronger current when the liquid is warm or cold. *(Note: Be sure the metals are washed and dried between each test. Students will find that almost any two dissimilar metals immersed in most chemical solutions will produce electricity.)*

Producing electricity from a "dry" cell. Insert copper and zinc strips about ½ inch (1 cm) apart in a potato. Attach the bared ends of two 1 foot (30 cm) lengths of insulated copper bell wire to the strips and to a current detector (see *352.15*). Students

03
Bb
Gc

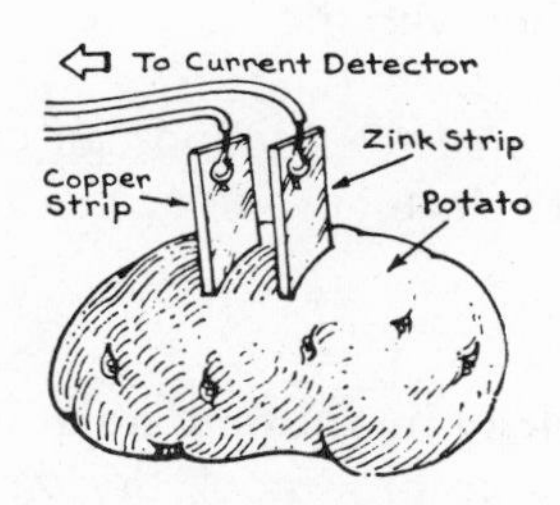

FIGURE 352.03

will see that a current is produced. (They can increase the number of potatoes to make the current stronger.) Explain that the potato is somewhat like a *dry cell* because, unlike "wet" cells, no chemicals can spill out regardless of how it is held. *(Note: A dry cell is not really dry inside.)*

Testing materials to make "dry" cells. Roll a lemon on a table until it feels soft and juicy inside. Bare about 1 inch (2.5 cm) at the ends of two 2 foot (60 cm) pieces of insulated copper bell wire. Wrap one end tightly around the head of an iron nail and insert the nail into the lemon. About ½ inch (1 cm) from the nail, insert one end of the second piece of wire. Use a current detector to see if electricity is being produced (see *352.15*). Now have students carry out various tests.

04
Bc
Ec
Gc

a. They can see what other metals, such as aluminum, gold, lead, pencil "lead" (carbon), steel, zinc, and so forth, will produce electricity. They can also test a dime and a penny. Keep records of the results to see which combinations produce the strongest current.
b. They can test to see what effect changing the distance between the two metals in the lemon has on the current. (The closer the metals are without touching, the greater the amount of current produced.)
c. They can test other fruits or vegetables in place of the lemon.
d. They can test nonmetal materials such as wood, plastic, or glass.

Examining a "dry" cell. Distribute "D" cells (standard flashlight cells) to students. Place the cells on newspapers and, using pliers and screwdrivers, pry open the top edges, and remove the top piece and the tarlike material beneath it. Twist the center black (carbon) rod carefully, loosen it, and slowly pull it out. Scrape out and examine the remaining materials. *(Note: The materials inside the cell are not harmful to the*

05
Aa

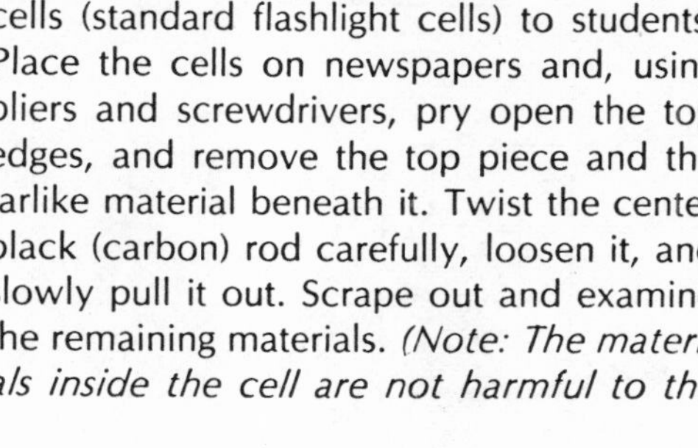

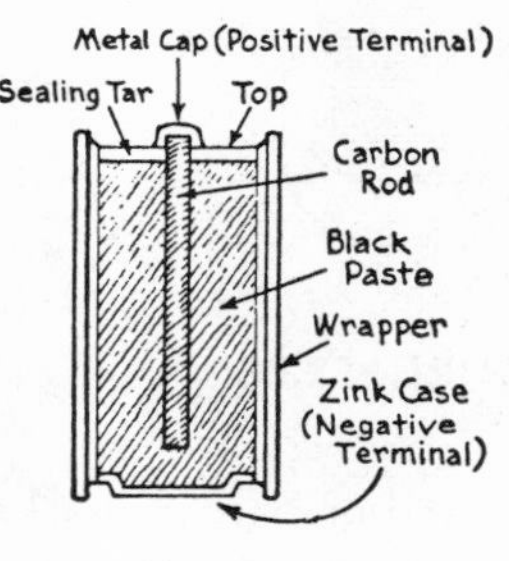

FIGURE 352.05

skin; however, they may stain clothing so some care should be taken.) As an additional experience, the cell can be attached to a bulb in a circuit to see at what stage of the dissection the cell stops lighting the bulb. Conversely, students can find out at what point a bulb begins to light as the black rod is pushed back into the cell. Students can test and list all the materials that make up the cell that do not conduct electricity. For study and display, a large cell can be cut vertically or horizontally with a hacksaw.

Making a battery. Polish a penny and a dime with steel wool. From blotter paper or paper toweling, cut three or four six-sided pieces that have a diameter no larger than a dime *(a)*. Sandwich the pieces between the **06 Bb Gc**

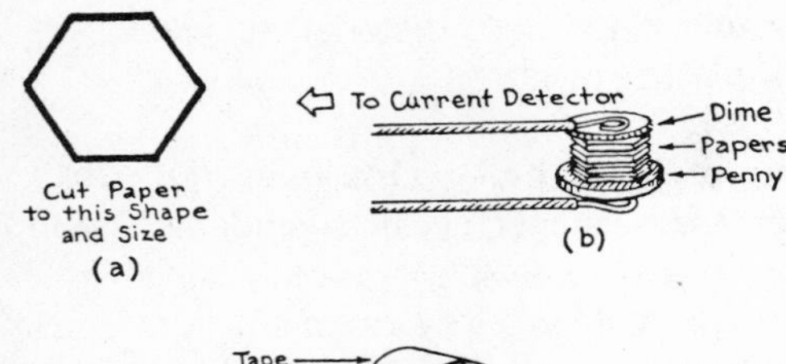

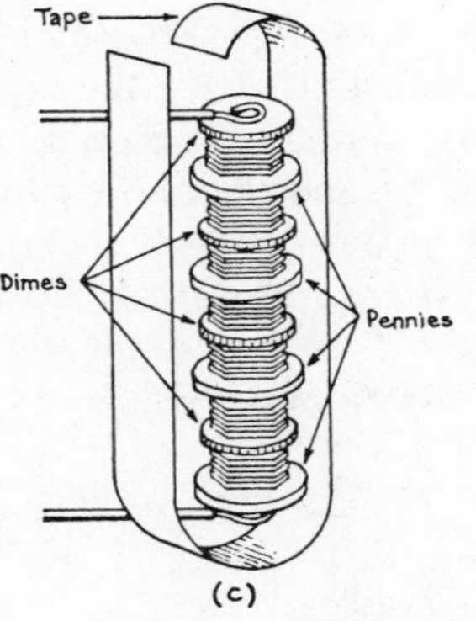

FIGURE 352.06

penny and dime, and then bare about 1 inch (2.5 cm) from the ends of two 2 foot (60 cm) lengths of insulated copper bell wire. Tape one end of one wire to the penny and one end of the other wire to the dime *(b)*. Use friction tape to hold the materials tightly together. With a medicine dropper filled with a salt water solution, moisten the papers between the coins and wipe any excess solution from the edges of the coins. The presence of an electrical current can be detected by connecting the free ends of the two wires to a current detector (see *352.15*). Tell students that they have made an *electric cell*. Tell them that when several are put together, the collection is called a *battery*. Encourage them to find out if the electric power increases as the number of cells increases. This can be done by polishing more pennies and dimes, cutting more blotter pieces, and sandwiching them between alternating pennies and dimes. The battery can be held together with friction tape *(c)*.

Contributing Idea B. Motion can be a source of electricity (mechanical propagation).

Producing electricity by rubbing objects. Darken the classroom on a cool, dry day. Stretch a pair of rubber bands at right angles to each other across an aluminum pie pan. Place a plastic bag on a table, rub it vigorously with wool, and hold it up vertically by one edge. Hold the pie pan by the rubber bands, press it against the bag, then touch it to one end of a burned-out fluorescent tube. *(Note: Be sure that fingers do not touch the pan.)* Repeat this activity several times, and let students touch the pan to their fingers, a nose, or other objects in the room. They will see, feel, and hear the electricity produced by the rubbing of the wool on the plastic. **07 Gc**

Producing electricity by moving a magnetic field. Loop fifty turns of insulated **08 Bb**

bell wire to make a coil that will pass easily over a strong bar magnet. Connect the free ends to a current detector (see *352.15*). Move the magnet in and out of the coil. Students will see a current being registered on the detector. Now see what happens when the magnet is held still while the coil is moved, when both are moved simultaneously, or when the opposite end of the magnet is used. Test other types of magnets. Tell students that the process by which a current is produced by moving the magnetic field near a wire is called *electromagnetic induction,* and that the principle is used in most telephones and in radio and television speakers.

Ec
Gc

Generalization II. Current electricity produces a magnetic field.

Contributing Idea A. An electric current carried by a wire produces a magnetic field around it.

Observing that an electric current produces a magnetic field. Remove the insulation from both ends of a 3 foot (1 m) length of copper wire. Lay the wire flat on a table, and put a pencil on each side of it. Place a sheet of glass or stiff cardboard on the pencils and sprinkle iron filings on it. With a switch in the circuit, connect the wire to an

09
Aa

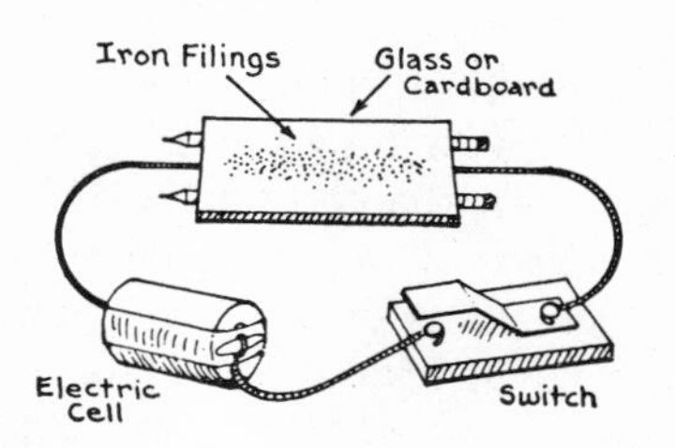

FIGURE 352.09

electric cell. Close the switch and tap the glass lightly. Students will see a pattern made by the magnetic field around the wire.

Comparing magnetic and electric fields. Suspend two needles from threads—one magnetized and one unmagnetized. Bring a charged comb near each end of each suspended needle. (The needles will be equally attracted by the electric field of the charged comb.) Now bring one pole of a strong magnet near each end. (Both ends of the unmagnetized needle will be attracted; one end of the magnetized needle will be attracted, while the other will be repelled.) From this experience, students will realize that magnetic and electric fields are not the same.

10
Ca
Fa

Testing the effect of magnetic and electric fields on different materials. Use lengths of thread to suspend, in turn, small objects of various materials (buttons, paper-

11
Bc
Ec

TABLE 352.11. Testing Magnetic and Electric Fields

Test Material	*Moved By Electric Field*	*Moved By Magnetic Field*
iron	yes	yes
steel	yes	yes
brass	yes	no
plastic	yes	no

clips, pins). Be sure the suspended objects can swing freely. Test each by first bringing to it a magnet, then a charged object such as a comb rubbed with wool. Students can record their observations on a table. They can deduce that magnetic and electric fields are not the same (all materials are attracted in electric fields; only certain materials are attracted by magnets).

Contributing Idea B. A compass is an instrument that can be used to detect magnetic fields.

12 Aa Ec

Detecting the magnetic field produced by an electric current. Have students connect a switch to a circuit. With the switch open, place a compass beside one of the wires so that the needle on the compass and the wire point in the same direction.

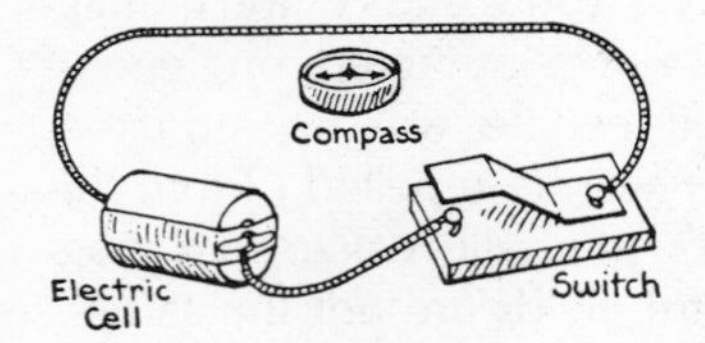

FIGURE 352.12

Close the switch. (The compass needle will be influenced by the magnetic field.) Repeat this activity by holding the compass first over the wire, then under it, to see what happens. Students will realize that a wire carrying an electric current seems to have a magnetic field around it and that this field can be detected by a compass.

Generalization III. Current electricity moves when there is a complete circuit.

Contributing Idea A. Current electricity can be detected.

13 Ae

Detecting electric currents. Have students touch the ends of wires from a flashlight cell a short distance apart on their tongues. If a current is produced, they will detect a metallic taste. When the current is stopped, the taste will disappear.

14 Ac

Using a radio to detect electric currents. Obtain a transistor radio, turn it on, and dial it so that it is between stations with the volume turned up. Now tap together two wires that are attached to an electric cell. Students will hear a clicking sound on the radio. This indicates an electric current is nearby.

15 Bb Gc

Building a galvanometer. Use transparent tape to fasten a small compass to a piece of cardboard *(a).* Cut two slots in the card-

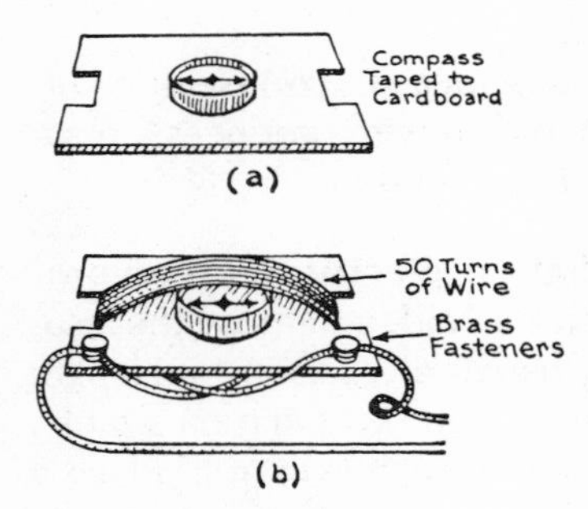

FIGURE 352.15

board at the North and South markings on the compass, then wind about fifty turns of fine, insulated copper bell wire through the slots and around the compass and cardboard *(b).* Leave about 2 feet (60 cm) of wire free at each end. Now put two brass paper fasteners through the cardboard and twist the free ends of the wire around them once or twice to keep the wire from unwinding. Scrape the insulation away from an inch of wire at each of the free ends and attach them to the terminals of an electric cell. (The needle of the compass will move, thus indicating that electricity is moving.) Tell students that the instrument is called a *galvanometer* and that it is used to detect electric currents. Now have them keep their eyes on the compass needle as you reverse the wires so that they touch the opposite terminals. Discuss what effect reversing the connections has on the compass needle. (The needle turns in the opposite direction.)

Contributing Idea B. A circuit is a continuous path through which electricity can move.

Arranging a circuit to light a bulb. Distribute to individuals or pairs of students, one "D" (flashlight) cell, one #48 (flashlight) bulb, and two 1 foot (30 cm) lengths of #20 bare copper wire. Insulated wire can be used, but the ends must be stripped or scraped so that contact can be made. Ask students to see if they can make the bulb light. *(Note: This makes a fine "discovery" experience, and students will find several ways to make the bulb light.)* Discuss similarities among different ways to light the bulb (the points of contact always remain the same). Next, challenge them to light the bulb using only one wire. As the students freely experiment, explain that an electric bulb lights when it is part of a continuous loop of materials through which electricity moves. Tell them that such a loop is called a *circuit*. Have them trace the circuit in each arrangement they make.

16
Bb
Ec

Making a holder for a "D" (flashlight) cell and preparing wires for use in a circuit. Cell holders can make it easier for students to connect two or more "D" cells in a circuit. Simply press a brass paper fastener into each end of a wide rubber band *(a),* and stretch the band over a "D" cell so that the fasteners make proper contact. Connecting wires can be wrapped around the fasteners.

17
Gc

Fahnestock clips, available at hardware stores, can be used instead of the fasteners, and wires can be easily inserted into their eyelets. Another type of cell holder can be made using paperclips and right angle brackets *(b). (Note: When the cells are put away after a class, be sure that the wires are disconnected to prevent the cells from becoming "worn out" by short circuits.)*

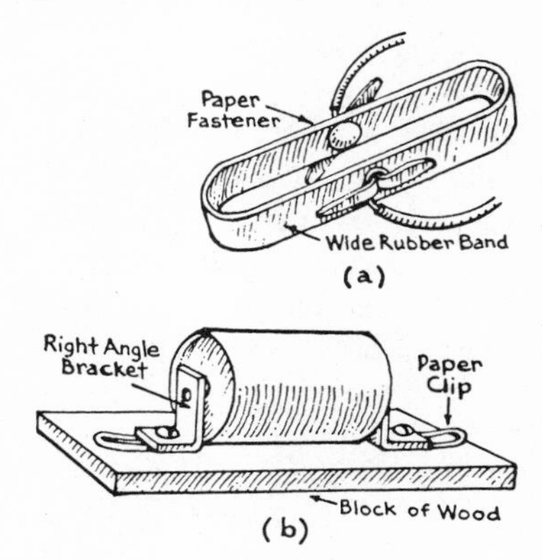

FIGURE 352.17a,b

If larger dry cells are used and if terminals are of the screw-type, it is best to prepare and use wires in a certain way. If their insulation is lacquered, students can scrape it off with a scissors or knife blade. If it is made of a thick material, cut the material down to

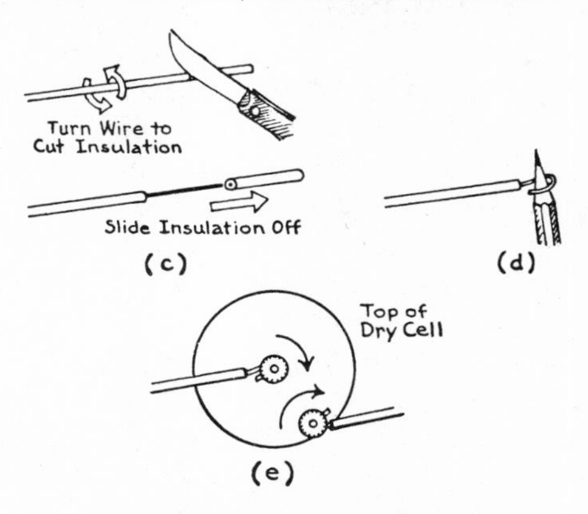

FIGURE 352.17c,d,e

the wire, and pull or whittle it off *(c)*. Be careful that the wire is not cut in the process. Generally a removal of about an inch (2.5 cm) of insulation is sufficient. To connect the wires to terminals, bend the bare end around a pencil point to form a hook *(d)*. Fasten the hook clockwise around a terminal (i.e., in the same direction that the screw turns) *(e)*. Wires can also be spliced easily together. Such wires should then be taped.

Making a holder for a small (flashlight) bulb. Bulb holders can make it easier to connect two or more bulbs in a circuit. Have students spread a paperclip slightly apart, then wrap a 1 inch (2.5 cm) piece of masking tape around the small arm of the clip *(a)*. 18 Gc

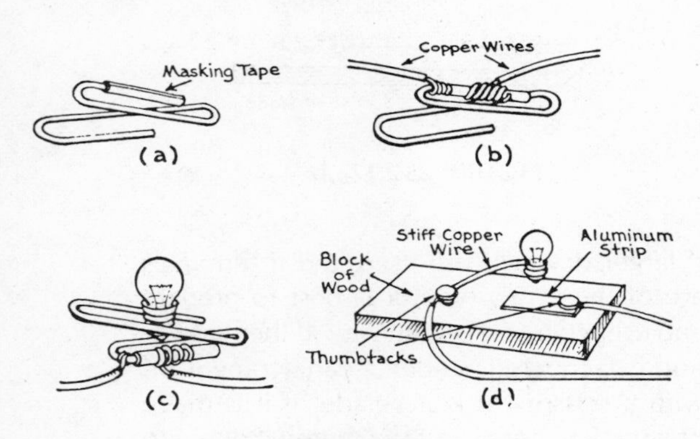

FIGURE 352.18

Remove about 1½ inches (4 cm) of insulation from a 1 foot (30 cm) length of copper wire. Wind the bare portion around the masking tape and fasten the free end to a "D" cell. Wind a second length of wire around the clip as close to the tape as possible *(b)*. Now turn the clip over and screw a small bulb into the large portion so that its tip touches the wire around the tape *(c)*. *(Note: It may be necessary to bend the large arm of the clip to ensure a tight fit and a good contact for the bulb.)* Another type of bulb holder can be made using paper clips, thumbtacks, and an aluminum strip *(d)*.

Exploring electric circuits. Provide small groups of students with two or more "D" (flashlight) cells with holders, two or more #41 bulbs with holders, and lengths of insulated copper wire. Allow them to explore different ways to put the materials together. They might try several bulbs and one cell, several cells and one bulb, or several bulbs and several cells. Whenever a bulb lights, have them trace the completed circuit. If they note differences in the brightness of the bulbs in their arrangements, challenge them to find out if there is any relationship between the brightness and the arrangement. Encourage further experimentation: they can find out how many cells it takes to light a household bulb (it takes about eight "D" cells to dimly light a 60-watt bulb); they can find out how many bulbs they can light with one cell; they can find out how brightly a bulb will shine when several cells are used. 19 Ec Fa

Making a circuit tester. Using insulated copper wire, have students arrange a bulb, bulb holder, "D" cell, and cell holder as shown. Various materials can now be touched by the two unattached bare ends to see whether or not the materials conduct electricity. Students will realize that if the electricity passes through a material, a circuit is completed, and the bulb will light. 20 Ec Gc

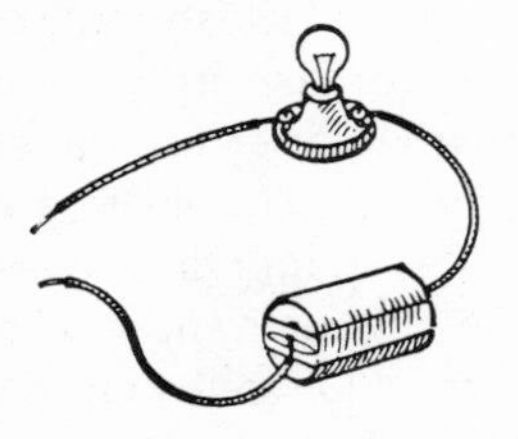

FIGURE 352.20

Conversely, if the bulb does not light, a circuit is not completed. The tester can also be used to test bulbs to see if they are "worn out" or not.

Locating hidden circuits. Students can use a circuit tester (see *352.20*) to locate hidden circuits, much as an electrician checks electrical circuits in a home or appliance. To do this, prepare some "mystery" circuits as follows. Put four brass paper fasteners through a 4 inches by 6 inches (10 cm by 15 cm) piece of heavy cardboard, and label each with a letter *(a)*. Connect lengths of insulated copper wire to the backs of the fasteners *(b)*. Cover the arrangement

21 Be Eb Ec

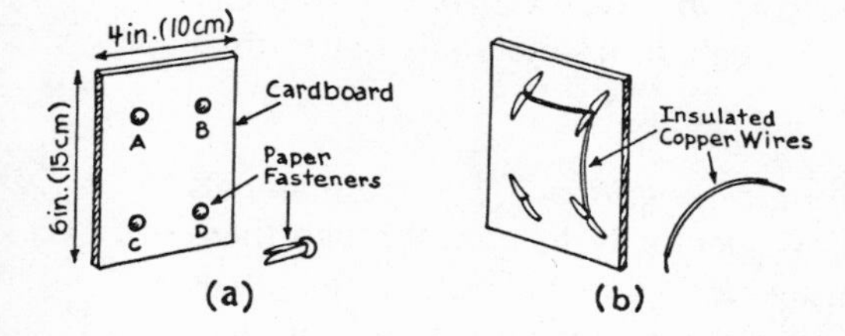

FIGURE 352.21

with another 4 inches by 6 inches (10 cm by 15 cm) piece of cardboard. Fasten the two pieces together with staples and/or tape. Many possible "mystery" circuits can be made. For each, have students infer the arrangement by testing for connections with a circuit tester. They can keep a record of their tests by drawing a picture of the arrangement. Compare drawings prepared independently by different students. *(Note: Some of the circuits can be connected several different ways.)* After this experience, students can create "mystery" circuits of their own.

Contributing Idea C. Current electricity takes the shortest and easiest circuit back to the place where it started.

Observing a short circuit. Demonstrate that electricity will take the easiest pathway back to its starting place by removing some insulation from two wires connected to a "D" cell and bulb. Touch the bare places together or place a length of thick copper wire across the bare places so that it makes a good contact. **(Caution: Do not touch the bare wires at the contact points because they get very hot.)** The bulb will go out. It is best to do this quickly so as not to greatly drain the cell's energy. Have students trace the circuit. Tell them that such a pathway is called a *short circuit. (Note: A short circuit may not always be the shortest path—electricity will follow the path of least resistance.)* Now place a length of #26 Nichrome wire across the bare places. Students will see that the bulb remains lit. Next, shorten the Nichrome wire bit by bit, placing it across the copper wires each time. The bulb will become dimmer and dimmer, thus indicating that the electricity takes the easiest as well as shortest path in a circuit. Let students use insulated copper wire to put together various short circuit arrangements. Discuss the fact that some fires begin in homes due to short circuits. (Insulated wires often become worn, and the two wires of a circuit may touch.)

22 Aa Bb

Observing a grounded circuit. Have students arrange a bulb and "D" cell on an aluminum pan as shown. Place a thick copper wire across the bare portion of one wire

23 Aa Bb

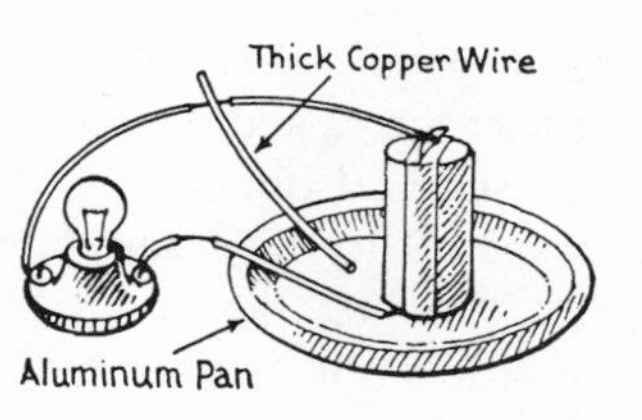

FIGURE 352.23

to the aluminum pan. Students will see the light go out. Trace the circuit for them. Tell them that such a pathway is called a grounded circuit because the electricity travels back to its source through the ground. Discuss with them the danger in turning on an electric appliance that has defective wiring while touching a radiator or other object that has pipes connected to the ground. Caution them that it is particularly dangerous if the body is wet (body salts and water make good conductors for electricity); electrocution is possible.

Contributing Idea D. Switches are used to open and close circuits.

Making a switch. Have students prepare a circuit tester (see *352.20*). Let them turn the bulb on and off by connecting and disconnecting the ends of the free wires. Tell them that a *switch* is a device designed to do this easily and safely. Now let them build a switch by inserting a screw through a rubber washer and fastening it to a block of wood. Next, bend a strip of copper or aluminum as shown and punch a hole in one end.

24
Bb
Gc

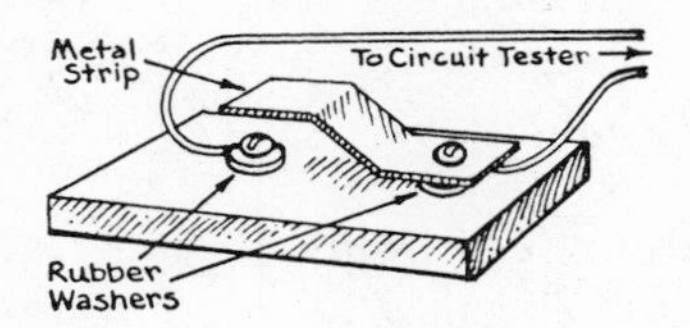

FIGURE 352.24a

Push a second screw through the hole, then through a rubber washer, then into the wooden block so that the raised end of the strip is over the first screw. Connect the end wires of the circuit tester to the screws and test the switch by pressing down on the raised end of the strip until it touches the screw underneath it *(a)*. Students can place

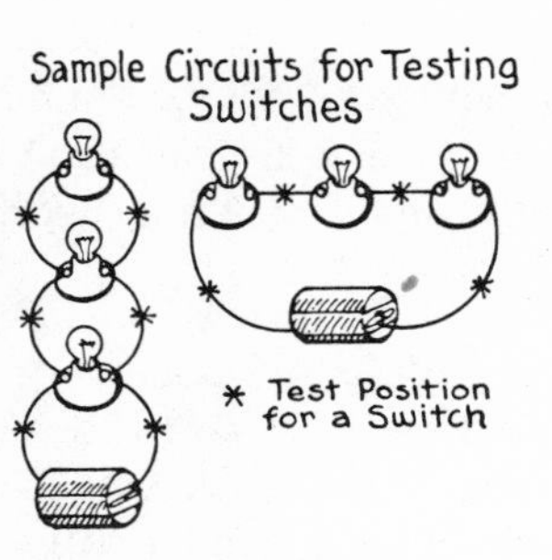

FIGURE 352.24b

their switch or combinations of switches in various parts of a circuit arrangement to see how switches function in turning appliances on or off *(b)*.

Generalization IV. Current electricity travels through some materials better than others.

Contributing Idea A. Some materials allow electricity to pass through them easily (conductors); other materials tend to block the movement of electricity (insulators).

Identifying conductors and insulators. Let students build a circuit tester (see *352.20*) to test several objects or materials to see which complete the circuit. They might test door hinges and knobs, coins, metal desks, faucets, paperclips, pins, or pipe cleaners. **(Caution: Pliers should be used to hold the wires, for they may heat up quickly.)** Results can be recorded on a table. Mention to students that materials that permit electricity to pass through them easily are called *conductors,* while those that do not permit easy passage are called *insulators.* Discuss why electrical wires, plugs, and switches are covered with insulating materials. As an extending experience, some students might try increasing the number of "D" cells in a par-

25
Bb
Bc
Ec

TABLE 352.25. Testing Conductivity

Lights the Bulb (conductors)	*Doesn't Light the Bulb (insulators)*
aluminum foil	rubber
carbon	paper
graphite (pencil "lead")	chalk
silver plated knife	glass
copper wire	wool

allel circuit to test the conductivity of liquids such as water, cooking oil, vinegar, coffee, or various solutions of salt, baking soda, and so forth. They might also try varying the distance between the wires in the liquids.

> *Contributing Idea B.* Thinner wires have more resistance than thicker wires of the same material; longer wires have more resistance than shorter wires of the same material.

Examining the influence of size factors on the conductivity of wires. Let students test various thick and thin wires (e.g., #22 and #36 copper wires) of equal length to see if thickness influences a circuit in any way. (Both wires will complete the circuit, but the thicker wire will cause the bulb to dim considerably.) *(Note: Students might check each test wire using other bulbs and cells to make sure the effect is caused by the wires and not a "worn out" cell.)* Now see if length makes any difference by placing a 2 foot (30 cm) length of #32 Nichrome wire, available at hardware stores, in a circuit. Test each time as the wire is shortened. Now prepare a different circuit using both

26
Ca
Ec

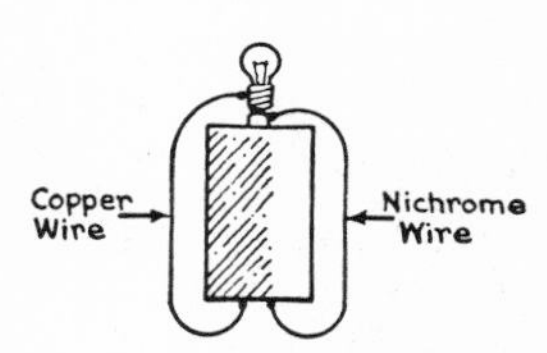

FIGURE 352.26a

kinds of wires *(a)*. Students should note that while the Nichrome wire should form a short circuit, it does not because its length provides too much resistance for the electricity to travel through it. Gradually shorten the Nichrome wire, testing each time. (As the length is shortened, the bulb goes out and the wire gets hot.) Repeat this experience using other wires (e.g., #26 Nichrome and #36 copper wires). The influence of wire length can also be tested by stringing

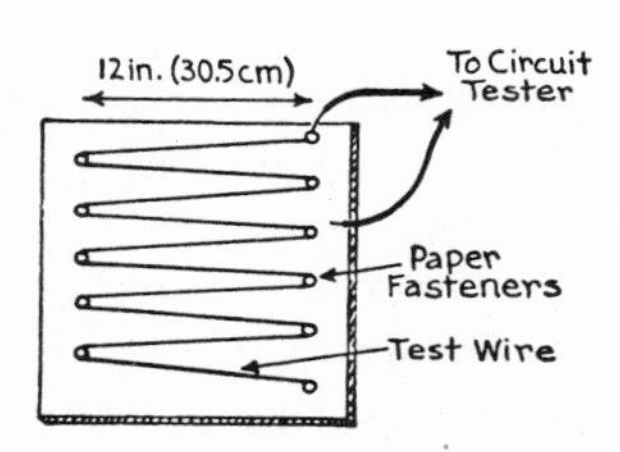

FIGURE 352.26b

a test wire around the heads of paper fasteners attached to a board or stiff cardboard *(b)*. Test by touching the wire at different distances along its length.

Making a rheostat. Prepare a 1 foot (30 cm) length of coat hanger wire by straightening it and sanding off any coating of paint on it. Now fasten one wire from a circuit tester to one end of the hanger wire. As the second tester wire is moved up and down along the hanger wire, students will see the brightness of the bulb change. They will realize that the electricity seems to move more easily when the distance traveled is shortened. Mention to them that the device they have made is called a *rheostat* and that it is used to control the amount of current.

27
Bb
Gc

Contributing Idea C. Some metals have more resistance to electricity than others.

Examining resistance in metals. Obtain equal lengths of iron picture wire and copper bell wire of equal diameters. Using a circuit tester (see *352.20*), compare the brightness of the bulb as the two different metals are tested. (Due to the resistance of the iron, the bulb will appear dimmer.)

28
Ca

Using resistance in metals to make a model light bulb. Place two pieces of #20 bare copper wire through some clay so that the wires resemble the interior of a bulb. Fasten various thin materials across the upright wires to see which ones resist the passage of electricity, get hot, and give off light. Students can test pieces of steel wool and very thin wires (e.g., #32 Nichrome wire, available at hardware stores, will get very hot and begin to glow). They can compare their results to observations of commercial light bulbs. Interested students might

29
Bd
Ec
Gc

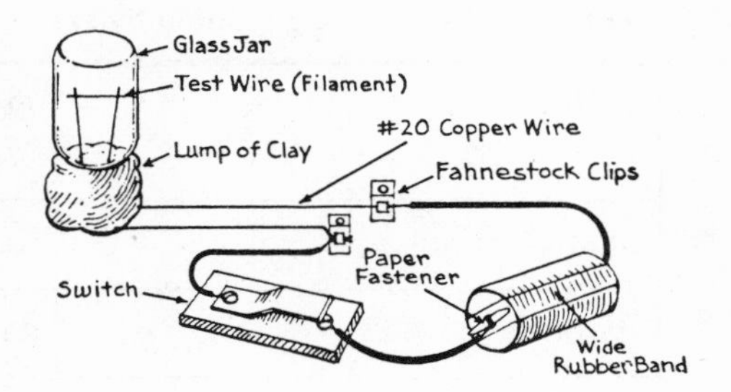

FIGURE 252.29

research the work Thomas Edison (1847–1931) did while inventing the first practical incandescent bulb.

Contributing Idea D. Overloading a circuit with too many resistors can overheat the wires.

Making a fuse. Have students cut a thin strip of foil from a chewing gum wrapper, and trim it into the shape as shown *(a)*. It is not necessary to remove the paper backing

30
Bb
Ec
Fc

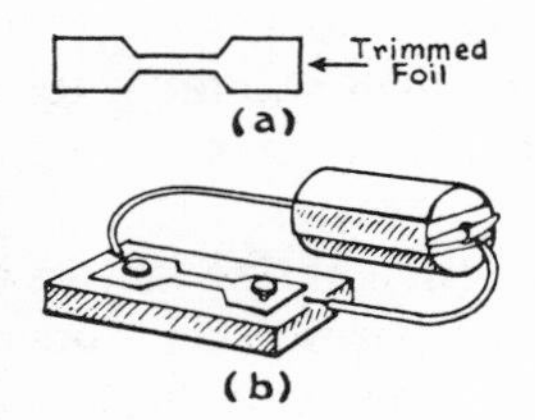

FIGURE 352.30

from the foil. Now place the strip on a block of wood and push a thumbtack through each end. Attach the bare ends of connecting wires from a "D" cell to the two thumbtacks and observe *(b)*. The foil strip will melt and break apart. Tell students that what they have made is called a *fuse*. Explain that when there is too much current in a circuit, the fuse overheats and melts. They will understand that a fuse acts like an automatic switch—it opens the circuit (to turn off the current) when the circuit is overloaded. Now let them experiment with various kinds and sizes of foil to see if they can also be made into fuses. *(Note: Household foils are usually such good conductors that they are difficult to melt unless they are trimmed very thin.)* Students can also examine and compare a new fuse to a burned-out fuse.

Using fuses. To show that attaching too many appliances to a house circuit can cause the wires to overheat, have students make a fuse and place it in a circuit with two "D" cells and a 2½-volt bulb as shown.

31
Ga

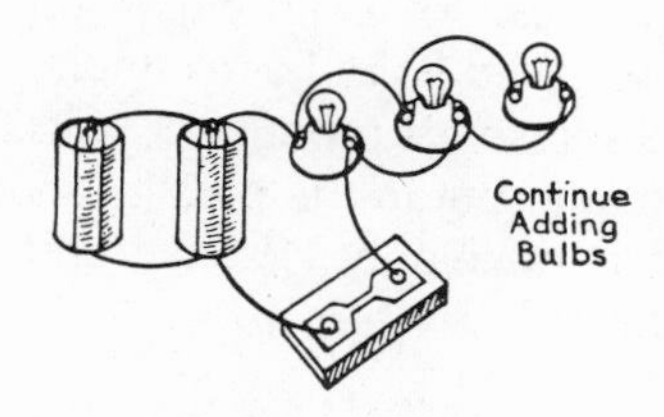

FIGURE 352.31

Overheat the circuit by adding more 2½-volt bulbs, one at a time, until the fuse melts. Students will realize that fuses keep the wires from overheating and causing fires. Now let them try trimming the center width of the fuse until it will melt according to a particular specification such as a "three bulb" fuse, a "four bulb" fuse, etc. Tell them that in a similar way, engineers design fuses that will melt at a specified temperature or amount of electrical current.

Generalization V. Current electricity can be controlled.

Contributing Idea A. Series circuits channel all the electricity through a single pathway; parallel circuits channel all the electricity through a main pathway and branching pathways.

Comparing series and parallel circuits. Cut five 1 foot (30 cm) lengths of insulated bell wire, remove the insulation from the ends, and connect them to three bulb holders, a switch, and a "D" cell *(a)*. Screw 1½-

32
Bb
Ca
Ec

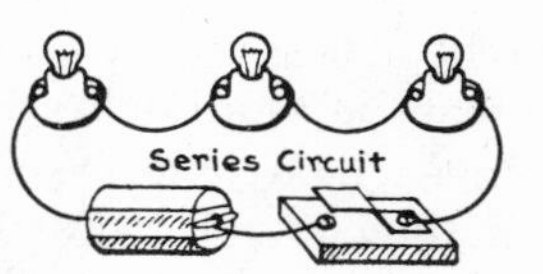

FIGURE 352.32a

volt bulbs into the holders and close the circuit. Students will see what happens when they unscrew one bulb in the circuit. (The other bulbs go out.) Trace the circuit. Tell the students that when the electric current from the cell moves through each bulb in turn, the arrangement is called a *series circuit.* Ask them to imagine what would happen if the appliances in their homes were arranged in series. (If one went out, all would go out.) Now challenge them to design a circuit that will keep the remaining lights on after one goes out. If allowed to explore freely, they will discover that this can be done by connecting the three bulbs as shown *(b)*. Trace the circuit for them, and

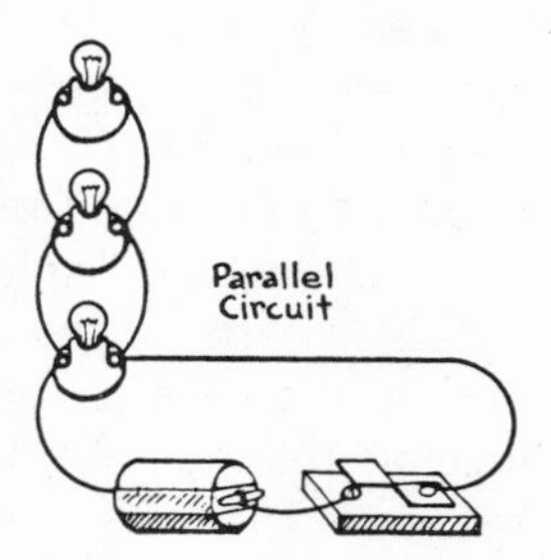

FIGURE 352.32b

indicate that not all the current goes through each bulb—it divides with some going down each branching pathway. Tell them that such an arrangement is called a *parallel circuit.* If possible, obtain strings of Christmas tree lights for students to examine.

Testing one electric cell and several bulbs. Connect two bulbs in series, then add more bulbs in the series, one at a time. Students will find that the more bulbs they connect, the dimmer each becomes. Explain that as the length of the circuit is increased, the resistance to the movement of the electricity is increased, and there is a corresponding cutdown in the current moving in the circuit. Now have them connect two bulbs in parallel, then add more bulbs in parallel, one at a time. Trace the circuit, and indicate that not all the current goes through each bulb, but divides along each branching pathway. As more bulbs are added, students will find that there is no cutdown in the current moving through any one bulb.

33
Ec

Contributing Idea B. Electric cells can be arranged to produce different voltages.

Testing several electric cells and one bulb. Have students hook up a 2½-volt bulb to a single "D" cell, and observe the bulb's brightness. Now arrange two cells in series and observe again. Students will find that increasing the number of cells in series increases the brightness of the bulb. Explain that increasing the cells in series increases the electrical pressure in the wires causing the bulb to burn brighter. Tell them that the increase in pressure is called *voltage.* Next, have them arrange the cells in parallel. They will see that there is a return to the original brightness. As an additional experience, students can substitute a galvanometer for the bulb. Ask them what they can conclude about the force of the current in each arrangement.

34
Bb
Ec

Testing several electric cells and several bulbs. If allowed to experiment freely with 2½-volt bulbs and "D" cells, students can discover that when connecting cells, the series arrangement gives more light, but when connecting bulbs, the parallel arrangement gives more. They can also discover that in the series arrangement of cells, the lifetime of the cells is no greater than a single cell, but in the parallel arrangement of two or more cells, the life-time is at least twice as long as a single cell. You might explain that the series arrangement of cells builds up electrical pressure called *volts.* For example, two 1½-volt cells in series produce 3 volts of electrical pressure, but in the parallel arrangement they produce only 1½ volts and last twice as long.

35
Bb
Ec
Fa

Contributing Idea C. Switches can be used to control circuits and voltages easily and safely.

Turning an appliance on and off from two or more switches. Have students recall where they have used switches that turn lights on by flipping them up *or* pushing them down. (Such switches are usually used at the bottom and top of a stairway or at opposite ends of a hallway.) Have them make several switches, and challenge them to design an arrangement that can turn a bulb on and off by either of two switches. Challenge students to design a way to turn an appliance off and on using three switches.

36
Ga
Gb
Gc

Making a reversable switch. Students can build a switch that can change the current direction quickly and easily. Such a switch can be used to reverse a motor's di-

37
Ga
Gb
Gc

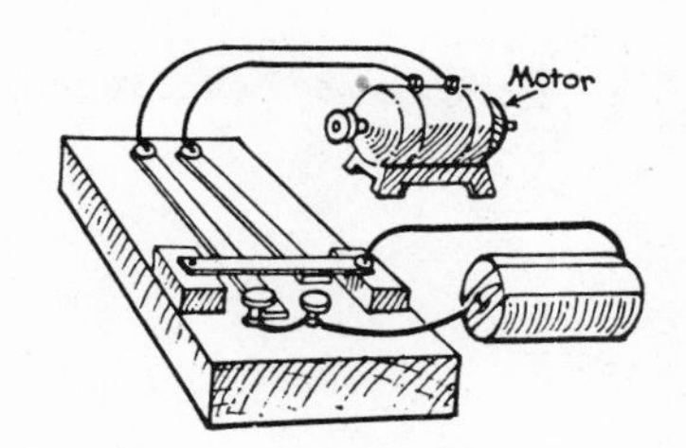

FIGURE 352.37

rection from time to time without removing and rearranging the wires.

Making a dimmer switch. Carve the wood from a pencil to expose the "lead" (graphite) inside. By sliding a contact wire from a circuit tester along the length of the "lead," the light bulb will become brighter or dimmer.

38
Gc

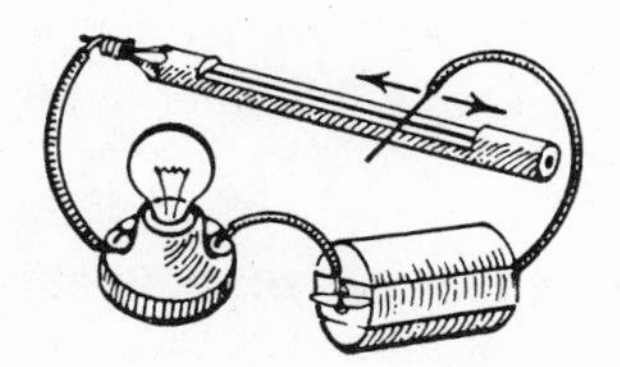

FIGURE 352.38

Listing safety precautions for use of electricity. Have students discuss the dangers related to electricity (it can hurt or kill living things, and it can set things on fire), then let them develop a listing of guidelines to follow when using electricity. The following are a few suggestions that might be listed.

39
Ba
Bc

1. Electric wires or cords that have worn or frayed insulation should not be used in an electric circuit.
2. Electric wires, cords, switches, fuses, or appliances that are connected in a circuit should never be touched with wet hands.
3. Switches should be repaired immediately if you get a shock when you use them.
4. Electric circuits should be disconnected from the source of electricity before any work is done to them.

360 magnetism

361 CHARACTERISTICS

Generalization I. Magnetism has identifiable characteristics.

Contributing Idea A. Magnetism exists as a magnetic field surrounding all magnets.

Observing magnetic fields. Have students place a sheet of cardboard, glass, or plastic over a bar magnet. Punch holes in the lid of an old jar and fill it with iron filings or cut up pieces of steel wool. From about 1 foot (30 cm) above the cardboard, sprinkle the filings lightly and evenly. Tap the cardboard gently to help the filings line up in the magnetic field. Students will see that some force moves the filings into a definite pattern. Tell them that the force is called a *magnetic field* and the place where the field is concentrated is called the *pole.* If they repeat this experience using magnets of different sizes and shapes, they will find that all magnets have a magnetic field surrounding them. **01 Aa Bb**

Feeling magnetic fields. Have students work in pairs to push the ends of two bar magnets together in various ways. They will be able to feel the force of attraction or repulsion between the two magnets depending upon which ends they bring together. Let them also test the middle and sides of the magnets. Tell them that what they feel are the *magnetic fields* that surround the magnets. **02 Ab Bb**

Contributing Idea B. Magnetism extends from one end to the other in the same magnet.

Making permanent pictures of magnetic fields. The patterns produced by magnetic fields can be preserved in several ways for display or for further study. If students compare the preserved examples, they will find that in each case the magnetic field extends from one end of the magnet to the other. **03 Aa**

a. Cover a magnet with cardboard or glass and place a sheet of waxed paper on top. After sprinkling iron filings over the paper, fix the patterns by warming the waxed paper near a heating unit. After the wax cools, the filings will stick well enough so that normal movement of the paper will not disturb the pattern.

b. Cover a magnet with a sheet of white cardboard and use an atomizer to spray the iron filings with a fine water mist. If the filings are kept damp for a short time (24–48 hours), they will rust, and the stains will leave a pattern on the cardboard.

c. Spread newspapers, place a magnet on

them, then cover the magnet with a sheet of cardboard. Using a clear lacquer in a commercial spray can, spray the iron filings. When dry, the patterns can be displayed. *(Note: Some practice may be necessary to properly use the spray can—if it is held too close, the force of the spray may disturb the pattern.)*

d. Place a magnet in position beneath blueprint paper, scatter iron filings over it, then produce a print by following the directions on the blueprint paper package. Generally, the procedure is to expose the paper to strong sunlight for several minutes as soon as the iron filings have arranged themselves along the lines of force, then carefully pour the filings into a container, rinse the blueprint paper in a pan of water for about five minutes, and pat dry. The patterns of white on the blue background can be easily studied. Similarly, students can make ozalid prints that produce dark lines of force on a white background.

Generalization II. Materials can be organized by the way they are influenced by magnetism.

Contributing Idea A. Magnetism attracts some materials.

Sorting objects by the way they are affected by magnetic fields. Collect a variety of small objects such as coins, wires, matchsticks, nails, papers, rubber bands, thumbtacks, tin cans, small pieces of aluminum foil, silk cloth, and woolen cloth. Touch the objects with the end of a bar magnet to see which are attracted to it. Let students sort the objects into groups—those which are attracted and those which are not. Tell them that the objects which are attracted to a magnet are called *magnetic,* while those that are not attracted are called *nonmagnetic.* The results of the sorting can be tabled.

04
Bb
Db

Contributing Idea B. Some materials can be magnetized.

Classifying materials according to which can and cannot be magnetized. Have students slowly stroke a large steel needle twenty to twenty-five times across one end of a bar magnet using the same end for each stroke, stroking the needle in the same direction, and lifting it completely away from the magnet at the end of each stroke. When finished, hold the needle by the other end, and stroke it slowly over the opposite pole of the bar magnet. Now lower the needle over some iron filings. (It should attract them.) Let students try making magnets out of other materials such as iron and steel nails, hacksaw blades, etc. By testing, they will find that magnets cannot be made out of any materials except iron, nickel, or combinations using these materials (e.g., steel). Students can classify the materials they test by the material's ability to become magnetized.

05
Dc
Ec

362 INTERACTIONS

Generalization I. Magnetism is a form of energy produced by different sources.

Contributing Idea A. The earth is a source of magnetism.

Detecting the earth's magnetism using a "horizontal lines of force indicator." Magnetize a large steel needle (see *362. 78–79*) and push it horizontally through the side of a small cork. Float the cork in a bowl of water. Be sure the needle is horizontal to

01
Aa
Gc

the water's surface. (Several drops of sudsless detergent will keep the cork from floating toward the side of the bowl.) The needle will align itself with the earth's magnetic field so that one end points north. Let students move and turn the cork, then move the bowl to different areas around the schoolgrounds. *(Note: If large metal objects or metal within the structure of buildings influences the position of the needle, do this activity outdoors.)* When students are satisfied that the needle always returns to the same position, mark the edge of the dish for north. Tell them that the magnetized needle is affected by the earth's magnetic lines of force. Now add the other points of the compass to the bowl, and use the floating needle to indicate directions. As an additional activity, set a compass so the indicator is directly in line with the north-south marks. Place a bar magnet to the side of the compass and adjust its distance so the compass indicator moves 45° toward the magnet. Measure and record the distance from the compass to the nearest end of the magnet. After moving the bar magnet closer to or farther away from the compass, students will develop some concept of the strength of the earth's magnetic field compared with the magnetic field of a small bar magnet.

Detecting the earth's magnetism using a "vertical lines of force indicator." Tell students that a compass only indicates the horizontal direction of the earth's lines of magnetic force. Explain that at the equator a compass indicator would be horizontal, but as it is carried closer to the magnetic poles, it would dip down more and more, and would be vertical over the pole. To measure the dipping, insert a large steel sewing needle through a large cork, then push two smaller needles into the cork from opposite sides. Balance the cork by the pro-

02 Aa Gc

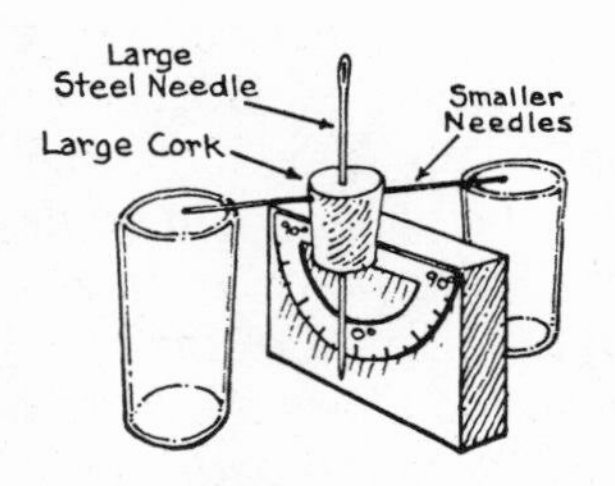

FIGURE 362.02

truding smaller needles between two glasses and place a scale beside the large needle so that it rests at 0° on a protractor. Now remove the large needle, magnetize it, then re-insert it in the cork. The angle at which the needle now dips can be measured by the protractor. Next, move the indicator to different parts of the school grounds to see if the angle remains the same. Students will realize that the angle of dip varies from place to place on the earth. As an additional experience, they can use the indicator to locate magnetic materials buried in some sand. Discuss what practical use could be made of such an indicator.

"Seeing" the earth's magnetic field. Unroll large sheets of butcher paper on the schoolgrounds. Set compasses at various places on the paper. When a compass needle stops moving, have a student put a dot on the paper at the North and South Poles of the indicator. Remove the compass, draw a line between the two points, then replace the compass so that the South Pole of the indicator is at the top of the line. Put a dot at the North Pole again, remove the compass, and connect the points with a line. Continue doing this for several feet for each compass. Students will see that none of the lines ever cross—they follow the same direction and seem to be parallel. Tell them that they have been marking lines that represent the lines of force of the earth's magnetic

03 Aa Be

field, and that such lines of force exist everywhere on earth. Explain that if the lines were continued in each direction, they would converge at the earth's magnetic North and South Poles.

Making a model to represent the earth's magnetic poles. Make a ball of clay, use a large ball of yarn, or obtain a rubber ball with a diameter the size of a bar magnet. Make an opening in the material and insert the magnet, then support the ball on a wooden meat skewer at an inclined angle that represents the axis of rotation of the earth. Tell students that the ball represents the earth, the skewer represents its axis of rotation, and the magnet represents its magnetic poles. Explain that the magnetic poles and the geographic poles are not located in the same place. Tell them that maps showing the magnetic poles must be periodically changed because the poles slowly change position. Students can use small compasses to study the lines of magnetic force surrounding the model.

04
Fc

Contributing Idea B. Certain natural and artificial materials can be sources of magnetism.

Examining natural magnets. Obtain several lodestones. Let students use them to pick up iron and steel objects such as pins, needles, paperclips, and bobby pins. Now set a lodestone on a board floating in a bowl of water. Students will find that the stone will always turn to face one direction no matter how the board is placed. Tell them that many years ago, sailors noticed that when this natural magnetic rock was placed on a floating board, it tended to line up so that one end always pointed north. Because of this, they called that end of the rock the *North Pole;* and since the rock was used to lead the sailors to distant destinations, the rock was called "lead stone" or *lodestone.* (*Note: Today geologists have decided that the end of any magnet that points north will be called* North.)

05
Aa
Bb

Examining artificial magnets. Have students work individually with artificial magnets of various sizes and shapes to test an array of objects for reactions. When they have ascertained that iron and steel are the materials attracted to the magnets, let them use the magnets to determine what objects in the classroom contain iron and steel. Prepare a list of the objects. Students can also move a magnet back and forth through a quantity of sand to see what kinds of particles the magnet removes. (It will remove the iron particles.)

06
Bc
Ec

Contributing Idea C. Electricity can be a source of magnetism.

"Seeing" the magnetic field produced by electricity. Have students draw a 4 inch (10 cm) diameter circle on an 8 inch (14 cm)

07
Aa

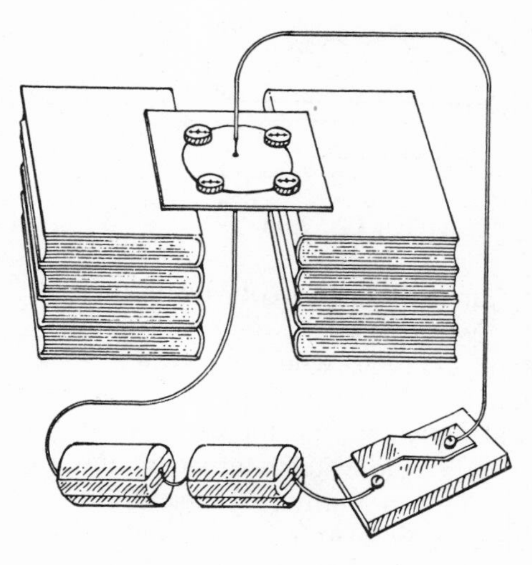

FIGURE 362.07

square of white cardboard and make a small hole in its center. Support the square on two tall stacks of books. Remove an inch (2.5 cm) of insulation from the middle of a 1 foot (30 cm) length of copper bell wire. Pull the wire vertically through the cardboard so that the uninsulated section is at the hole. Now place four small compasses on the circle. Attach the length of wire to a switch and two dry cells that are connected in series. When the circuit is closed, students will see that the four compass indicators outline the circle. Reverse the connections to the dry cells and repeat. (The compass indicators will reverse positions.) As a supplementary experience, sprinkle iron filings on the circle around the wire in place of the compasses. When the current is turned on and the cardboard is tapped gently, the filings will form a circular pattern. From these experiences students will realize that a magnetic field surrounds a wire when a current moves through it.

Using electricity to produce magnetism. Remove about one inch (2.5 cm) of insulation from a 1 foot (30 cm) length of copper bell wire. Connect the length of wire to a dry cell and a switch. Sprinkle iron filings on a sheet of paper, then, when the switch is closed, touch the bare part of the wire to the filings. Students will see that the wire attracts the filings. When the switch is turned off, the filings will fall back onto the paper. (**Caution: The bare wire gets hot enough to burn.**)

08
Ga

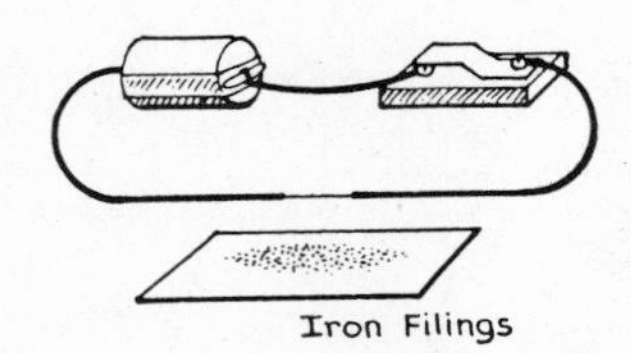

FIGURE 362.08

Generalization II. All magnetic sources are surrounded by a magnetic field.

Contributing Idea A. Magnetic fields extend in unbroken lines from one pole of a magnetic source to the opposite pole of another magnetic source; magnetic fields from one pole of a magnetic source turn away from the same pole of another magnetic source.

Observing interactions between magnetic fields. Obtain a transparent, flat-sided bottle such as a hip flask. Fill it with clear glucose, and add a spoonful of iron filings. Cap the bottle, then shake it to distribute the filings throughout the liquid. Set the bottle on a table and bring one end of a bar magnet to one side. (The filings will arrange themselves into a three-dimensional representation of the magnetic field around the magnet.) Shake the bottle again, and repeat using like or unlike poles on opposite sides of the bottle, using magnets of various shapes, or by suspending an iron washer from the cap of the bottle to see the magnetic field induced by an approaching magnet. Students can also test other liquids in the bottle to see if they can improve the visual effect (e.g., clear molasses, white syrup).

09
Aa
Ec

Determining the polarity of magnets. Construct a wire support to hold a bar magnet. Tie a string to the support, and suspend a magnet from it so that the magnet is balanced. When the magnet stops turning, compare the direction it points to the alignment of a compass indicator. Explain that the magnet, just like the compass indicator, lines up with the magnetic poles of the earth. Tell students that the end that points north is labeled the *N* or *North Pole* of the magnet while the end that points south is labeled the *S* or *South Pole*. Now magnetize some iron

10
Aa
Bb

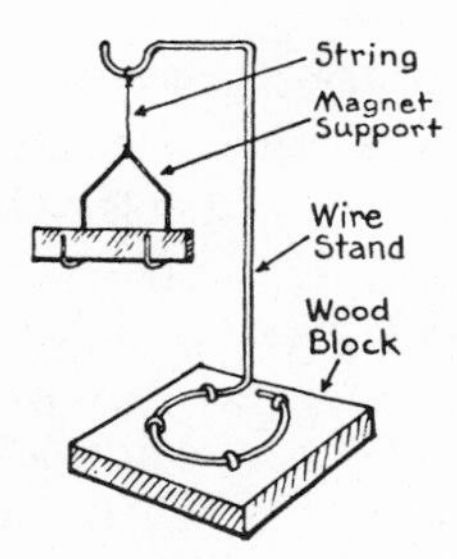

FIGURE 362.10

nails or steel needles and determine the poles of each by suspending them from threads and observing how they line up with a compass. Red nail polish can be used to mark the North Pole end of each magnetized object.

Examining polarity. Have students bring the North Pole of a bar magnet to the South Pole of a suspended bar magnet. Determine how far away from each other the poles are when the suspended magnet begins to move. Compare the strengths of various magnets by bringing each toward the suspended one and measuring the distance at which it begins to move. Now place two other bar magnets beneath a sheet of white cardboard, glass, or plastic so that the North Pole of one and the South Pole of the other are about an inch (2.5 cm) apart. Sprinkle iron filings on the cardboard over the magnets, tap the cardboard, and observe the pattern that appears. Students will note that the lines of force indicate an attraction between the unlike poles. Next, bring the North Pole of the first bar magnet toward the North Pole of the suspended magnet. Students will find that the poles repel each other. Likewise, South Poles will repel each other. Now place two like poles about an inch (2.5 cm) apart beneath the sheet of cardboard. Sprinkle iron filings over them and tap the sheets. Students will find that there seems to be a wall between the like poles—the lines of force deflect away from the like poles, indicating that these poles repel each other. Have them devise a rule that explains what happens when various poles are brought together (unlike poles attract each other, like poles repel each other).

11
Cd
Ec
Fa

Determining the polarity of unmarked magnets. Set one bar magnet on a support, suspend it from a string, and bring another bar magnet close to it to test different combinations of poles. From the results, students will be able to generalize that like poles repel and unlike poles attract. Now tape markings on several magnets so that students cannot identify the poles. Using a marked magnet, let them test to identify the poles of unmarked magnets. The tape can be removed to check answers. If unmagnetized iron or steel bars can be obtained, tape the ends and mix them with the bar magnets. Students will find that although the bars are attracted to another magnet, they are never repelled. They will realize that repulsion, rather than attraction, is the best test of whether or not a material is magnetic. Since large, stationary steel objects, such as cabinets or lockers, usually have magnetic poles, students can use their gained knowledge to locate and identify these poles. (They can bring a compass very close to all parts of the object and note which way the indicator points.)

12
Bc
Cc
Ec

Differentiating among magnetic, magnetized, and nonmagnetic objects. Tape each of various small objects between two squares of cardboard so that they cannot be seen. The cards should contain a variety of magnetic (iron washer, Canadian nickel), magnetized (small magnets), and nonmagnetic objects (pieces of plastic and wood). Students can test each card using a marked

13
Db

magnet and sort the cards into groups. (They may first sort them into two groups—magnetic and non-magnetic, but if challenged to determine whether or not there is a magnet inside a card, they will find that there are two kinds of magnetic objects.)

Contributing Idea B. Magnetic fields can be induced in some materials.

Inducing magnetic fields. Sprinkle some thumbtacks, paperclips, or bobby pins on a sheet of paper. Test the point of an iron nail to see if it will pick up any of the objects. If it doesn't, place one end of a bar magnet against the head of the nail, and test again. (The objects will be picked up by the nail.) **14 Bb Bc Ec**

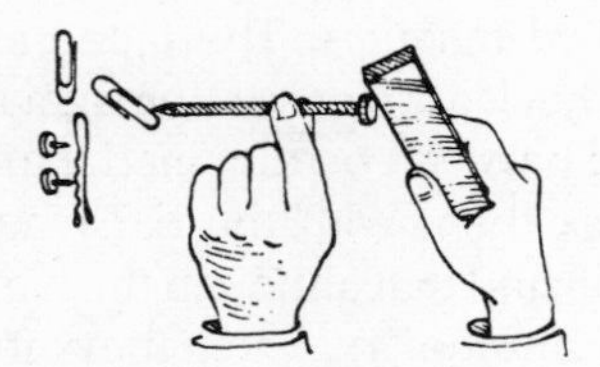

FIGURE 362.14

Remove the magnet. (The objects will drop back on the paper.) Explain that the magnetic field surrounding a magnet can temporarily magnetize certain other objects. Tell students that when this happens, the object is said to be magnetized by *induction* and is called a *temporary magnet.* Now let them test other materials and keep a record to see which can be temporarily magnetized by induction. They will find that magnetic fields can be induced only in materials made of iron or steel.

Contributing Idea C. The compass is an instrument used to detect magnetic fields, to determine the polarity of a magnetic source, and to indicate directions on earth.

Making a compass. From the following experiences, students will find that a compass is simply a small magnet that is allowed to turn freely. It can be used to detect magnetic fields or to determine the polarity of a magnetic source; however, it is usually used to indicate directions on earth. Simple compasses can be made in several ways. *(Note: For proper functioning, be sure to use the compasses away from large metal objects and structures containing iron or steel.)* The following compasses can be carried around and used outdoors even on windy days. **15 Gc**

a. Cut out two sides of a quart (liter) milk carton and suspend a small bar magnet from a string so that it is free to turn in

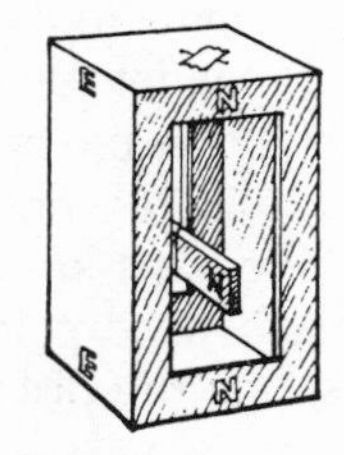

FIGURE 362.15a

the carton. Let the bar magnet align itself, and mark the four compass directions on the carton.

b. Magnetize a large steel needle, fold in half a length of cardboard that is just a

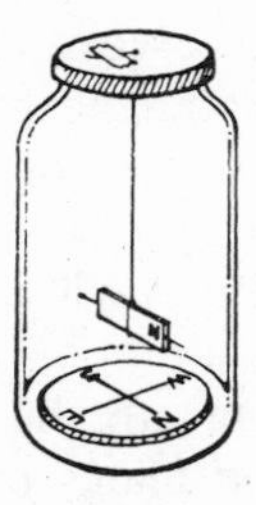

FIGURE 362.15b

little shorter than the needle, push the needle through the cardboard, and suspend it from a thread within a wide-mouth jar. A protractor can be used to prepare a circular directional disk that can be glued to the inside bottom of the jar.

c. Place a magnetized needle in a cork and float the cork in a bowl of water. Add a few drops of sudsless detergent to keep the cork from being drawn to the sides of the bowl. You might tell students that Christopher Columbus used a floating compass when he arrived in North America.

d. Tie a thread to the center of a magnetized bobby pin, and suspend it in a glass jar to shield it from wind. Test to see if this compass will work as well when the jar is filled with water. You might tell students that airplane and ship compasses are liquid-filled to keep the indicator from swinging too much.

e. Push a wire through a hole in the lid of a wide-mouth jar, make a hook at the lower end, and tape it securely to the lid at the upper end. Tie one end of a thread to the hook and the other end to the middle of a U-magnet. Place the lid on the jar, and mark the points of the compass on the lid when the magnet lines up in a north-south position.

Discovering that not all compasses and magnets are correctly marked. When compass indicators line up in a north-south position, students can infer that the North Pole of a compass is really the "north-seeking" pole, however, if they use a marked magnet to test the polarity of several inexpensive compasses from different companies, they will be likely to find that some of the compasses are incorrectly marked. (There is no agreement among companies on which pole points north; and some companies try to avoid confusion by coloring one end instead of imprinting the words North and South. Thus, in some compasses, the colored end points north, and in others it points south.)

16 Ec Fb

Using compasses. Have students work in groups of three. Provide each group with two large cardboard boxes with open tops and one compass. In an outdoor area, have two members from each group place the boxes over their heads. Give the compass to one of the two students, then have them both walk in a straight line for 300 steps, then retrace their steps. The observing students will see that the student with the compass is able to come closest to the starting point by following the compass indicator in one direction, then in the opposite direction after turning around. You may wish to introduce map reading to the students and show them how to orient a map using a compass.

17 Ga

Generalization III. Magnetism can be transferred to some materials.

Contributing Idea A. Electricity or another magnet can be used to make a magnet.

Making a magnet using another magnet. Spread apart the prongs of a bobby pin so that they are in a straight line except for the small bend in the middle. Hold the bobby pin by the bend, and stroke its entire length about fifty times across one end of a strong bar magnet, always in the same direction. Test the bobby pin to see if it will pick up paperclips or thumbtacks. (If it doesn't, repeat the procedure.) Similarly, hold a steel needle by its eye end or an iron nail by its head, and stroke its entire length across one

18 Gc

end of the magnet. Suspend the magnetized object from a thread or float it on a piece of aluminum foil in a bowl of water. Students can try turning the object to see if it always returns to a north-south position (the position can be checked with a compass). Mark the north-seeking pole of the object with some red nail polish. If several needles and nails are made into magnets and tested, students may find that the north-seeking pole is not always at the same end. (If the object is held by the point and stroked, or if the opposite pole of the magnet is used, the polarity of the object will be reversed.) Students can also make temporary magnets by inducing magnetism from a permanent magnet. To do this, touch an iron nail to one pole of a magnet and try to pick up a paperclip with it. Students will find that as long as the nail is in contact with the magnet, it also becomes a magnet.

Making a magnet using electricity. Roll a strip of light cardboard to make a 4 inch (10 cm) long tube with a diameter of about ½ inch (2 cm). Hold the tube together with tape. Wind 100 turns of #22 or #26 insulated copper wire around the tube in a clockwise direction leaving about 1 foot (30 cm) free at each end. Strip an inch (2.5 cm) of insulation from the ends, and mark the beginning end of the coil with adhesive tape *(a).* Attach the coil to a switch and two dry cells (in series) with the beginning end on the negative terminal. Now insert an entire bobby pin, steel needle, or iron nail into the tube from this end, head end first *(b).* Close the switch for five seconds. Open the switch and remove the bobby pin. Students can use a marked magnet to be sure the head end of the bobby pin is the north-seeking pole. Now magnetize a second bobby pin in the same way, mix the two magnetized bobby pins with some unmagnetized ones, and challenge students to pick them out. Students can use the coil to magnetize other objects such as iron curtain rods, hacksaw blades, and screwdrivers.

19
Ga
Gc

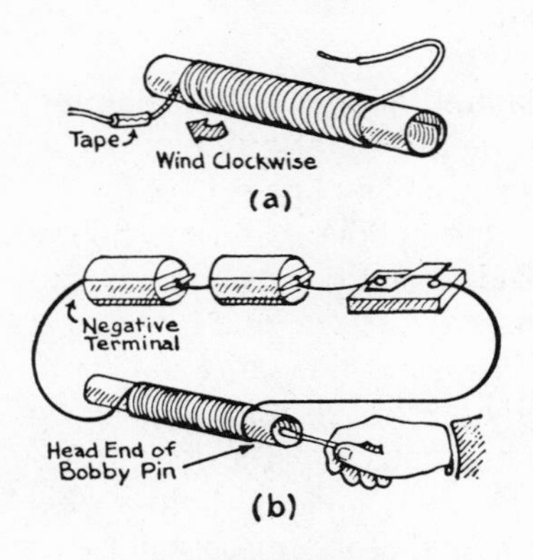

FIGURE 362.19

Contributing Idea B. Permanent magnets retain their magnetism longer than temporary magnets.

Comparing a permanent magnet to a temporary magnet. Using the same number of strokes, have students magnetize a large steel needle and an iron nail of about the same size. Keep a record of how many steel thumbtacks or paperclips each magnetized object will attract. After the magnets have been put aside for three days, test again, and compare how many tacks each now attracts. Explain that although any material attracted by a magnet can be made into a magnet, materials differ in how permanently they can be magnetized (e.g., iron loses its magnetism rapidly; steel retains its magnetism for a longer period of time). Now make magnets out of various other objects, and test each after several days to see which retain their magnetism the longest. Tell students that magnets made of materials that lose magnetism rapidly are called *temporary magnets,* while magnets that hold mag-

20
Bb
Ca

netism for a long time are called *permanent magnets.* (*Note: Permanent magnets are not really permanent. They can lose their magnetism in a number of ways—see* 362.21.)

Observing that a magnet can be demagnetized. Have students make several magnets out of steel needles, bobby pins, or iron nails. Test each to be sure it attracts paperclips or thumbtacks. Now place one of the magnets cross-wise to a coil of wire and turn on the current for five seconds. Test the magnet for magnetism again. Hold a second magnet with a clothespin or pair of pliers, and heat one end in a flame. When cool, test it. Drop a third magnet several times on a hard surface or strike it several times with a hammer, then test it. Students will find that magnets can lose magnetism by influencing them with a strong magnetic field, by heating them, or by giving them a sharp blow. **21 Aa**

Generalization IV. Magnetism travels through some materials better than others.

Contributing Idea A. Magnetic fields travel through space.

Observing and comparing magnetic fields through space. Draw lines through a 36 inch (1 m) square of paper to divide it **22 Ca Cd**

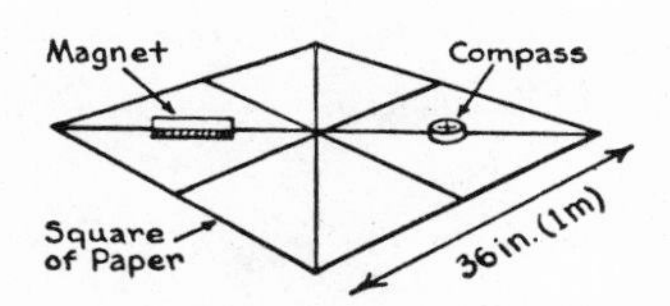

FIGURE 362.22a

into quarters. Next, draw diagonal lines to divide it into eighths. Place a compass on one of the diagonal lines near the center of one of the quarters. Students can notice how the compass indicator moves when a magnet is pushed back and forth along the line on which the compass rests *(a).* If a "vertical lines of force indicator" is constructed (see *362.02*), students can move the magnet vertically to see how it affects the indicator. Students will find that the magnetic field becomes weaker as the space between the compass and the magnet is increased. Now mark the place on the line under the compass where the magnet's field no longer disturbs the compass indicator. If the distance is recorded, other magnets can be tested in the same way and compared in terms of the distance the magnetic field extends through space. Next, set the magnet upright in the center of the paper and move the compass along each line toward it. Mark the place on each line under the compass to indicate where the magnet's field first disturbs the compass indicator, then join the marks with a heavy line *(b).* On identical 36 inch (1 m)

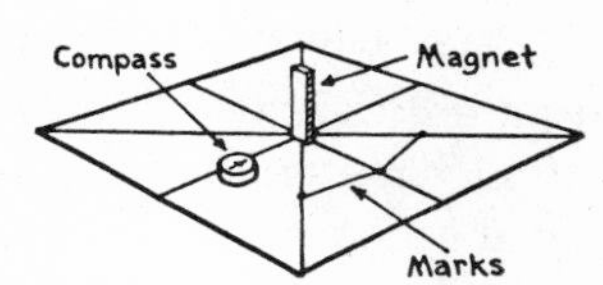

FIGURE 362.22b

squares of paper, students can plot, then compare the extent of the magnetic fields around the poles of other magnets. They can also try plotting the fields when the magnet is on its side or when magnets of other shapes are used. As a supplementary activity, suspend a U-magnet by a string over a compass indicator. Spin the magnet, and observe what happens to the indicator. Students will realize that the magnetic field travels through space and induces the spinning motion in the compass indicator.

Contributing Idea B. Magnetic fields pass through some materials, but are blocked by others.

Testing and comparing the penetration of magnetic fields through various materials. Suspend a magnet from a support. Fasten a string to a block of wood so that a paperclip tied on its end nearly reaches the magnet. Let students determine the greatest distance the magnet can be moved away and stili hold the string and paperclip in the air. Now insert various materials between the paperclip and the magnet (e.g., aluminum foil, copper, glass, iron, paper, plastic, steel, wood). Students will find that all materials are transparent to magnetism except iron and steel. (Such materials block the magnetic field, and the paperclip falls.) By testing with the magnet, they will also find that any material that blocks the magnetic field *is* attracted by the magnet, while any material that is transparent to the magnetic field *is not* attracted.

23
Ca
Ec

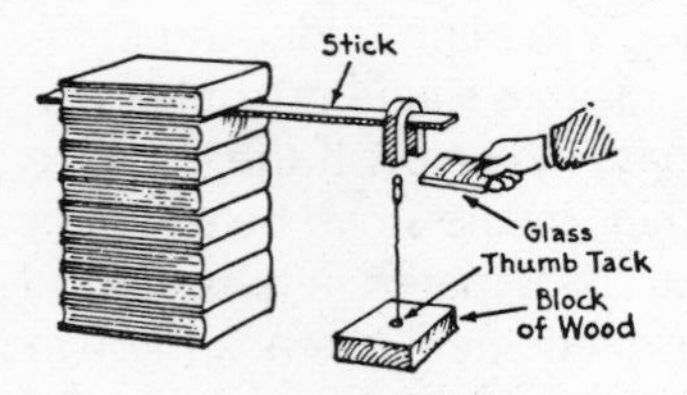

FIGURE 362.23

Measuring and comparing the penetration of magnetic fields through various materials. Prepare a variety of barriers to place between an end of a bar magnet and some magnetic objects. Barriers can be made from iron coat hangers, Canadian nickels, wooden rulers, strips of corrugated cardboard, etc. Each barrier should be made about the same size so that comparisons can be made. Let students place paperclips, thumbtacks, bobby pins, or iron filings on one side of the barrier, and slide the magnet toward the barrier from the other side. For each barrier, measure and record the distance between the paperclips and the magnet when the paperclips first move. From their tests, students will find that materials have no noticeable affect on the penetration of the magnetic field, except those that can be magnetized (e.g., iron, nickel, and their alloys). *(Note: Canadian nickels are a good source of nickel as the coins are 100 percent nickel, while United States nickels are 75 percent copper and 25 percent nickel.)*

24
Ca
Cd

Observing that magnetic fields can penetrate liquids. Submerge a magnet in an aluminum pan or heat-resistant dish of water. Put steel thumbtacks in a cork and float it on the water. Turn the cork in various ways to see if the magnetic field penetrates the water and influences the metal on the cork. Students can test other liquids in a similar way.

25
Aa

Contributing idea C. The strength of a magnetic field can be measured.

Measuring the strength of a magnetic field. Set a bar magnet on a glass slide bridge, supported between two stacks of books. Open a paperclip to make a hook.

26
Bc
Bf
Cc

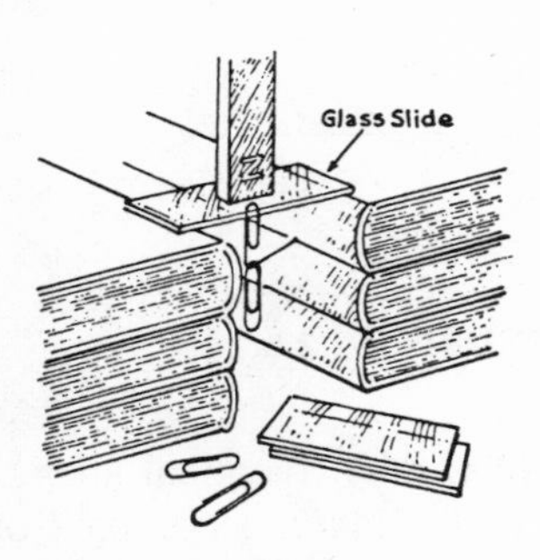

FIGURE 362.26

Touch the paperclip to the underside of the glass slide so that it is held up by the magnetic field of the magnet above it. Now hang other paperclips, one at a time, on the first one until their weight pulls them away from the glass slide. Record the number of paperclips held up by the magnet. Next, raise the magnet so that it rests on two glass slides, and repeat the tests. Continue adding slides, one at a time, each time recording the number of paperclips supported. Students can table and graph the results.

TABLE 362.26. Measuring the Strength of Magnetic Fields

Number of Glass Slides	*Number of Paper Clips*
1	8
2	6
3	4

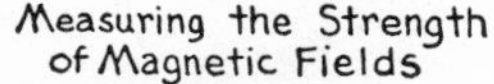

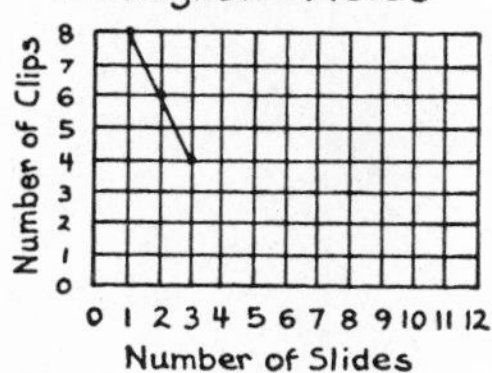

GRAPH 362.26

Testing the strength of magnetic fields through various materials. Measure the strength of magnetic fields through materials such as paper, cardboard, cloth, aluminum foil, or plastic wrap by stacking sheets of the material into sets of 1, 2, 3, 7, 10, 20, 30, 70, and 100. A magnet can be placed on one side of a set and a paperclip on the other side. The sets can be combined to increase the number of pieces between the magnet and the paperclip. Continue increasing the number of pieces until the paperclip no longer hangs from the material. The data can be tabled and graphed.

27
Bf
Cc

Measuring and comparing the strength of magnetic fields. Place a bar magnet on a sheet of lined paper so that one end of the magnet is on a line of paper. Push a paperclip slowly toward the end of the magnet, and count the number of spaces remaining at the point when the magnet moves the paperclip. If this is repeated using different magnets, students can order them by strength, from those that affect the paperclip from the greatest distance to those that affect it from the shortest distance.

28
Cc
Cd
Da

400-499
Instructional Apparatus, Materials, and Systems

410 safety precautions

411 GUIDELINES

Generalization. Science learning requires active participation with equipment, thus safety practices are essential.

> *Contributing Idea A.* Some general precautions are important when using equipment in the classroom.

Practicing safety in the classroom. Various precautions must be considered by the teacher prior to permitting students to use equipment in science experiences. The teacher should: **01 Ga**

a. Handle the equipment and materials prior to student usage to check out possible hazards.
b. Organize the classroom so that there can be constant supervision during individual or group work with equipment.
c. Limit the number of students in groups when there is a limited amount of equipment, because large groups can cause confusion among students and result in accidents.
d. Allow sufficient time for students to carry out experiments because haste can cause accidents.
e. Safety glasses should be considered for certain activities. In many school districts they are mandatory.
f. Long hair should be tied back or kept in a hair net when students are working near an open flame.

> *Contributing Idea B.* Some precautions are important when using heat sources.

Practicing safety when using heat. Various precautions should be taken before using heat in a classroom science experience. The teacher should: **02 Ga**

a. Become familiar with the school's regulations regarding fire, personally practice general safety procedures when using heat, and instruct students in the same procedures.
b. Take extreme care to see that volatile or flammable liquids are at a safe distance from heat sources.
c. Always place heat sources on flat surfaces away from student traffic and be sure electrical cords cannot tip over equipment by looping them around the legs of sturdy furniture.
d. Use an oven mitt, pot holder, or other device when handling hot objects.
e. Set hot objects on asbestos sheets or other insulated pads to protect surfaces.
f. Provide small containers of water for matches after they have been lit.

Contributing Idea C. Some precautions are important when using electricity.

Practicing safety with electricity. Various precautions should be taken when studying electricity. The teacher should:

03 Ga

a. Tell students not to experiment with electric current at home or at school.
b. Be sure students do not handle electric devices immediately after use because such devices sometimes retain a high temperature.
c. Instruct students on the proper way to remove electric plugs from sockets by pulling the plug and not the cord.

Contributing Idea D. Some precautions are important when using glassware.

Practicing safety when using glassware. Various precautions should be taken when using glassware. The teacher should:

04 Ga

a. Label all bottles to identify their contents.
b. Tell students that only heat-treated glassware can be heated.
c. Use soap or petroleum jelly for lubrication when inserting tubing into a stopper. The tubing should be wrapped with several layers of cloth, held as close to the stopper as possible, and inserted with a twisting motion. Tubing that is stuck in stoppers can be removed by splitting the stopper with a single-edge razor blade. The stopper can be reclosed with glue or rubber cement.
d. Tape sharp or chipped edges of mirrors and sheets of glass.
e. Wrap tape around the base of glass objects that might break when dropped. The tape will keep fragments from scattering.
f. Never pick up broken glass with fingers. A whisk broom and dustpan should be used for large pieces and wet cotton should be used to pick up small pieces.
g. Be sure that broken glassware is disposed in a container marked **BROKEN GLASS.**

Contributing Idea E. Some precautions are important when working with animals and plants.

Practicing safety when studying animals. Various precautions should be taken when animals are brought into the classroom for study. The teacher should:

05 Ga

a. Be sure proper habitat and food are available for an animal before it is brought into the classroom. The living quarters should be kept clean, and plans should be made for care over weekends and holidays.
b. Be sure all mammals brought to the classroom have been inoculated for rabies. Scratches or bites should be reported immediately to the school nurse. Check the school's policy on having animals in the classroom.
c. Caution students against teasing animals and inserting fingers or objects into cages.
d. See that students handle animals only when necessary. Handling should be done properly according to the particular animal. Special handling is important if the animal is nervous, pregnant, or with its young.
e. Be sure students wash their hands after handling fish, amphibians, and reptiles. Water from their habitats should be disposed of carefully because of bacterial growth.

Practicing safety when studying plants. Various precautions should be taken when plants are studied. The teacher should:

06
Ga

a. Help children learn, before going on field trips, to identify plants that can produce poisonous effects.

b. Be careful not to use certain flowers and molds for study if some children are allergic to them.

c. When growing plants in the classroom, be aware of and caution children against, any parts of the plants that may be poisonous.

421 HEAT SOURCES

Generalization. Heat sources are essential for many science experiences.

Contributing Idea A. Some heating equipment is inexpensive and easily obtained.

Identifying inexpensive heating sources. Heat is essential for many science experiences. It can be obtained from candles, electric light bulbs, electric hot plates, alcohol lamps, and canned-heat cooking fuels. Liquified petroleum burners are valuable when very hot temperatures are required. **01 Ga**

Contributing Idea B. Some heating equipment can be constructed.

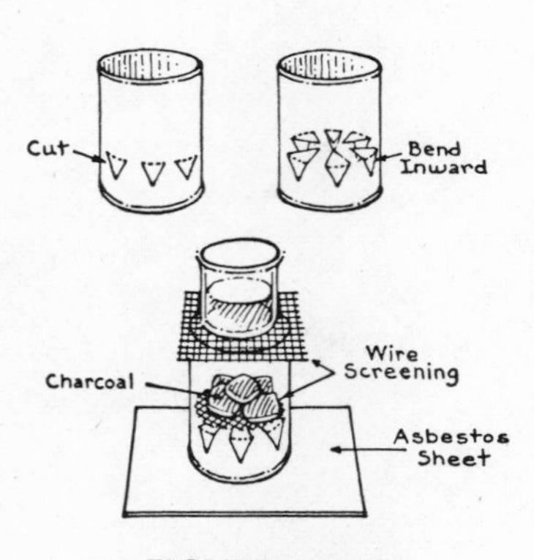

FIGURE 421.02

Constructing a charcoal burner. Use tin snips to cut triangular openings around the side of a large tin can about 1 or 2 inches (2.5 or 5 cm) up from the base. When the triangular cuts are bent inward, they will form a base for charcoal. (Wire screening can be used for additional support.) The openings allow for proper air circulation. Place the burner on an asbestos pad and heat objects over the top of the can. **02 Gc**

Constructing an alcohol burner. Obtain a small bottle, such as an ink bottle. Punch a hole in the metal screw top, insert a small metal tube, and solder it in place. Prepare a wick from twisted strands of cotton string or a rolled section of cotton bath towel. Thread it through the tube. Be sure it is long **03 Gc**

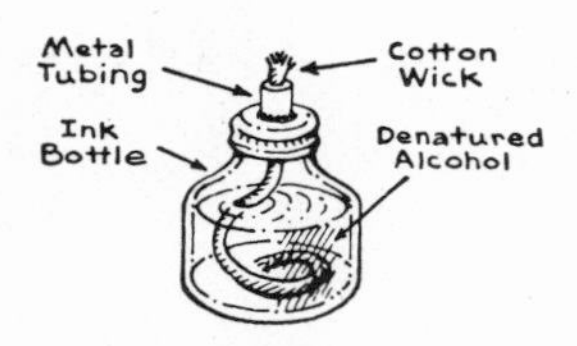

FIGURE 421.03

enough to rest on the bottom of the bottle. Partially fill with denatured alcohol. **(Caution: There is no need to totally fill a container—partial filling reduces the risk of spillage and fire.)** A cap over the wick will keep the alcohol from evaporating when the burner is not in use.

422 SUPPORT STANDS

Generalization. Support stands are useful for many science experiences.

Contributing Idea A. Some support stands can be constructed from inexpensive materials.

Constructing support stands. Many science experiences require support stands to hold and suspend objects. Such stands can be easily constructed from scrap materials. **01 Gc**

a. *Wire stands.* Insert coathanger wire into holes drilled into blocks of wood. Be sure the drilled holes are slightly smaller than the diameter of the wire so that the wire will fit snugly. Make the base of each stand large enough so that it cannot be tipped over easily when used.

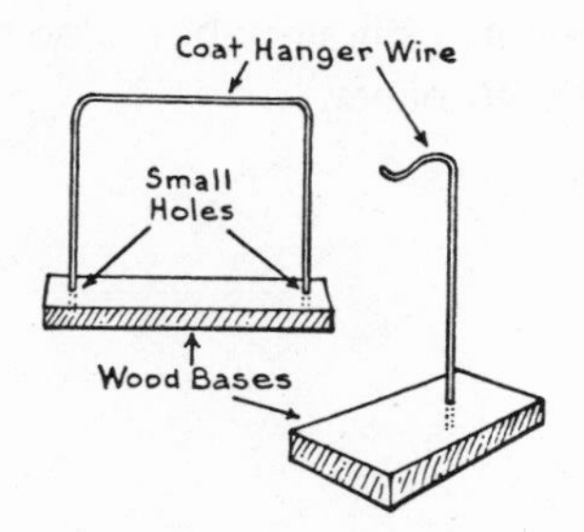

FIGURE 422.01a

b. *Wood stands.* Drill a hole through a small block of wood just large enough to fit a heavy dowel through it. Screw the block to a larger one, then fasten a cross beam to the dowel. Clothespins can be used to suspend objects from or hold objects to the stand.

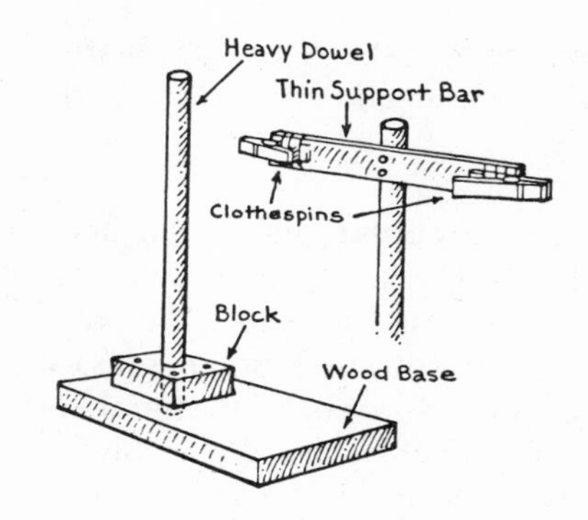

FIGURE 422.01b

c. *Test tube stands.* Wooden blocks with 3/4 inch (2 cm) holes drilled into them serve well as individual test tube holders. Stands for several test tubes can also be constructed.

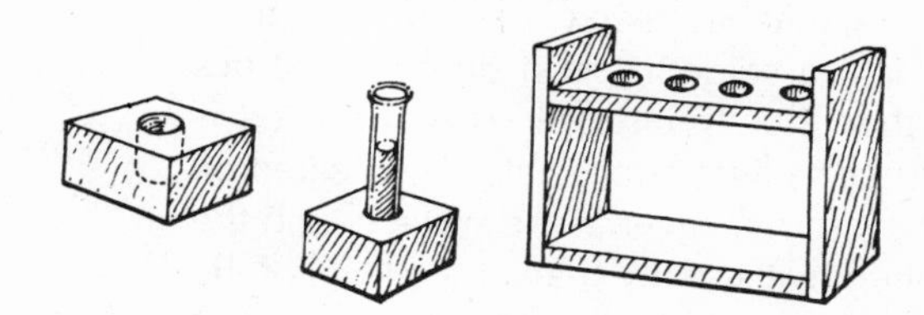

FIGURE 422.01c

Contributing Idea B. Special stands are necessary to support heated objects.

Constructing a heating stand. Cut arch-shaped holes from the sides of a tin can to form three or four legs. A heat source can be placed beneath the can and objects can be heated on its tin top. Differently sized cans can be made to fit differently sized heating equipment. **02 Gc**

423 GLASSWARE

Generalization. Various glasswares are useful for many science experiences.

Contributing Idea A. Some glassware can be constructed from discarded materials.

Constructing glass containers. Jugs, jars, and bottles that are generally discarded can be saved and converted for use as laboratory containers. **(Caution: Not all glassware is usable to heat materials.)** Preparation of containers usually involves the application of heat at the level of a jar or bottle where it is hoped the glass will fracture smoothly. For example, a piece of twine that has been saturated with alcohol or other inflammable material and tied about the bottle just below the cutting level can be ignited. (The bottle will usually break above the line.) Another method is to fill the container to the desired level with ordinary motor oil that has been cooled in a refrigerator. Heat a soldering iron until it becomes red, then slowly insert it so that the oil is disturbed as little as possible. The heated oil will rise, form a layer at the surface, and the container will usually snap at the level of the oil. If is does not, a hotter or larger iron must be used. This latter technique works satisfactorily with cylindrical or rectangular containers ranging in diameter from 2 to 8 inches (5 cm to 20 cm). Unsatisfactory results occur when the wall of the jar or bottle varies greatly in thickness around its perimeter. **01 Gc**

Cutting glass tubing. To cut tubing, draw a triangular file across one side of the glass where it is to be cut. With leather gloves on, hold tubing at arms length with thumbs together behind the cut. Push toward the cut and the tubing should snap cleanly. If it does not, draw the file across the cut again and repeat. Next, hold the cut end in a flame to smooth the edges. **02 Gc**

Contributing Idea B. Glass tubing can be shaped for specific purposes.

Bending glass tubing. To bend, hold tubing horizontally in a flame at the point to be bent. Rotate it slowly and allow the tube to bend into a right angle by gravity. (Such natural bending will keep the corner from flattening out.) Other angles can be achieved by holding the tubing in different positions. To make a U-bend, set tubing that has been bent at a right angle onto an asbestos sheet while it is still very hot. Slowly curve the angle with pliers. **03 Gc**

430 measuring systems and instruments

431 MEASURING SYSTEMS

Generalization I. Measurements can be estimated and/or depicted in various ways.

Contributing Idea A. Measurements can be estimated.

Estimating. Obtain six identical quart (liter) glass jars. Fill five of them, each with a different material that can be counted such as marbles, breakfast cereal, peas, raisins, rice, beans, candies. Fill the sixth container with salt or sand. Let students quickly guess how many individual pieces are in each jar, then have them think of ways to find out how accurate their guesses were. **01 Cb**

a. They might count each item one by one (the most accurate, but people often lose count or give up if there are many items).
b. They can count how many it takes to fill a small paper cup, then see how many cupsful are in the container.
c. They can weigh ten objects from a jar one at a time, to get the average weight for one object. Next, they can weigh the jar with the objects inside, then again when it is empty. By subtracting the weight of the jar from the total weight and dividing by the average weight of one object, they can tell how many objects the jar holds.
d. They can pour some objects from one jar into a square baking pan, and distribute the objects evenly to form one layer. By counting the number along one side, then squaring the number they can determine how many there are in one layer.

When finished, students can compare their original guesses with their estimates. Let them try estimating other measurable objects such as the number of books in the school library, letters on the page of a book, students in the school, people in a photograph. With practice, they can become very skilled at estimating numbers of objects.

Contributing Idea B. Measurements can be depicted pictorially.

Graphing. Perhaps the most visual means of presenting quantitative data is through graphing. Graphs enable learners to see relationships among quantities through a pictorial form, thus offering a clearer presentation of data. **02 Bf**

a. *Histograms.* A histogram is a form of graph that is made up of tallied observations showing relative frequencies. Students can count the number of seeds in a sample of pods and tally them on a histogram. The results provide a pictorial representation of the natural distribution of seeds in the pods of a particular plant. Other histograms can be produced by counting the number raisins in slices of bread in a loaf, paperclips a magnet can pick up, drops squeezed from a medicine dropper with one squeeze, students absent each day for a month, words on a

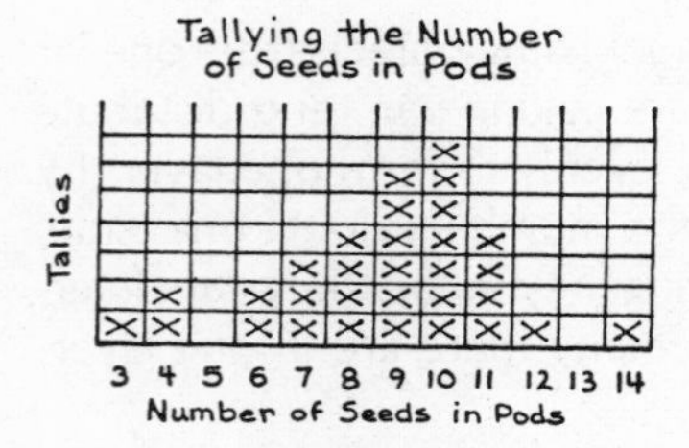

HISTOGRAM 431.02

page or letters in a line, seeds in a specific kind of fruit, different makes of cars seen in thirty minutes, seeds that sprout in a germination test, and so on. Anything that can be counted can be developed into a histogram.

b. *Bar Graphs.* A bar graph uses a solid bar to represent some measured observation. The growth of plants or heights of students can be converted into bar graphs.

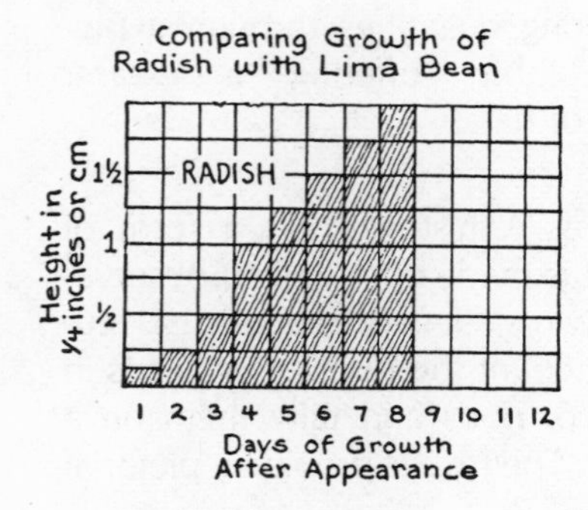

GRAPH 431.02a

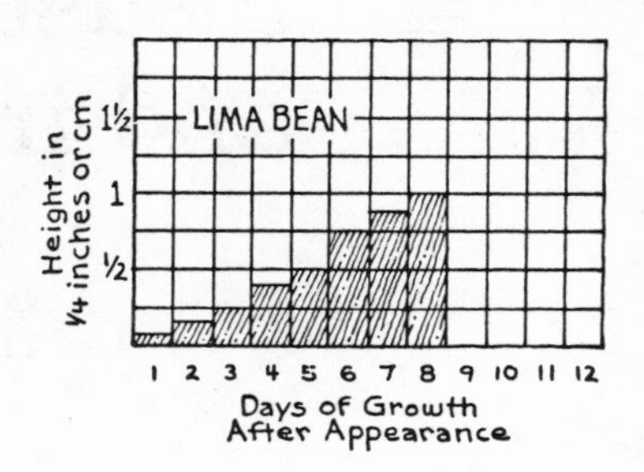

GRAPH 431.02b

c. *Line Graphs.* Intermittent measurements plotted on a grid produce a line graph. When the measurements are connected by lines, inferences can be made about changes between points and in some cases predictions can be made. To prepare such graphs, horizontal and vertical axes should correspond with the columns of data on a table; thus, each tabled recording represents a point on the graph.

1. Some tabled data produce straight-line graphs. Predictions can be made from such graphs. For example, if students burn a candle, record the time it takes to burn each ¼ inch for several minutes, then plot the data on a graph, they can make a prediction telling when the candle would burn out (if the conditions remain the same), because the data produce a straight

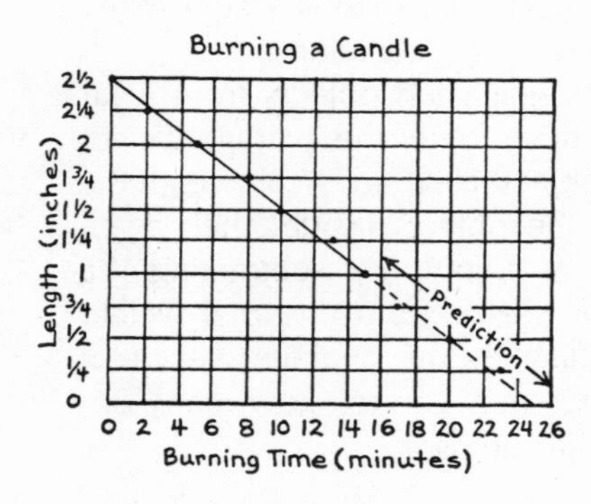

GRAPH 431.02c

TABLE 431.02. Burning a Candle (Diameter of candle = ¼ inch)

Length (inches)	*Time of Day*	*Burning Time (minutes)*	*Remarks*
2½	1:10	0	lit candle
2¼	1:12	2	
2	1:15	5	
1¾	1:18	8	
1½	1:20	10	
1¼	1:23	13	
1	1:25	15	blew out candle
1	1:37	15	lit candle again
¾	1:39	17	
½	1:42	20	
¼	1:45	23	
0	1:46	24	candle burned out

line. Similarly, students can predict how long it will take a liquid to evaporate by weighing cups of each liquid on a daily basis for a week, then graphing the data. They can make predictions about the flow of water by measuring at ten-second intervals the volume or weight of water collected in a container filled at a drinking fountain.

2. Some tabled data do not produce straight line graphs. Most graphs of this type cannot be used for predictive purposes, but they do produce interesting information that can be discussed. For example, students can heat water, record the temperature change at one-minute intervals, then graph the data after boiling point is reached. The graph can be compared with graphs made after boiling salt water or sugar water. Plant and animal growth also produce irregular curves on graphs.

Generalization II. Scientists throughout the world measure lengths, weights, and other values by a standard system.

Contributing Idea A. Units of measure are arbitrary.

Creating measuring systems. Students can develop and use their own systems of measurement. To measure lengths they might use unsharpened pencils, crayons, erasers, or edges of cards. To measure weights, they might use nails, pins, sequins, or other objects on a balance scale. To measure liquid capacities, they might use unmarked containers, paper cups of different sizes, or thimbles. As they measure various aspects of their environment, encourage them to make up words for each unit they devise, then have them describe the sizes of desk tops, tables, chairs, their own heights, weights, and so forth. You can guide them toward realizing that the different measuring systems used in the classroom are analogous to different, widely accepted systems developed by people. Students will discover that some systems are more accu-

03
Cc
Gb

TABLE 431.04a. Tables of the Metric System

	Unit	Symbol		Amount	
Units of Length	A kilometer	(km)	=	1,000	meters (m)
	A hectometer	(hm)	=	100	meters
	A decameter	(dm)	=	10	meters
	A METER				
	A decimeter	(dm)	=	0.1	of a meter
	A centimeter	(cm)	=	0.01	of a meter
	A millimeter	(mm)	=	0.001	of a meter
Units of Liquid Capacity	A hectoliter	(hl)	=	100	liters (L)
	A decaliter	(dal)	=	10	liters
	A LITER				
	A deciliter	(dl)	=	0.1	of a liter
	A centiliter	(cl)	=	0.01	of a liter
	A milliliter	(ml)	=	0.001	of a liter
Units of Weight	A kilogram	(kg)	=	1,000	grams (g)
	A hectogram	(hg)	=	100	grams
	A decagram	(dag)	=	10	grams
	A GRAM				
	A decigram	(dg)	=	0.1	of a gram
	A centigram	(cg)	=	0.01	of a gram
	A milligram	(mg)	=	0.001	of a gram
Units of Volume	A cubic hectometer	(hl^3)	=	1,000,000	cubic meters (m^3)
	A cubic decameter	(dal^3)	=	1,000	cubic meters
	A CUBIC METER				
	A cubic decimeter	(dm^3)	=	.001	of a cubic meter
	A cubic centimeter	(cm^3)	=	.000,001	of a cubic meter
	A cubic millimeter	(mm^3)	=	.000,000,001	of a cubic meter

rate than others, and that some standards are necessary if people are to communicate with each other about measurements. You might explain that the metric system is designed to be a single system for all people.

Contributing Idea B. No one system of units is inherently more accurate than another, but some, such as the metric system, are more convenient to use.

Converting units of measure to different systems. The following tables can be used to convert the more frequently used measurements in the United States system to the

04
Bd
Ga

TABLE 431.04b. U.S. → Metric Metric → U.S. Conversion Tables

	U.S. System to Metric Conversion Table		*Metric to U.S. System Conversion Table*	
Units of Length	1 inch	25.4 millimeters (mm) 2.54 centimeters (cm)	1 millimeter 1 centimeter	0.039 inch 0.394 inch
	1 foot	0.305 meter (m)	1 meter	39.37 inches 3.281 feet 1.093 yards
	1 yard	0.914 meter		
	1 mile	1609.347 m 1.609 kilometers (km)	1 kilometer	0.621 mile
Units of Liquid Capacity	1 fluid ounce	29.573 milliliters (ml) 0.03 liters (L)	1 milliliter	0.034 fluid ounce
	1 pint	473.167 ml=0.473 L	1 liter	2.113 pints 1.057 quarts 0.264 gallon
	1 quart	946.33 ml=0.946 L		
	1 gallon	3785.33 ml=3.785 L		
Units of Weight	1 ounce	28349.527 milligrams (mg) 28.35 grams (g)	1 gram	0.035 ounce
	1 pound	453.592 g 0.454 kilogram (kg)	1 kilogram	35.274 ounces 2.205 pounds
	1 ton	907.185 kg 0.907 metric ton	1 metric ton	2204.62 pounds 1.102 tons
Units of Volume	1 cubic inch	16.387 cubic centimeters (cm³) 0.016 cubic decimeter (dm³)	1 cubic centimeter	0.061 cubic inch
			1 cubic decimeter	61.023 cubic inches 0.035 cubic foot
	1 cubic foot	28.317 dm³ 0.028 cubic meter (m³)	1 cubic meter	35.314 cubic feet 1.308 cubic yards
	1 cubic yard	764.559 dm³ 0.765 m³		

Metric System. Some students might be interested in researching the development and use of metric units.

432 MEASURING INSTRUMENTS

Generalization. Measuring instruments can be used to compare any object or event with others of known or accepted dimensions.

Contributing Idea A. Some instruments are designed to measure distances.

Constructing a sextant. Attach two blocks of wood on the ends of a 1 foot (30 cm) long ruler or stick. (The blocks can be used as handles.) Tape a plastic straw to the top of the ruler and fasten a protractor to its side with a thumbtack. Hang a weight on a length of thread and suspend it from the thumbtack. Be sure that the thread hangs freely and parallel to the protractor's base.

01
Bb
Cd
Gc

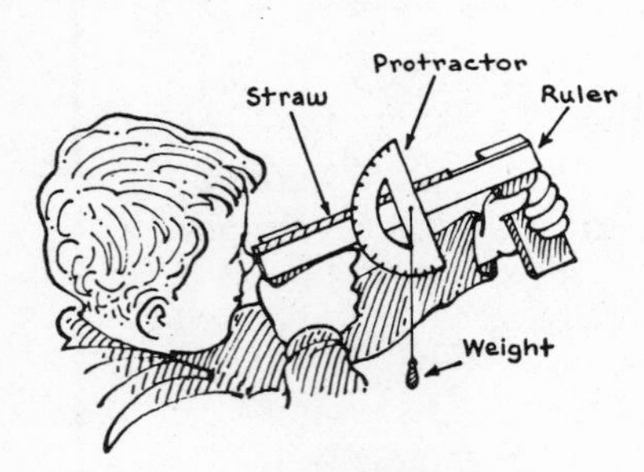

FIGURE 432.01a

When the sextant is tilted upward, the weighted thread will indicate the angle in relation to the ground *(a).* Students can use this instrument to measure the heights of various structures by triangulation. For example, they can look at the top of a flagpole through the straw on the sextant, determine the angle, then measure the distance (baseline) they are standing from the pole. Next, draw on a piece of graph paper, the angle found and add the baseline distance in a scale that fits on the paper. On the other end of the baseline, draw a perpendicular line to represent the pole. The height of the pole (in scale) is determined by the point where the angle line intersects the perpendicular line *(b). (Note: Students must add their own heights to this distance since they sighted on the pole several feet above the baseline.)* This sextant can be used to determine

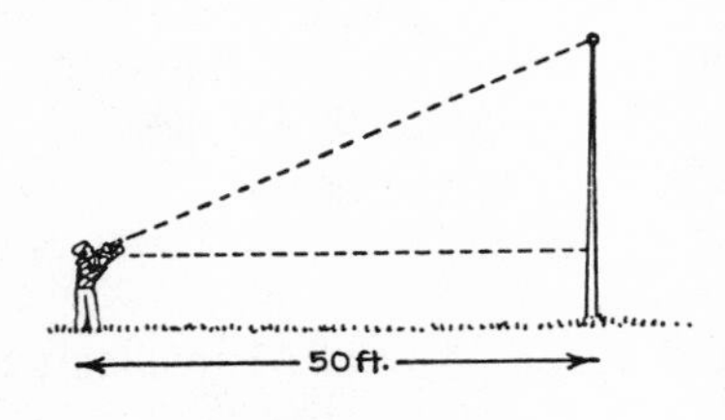

FIGURE 432.01b

the heights of structures whenever a baseline (the distance to the flagpole) and two angles are known (the angle of the sextant and the angle of the flagpole—90°—in relation to the ground), simply by drawing the imaginary triangle to scale. You might tell them that the instrument they constructed is called a *sextant.*

Constructing a range finder. Fasten two small protractors and plastic straws at the 3 inch and 33 inch marks on a yardstick (5 cm and 95 cm on a meterstick). The distance between the straws represents a baseline of 30 inches (90 cm) *(a).* Set the stick on the ground and look through the straws at a tree or other object some distance away *(b).* Mark a corresponding 3 inch (7 cm) baseline

02
Bb
Cd
Gc

on graph paper and draw the angles at each end. When the two lines are intersected, measure the vertical distance from the baseline to the intersection. In scale, this is the distance to the viewed object. Students can

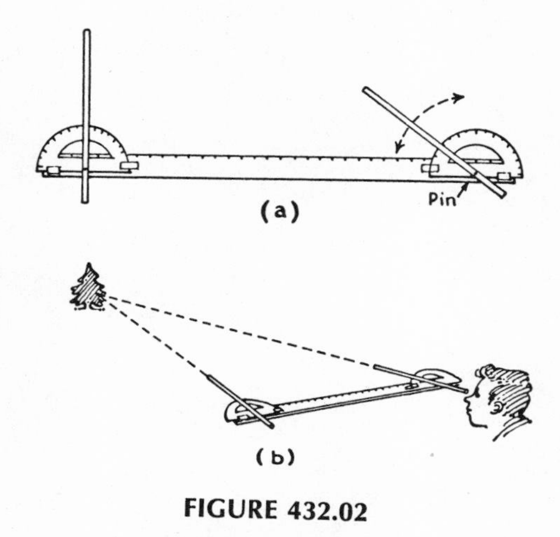

FIGURE 432.02

use a tape measure to check. You might tell them that such an instrument is called a *range finder*.

Contributing Idea B. Some instruments are designed to measure liquid capacity.

Making a graduated cylinder. Cut a strip of paper about ½ inch (2 cm) wide and fasten it to the side of a test tube or tall, thin jar such as an olive jar. Use a measuring cup or laboratory graduated cylinder to pour equal volumes of water into the container. Mark the water level with a fine pen and waterproof ink for each amount. The space between each mark can be divided into ten equal divisions. When the scale is finished, cover it completely with waterproof cellulose tape. **03 Cg Gc**

Contributing Idea C. Some instruments are designed to measure weight.

Constructing a spring scale. Suspend a spring from a support stand and hang a cup or jar lid from its base. Fasten a short plastic ruler to the support with rubber bands. Adjust the ruler so that the bottom of the spring is opposite the ¼ inch (1 cm) mark. (A wire pointer can be attached to the base of the spring to make readings of the scale easier.) Objects can now be placed in the lid and the weight read on the ruler (subtract 1 from the reading and record the weight in ¼ inches or centimeters). To calibrate the scale **04 Cf Gc**

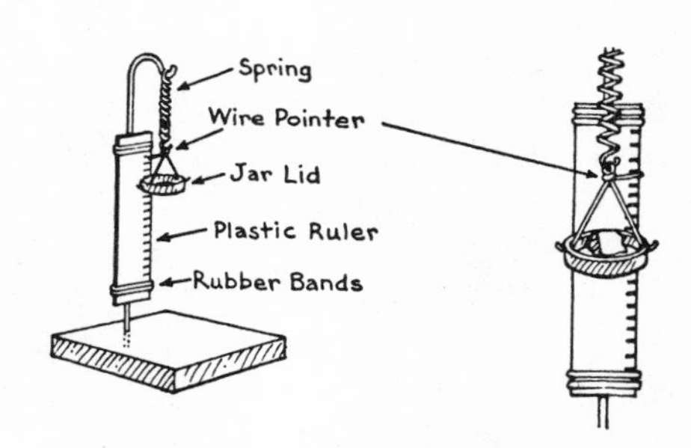

FIGURE 432.04

so that the readings can be converted into grams, place gram weights on the lid until the bottom of the spring reaches the 11-centimeter mark. Subtract 1 and divide the remainder (10) into the number of grams on the scale. The result will be the number of grams per centimeter. This number can be printed at the base of the scale. To change any readings to grams, simply multiply the number of centimeters the pointer moves by this number.

Constructing a balance scale. Balance scales can be made in numerous ways. Many variations can be created from the following. **05 Cf Gc**

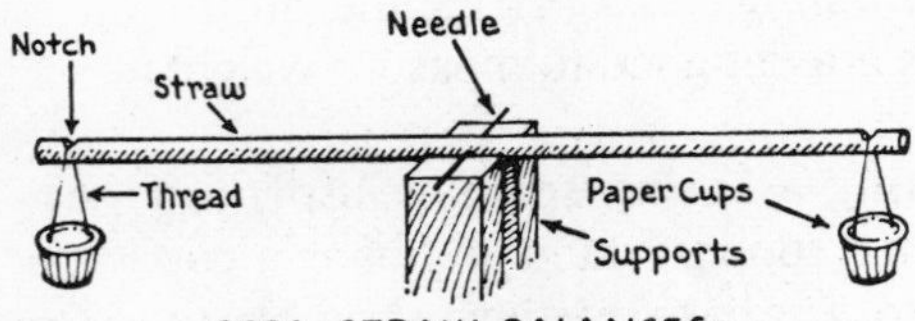

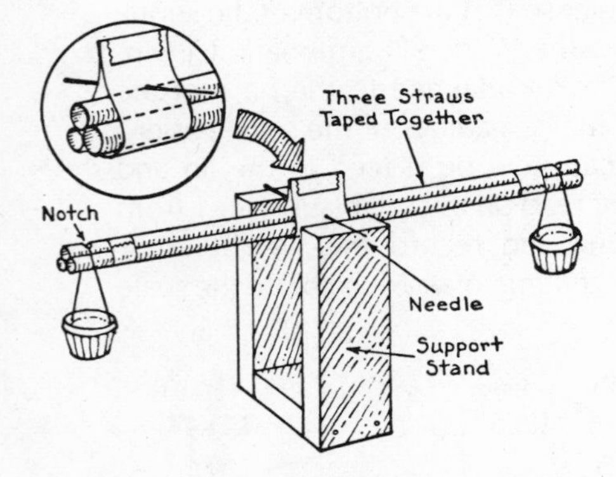

FIGURE 432.05a

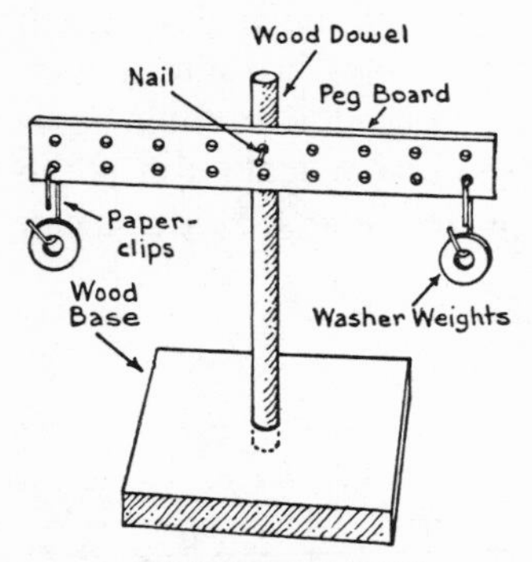

FIGURE 432.05c

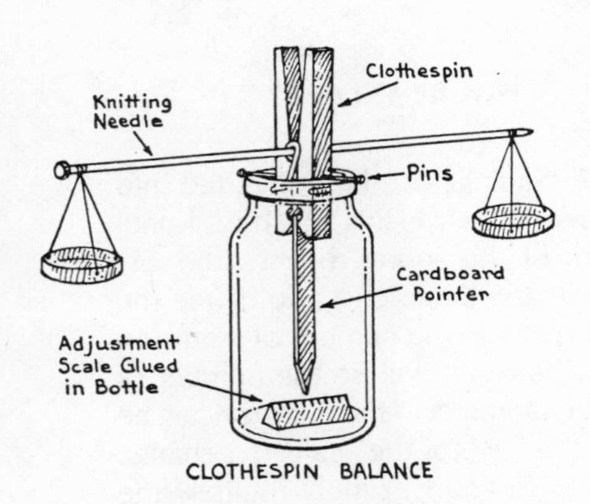

FIGURE 432.05b

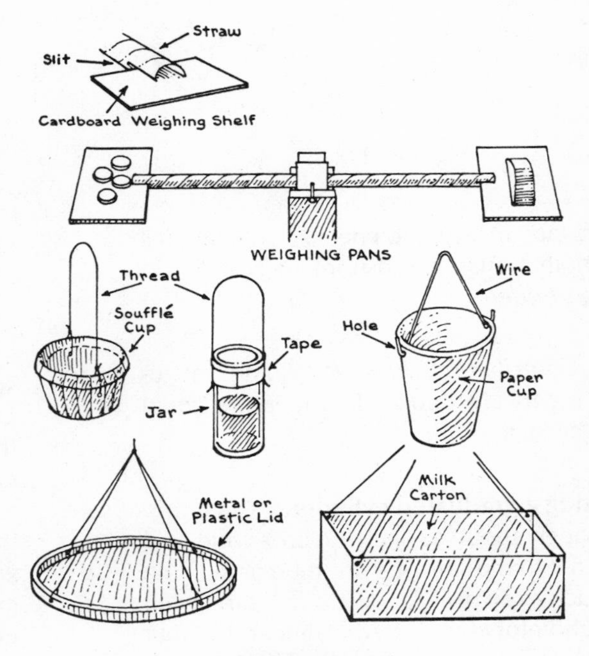

FIGURE 432.05d

Using a balance scale. Balance scales are used to compare unknown weights to known weights. To prepare a set of known weights, obtain a set of gram weights (e.g., grams, decigrams, centigrams). Let students balance each weight by cutting segments of wire or rolling tinfoil together until balance is attained. The wires or tinfoil can then be used in place of the gram weights. Various measurements can be made using these units of measure.

06
Bc
Bf
Cf

a. Very small dry particles such as salt, sugar, talcum, or soap powder can be weighed with the balance scale. To do this, cut two square pieces of waxed paper or aluminum foil the same size. Fold each from corner to corner twice. Place one on each balance pan and balance the beam. When small particles are poured, the weighing paper makes it easy to remove them from the pans.

b. Objects can be counted with a balance scale. To do this, obtain 100 small nails, pins, paperclips, and staples. Balance the scale, place a 1-gram weight in one pan, then weigh out 1 gram of the nails or pins in the other. Count the number of nails or pins in 1 gram. Now place an unknown number of nails or pins in one pan and weigh them. Without counting, the number can be calculated by multiplying the number of objects in 1 gram by the number of grams.

c. The liquid content of objects can be measured with a balance scale. To do this, obtain different foods such as bread, apples, pears, lemons, meat, peas, crackers, and so on. Put 10 grams of fresh bread or other food on a weighing paper in one of the pans. Do the same for other foods, and use a fresh piece of paper each time. Record the weights on a table, put the papers of food in the sun to dry for a few days; then, when dried, weigh them again. The difference between the fresh weight and the dry weight indicates the amount of moisture (liquid) in each food. By starting with 10 grams of each food, percentages are easy to figure.

d. Evaporation rates can be measured with a balance scale by placing the scale in a warm airy place where it will not be disturbed. Prepare two weighing papers and put them on the balance pans. Set a small piece of wet sponge on one dish. Weigh the sponge and record the weight and the time of day. Check the weight every thirty minutes, and record it on a table.

e. Amounts of absorption can be measured with a balance scale. To do this, obtain pieces of different materials such as sponges, blotters, cloths, and paper towels. Make weighing papers, and put one in each balance pan. Balance the scale, then weigh each of the sample materials and record the dry weight on a table. Now saturate each material with water. Shake off any excess water and weigh each when wet. Compare the dry weight with the wet weight. Students can compare different brands of paper towels to see which absorb more than others. They can also test other materials such as napkins and facial tissues.

Contributing Idea D. Some instruments are designed to measure temperature.

Making a sensitive thermometer. Obtain a dowel the same size as a thermometer bulb and a 1½ inch by 3 inch (3 cm by 6 cm) strip of copper. Bend the strip of copper tightly around the dowel *(a),* and fold the ends back to form a flat surface *(b).* Slide the

07
Ce
Gc

copper under the bulb of the thermometer and gently pinch it around the bulb *(c).* Set it in the shade for several minutes, then place it in sunlight. The copper should absorb enough heat energy to quickly affect the thermometer. The instrument can be improved by removing the copper strip, blackening it with soot from a candle flame *(d),* and attaching it to a stand so that the black surface faces outward *(e).* Use the instrument to measure the heat from electric bulbs by setting it so the blackened surface is about 1 foot (20 cm) from a single 100-watt

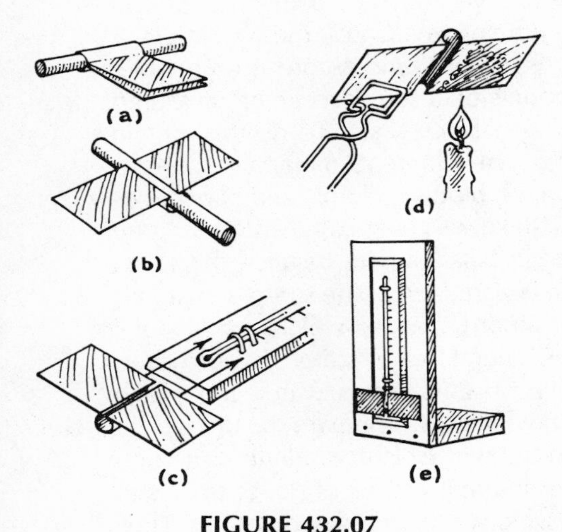

FIGURE 432.07

bulb. After some time, the temperature of the thermometer will stop rising. Now let students test different factors that affect the temperature such as the length of time the lamp is on, the area of blackened surface, the distance the surface is from the bulb, and the wattage of the bulb. They might also try other metals in place of the copper (aluminum foil, tin foil), or determine what happens to the energy output of a test bulb when others are added in either parallel or series circuits.

Contributing Idea E. Some instruments are designed to measure time.

Making a water clock. Use a pin to punch a hole in the bottom of a quart (liter) milk carton. Fill it with water, and set it on top of a quart (liter) glass jar that has straight sides. See how long it takes for the water to drip out by timing it with a clock that has a second hand. Now use a safety pin to punch a larger hole in another carton. Time the flow to see how long it runs. Tell students that they have made simple *water clocks.* Challenge them to make clocks that will run for five minutes, ten minutes, or thirty minutes. To calibrate a clock, attach a strip of tape that can be written on to the side of the jar. Hold a finger over the hole in the carton and fill it almost to the top. Put a mark in the carton at this level so that it can always be filled with the same amount of water. Hold the carton above the jar, and when the second hand of the clock reaches 12, start the clock running. At one-minute intervals, use a fine tipped, waterproof pen to mark the water level on the tape. Students will discover that as the water level gets lower, the water comes out slower and the spacing between the marks changes. *(Note: An unclear band will be seen at the water level because the water tends to climb part way up the side of the jar.)* Place a mark at the lower edge of the band because it is easier to see than the top level.

08
Bb
Ch
Gc

Making a pendulum clock. Tie a piece of thread to a metal weight such as a fishing weight or washer. Fasten the free end to a pin pushed into a block of wood. Set the block at the edge of a table so that the thread swings freely from the side *(a).* Pull the weight to one side and allow it to swing for one minute. Check this several times. If the pendulum does not swing exactly sixty times

09
Ch
Gc

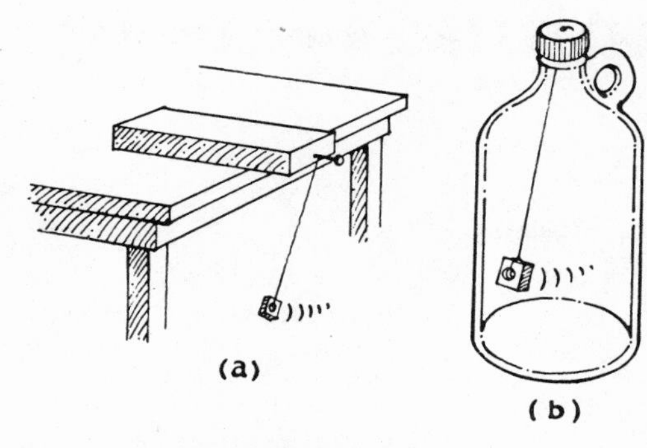

FIGURE 432.09

in one minute, challenge the students to change it until it does (the thread can be lengthened or shortened). When the pendulum is adjusted, each swing (back and forth motion) will represent one second of time. A similar timer can be made by suspending the thread and weight from the cap on a transparent jug or bottle. The pendulum swinging inside will not be influenced by passing breezes *(b).*

440 plant and animal containers

441 PLANTERS

Generalization. Planters can be used to control factors that influence the growth of plants.

> *Contributing Idea A.* Soilless planters are designed for observations of growth of all parts of a plant for short periods of time.

Constructing soilless containers. Growing plants in soilless containers has the advantage of keeping the seed and the germinating process visible for observations and measurements. Seeds sprouted in such gardens, however, will not grow long under the conditions. During the brief growth period, students can observe how seeds swell and crack, what happens to the seed coat as the embryo pushes through, how the roots and stem begin, how the first leaves develop, and what happens to the seed itself. Generally it is best to soak seeds overnight before placing them in a soilless container. Each of the following methods require only water and a means of transmitting it to the seed. Other creative variations can be made from these examples.

01
Gc

a. Obtain a strip of gauze or sheeting about 6 inches by 8 inches (15 cm by 20 cm). Place seeds upon the strip about 1 inch (2.5 cm) apart. Cover the strip with another dampened strip. Carefully roll the strips and fasten them at the ends with rubber bands. Set the rolled strips in a shallow dish of water. Unroll them every few days to observe.

b. Set a wet sponge in a saucer of water and sprinkle some seeds such as grass seed, radish seed, bird seed, or any individual seeds on it. If the sponge is set in a deep bowl, the bowl can be covered with plastic sheeting to create a greenhouse effect. A porous brick or wad of cotton can be used in place of the sponge.

c. Set a block of wood across a pan of water. Fold a blotter so that it rests across the block like a saddle and so the ends dip into the water in the pan. Place seeds on the blotter. A greenhouse effect can be obtained by inverting a glass jar over the seeds.

d. Place a blotter or layer of cotton on a sheet of glass such as a lantern slide glass. Set seeds on the blotter about 2 inches (5 cm) up from a long side base. Place another sheet of glass on top and fasten the pieces together with several rubber bands. If the selected seeds are large, small strips of wood along the edges will keep the pieces of glass separated. Stand the materials in a pan or tray of water to keep the center moist.

e. Roll a blotter, paper toweling, or cloth and slip it inside a jar or other glass container. Fill the center of the container with peat moss, cotton, excelsior, sawdust, sand, or some similar material. *(Note: Some materials, such as sawdust, tend to discolor the water and blotter.)* Place seeds between the glass and the blotter. Space the seeds evenly about a quarter of the way down the side. Moisten the material in the center of the glass. A little water can be kept in the bottom of the glass, but do not let the water touch the seeds directly. Occasionally, it might be necessary to replace the water that is lost by evaporation.

f. Tie a piece of cloth over the mouth of a small jar. Set the jar in a larger jar or pan, and let the extra cloth length hang down the sides into an inch or two (3–5 cm) of water. A sheet of glass can be placed across the large jar to keep the inside air moist. Scatter seeds onto the part of the cloth that covers the mouth of the small jar.

Contributing Idea B. Soil planters are designed to allow for observations of growth of plants over long periods of time.

Preparing small soil containers. Success in seed planting and germination in soil containers depends upon several variables. The following are suggestions that have provided a great amount of success in classroom experiences. *(Note: Seeds will germinate and grow with less care than described here; however, an oversight of any one of the following factors can hinder germination and growth.)* Many experiments can be designed by controlling for all factors except the one being tested.

02
Gc

a. *Containers.* Seeds can be grown in many kinds of common containers such as waxed paper cups, milk or cream cartons, cigar boxes, wooden cheese boxes, flower pots, and seed flats. Most containers, especially milk and cream cartons, should be thoroughly cleaned before using. If clay pots are used, it is helpful to soak them in water just before use or they might draw much moisture from the soil and away from the seeds.

b. *Drainage.* Containers should have proper drainage for good plant growth. A hole should be punched in containers that do not have holes so that excess water will drain out. The hole should be covered, but not blocked, with a small rock or piece of pottery so that the soil will not wash out. A drainage saucer such as a pie or cake tin or an aluminum foil pan should be placed underneath the container. A coarse-gravel layer in the saucer will aid in the drainage and diffusion of excess water. Because proper drainage is important, do not use glass or other sealed containers. Seeds planted in such containers usually do not thrive, for it is difficult to keep the soil from becoming too soggy or too dry. Soggy soil keeps air from the seeds, while dry soil will not bring enough moisture. If for some reason a container without a hole must be used, drainage can be aided by placing a layer of coarse pebbles about 1 inch (2.5 cm) deep in the bottom of the container.

c. *Soil.* Unless a special soil experiment is being attempted, several layers of soil should be placed in the container.

1. The first layer, about ½ inch (1 cm) deep, should be gravel.
2. The second layer should be of loam or good sandy garden soil. The loam

or garden soil should be sifted to remove rocks and other debris. Loam is composed of sand (which allows water to drain through), organic matter (dead plant or animal matter), and clay. The layer should nearly fill the container. Tap the bottom of the container until the particles settle to within ½ inch (1 cm) of the top. If the layer is not high enough, the plants might get too much water and/or be too shaded. If it is too high, the topsoil might wash away, and the plants might not get enough water.

3. The third or top layer should be a sprinkling of fine soil or sand to cover the surface. Vermiculite or other commercial materials can be used in place of soil. Such materials can be obtained from local nurseries or florists. Vermiculite is a material made from mica, and it holds water better than most soils. Because of this, germination is generally speeded up when it is used. It does not, however, contain nutrients needed for sustained plant growth.

d. *Seeds.* Germination will be more rapid if seeds are soaked overnight before planting. There are several ways in which seeds can be planted.

1. Seeds can be added to the top of the layer of loam, then covered with a ½ inch (1 cm) layer of fine soil or sand. The soil should be pressed firmly with the bottom of a milk bottle or similar implement. Pressing with fingers tends to compact the soil too firmly about the seeds.
2. Seeds can be inserted into the soil. The depth for planting should be in accordance with the depth prescribed on seed packages. If the proper depth for planting is not known, a general rule of thumb is to plant a seed to a depth of 1½ times its length. A mark of the depth to be planted can be made on an unsharpened pencil or stick. The pencil can be pressed into the soil to the mark. Seeds can be dropped into the holes and covered with humus. If seeds are sown onto the soil rather than placed individually, overcrowding might induce rot.

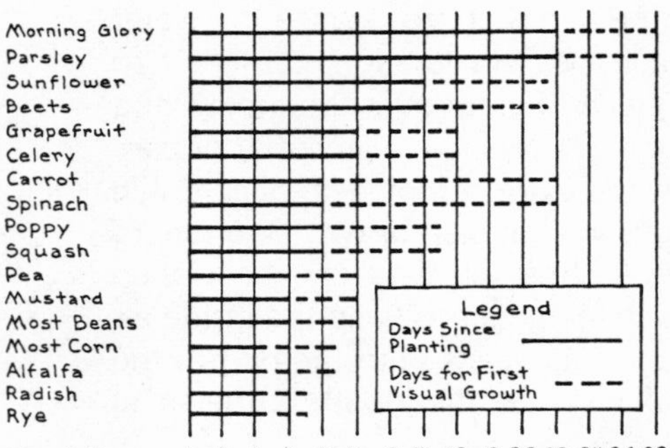

NOTE: Most seeds will germinate over a wide temperature range. Generally seeds take longer to germinate at low temperatures. The germination periods represented in this table are for a temperature range of 65°F–75°F (18°–24°C).

TABLE 441.02

On the other hand, too few seeds in an activity will not allow for casualties or nongerminating seeds. As soon as seeds sprout, place the container in a sunny place. Plants should have as much direct sun as possible unless an experiment measuring the effect of sun energy is being conducted. Lack of sunlight will make seedlings weak and spindly.

e. *Watering.* Soil can be moistened before or after it has been placed in a container.

1. To moisten the soil in advance, knead enough water into the soil to form a soft, but not muddy, ball that will hold its shape in the palm of your hand. This consistency will be right for planting.
2. To moisten the soil after placement, watering can be done from the top before the seeds are planted. Enough water should be added so that it drains from the base. Subsequent waterings are best done by adding water to the drainage saucer before all moisture is gone from it. Watering from the base will not disturb seeds or plants, for the water will move up through the hole in the container and through soil by capillary attraction. If, however, the container must be watered from the top, it is best to water thoroughly every four or five days rather than a little every day. Frequent light waterings tend to draw the moisture out of the soil by capillary attraction and evaporation. Water added to the top of the container should be carefully poured against the side so that the seeds or plants are not disturbed.

f. *Greenhouses.* In the classroom, weekends and vacations can pose a watering problem. Greenhouses will help retain moisture over such periods. Plastic or sandwich-wrap sheets and glass covers can be placed over some of the previously suggested containers to obtain a greenhouse or hotbed effect. With such an effect, temperatures will be raised and evaporation will be reduced.

Preparing large soil containers. Large planters, called Forsythe pans, can be obtained commercially. They can also be made from a large flower pot, 8 inches (20 cm) or more in width at the top, and a small flower pot, 3 inches (8 cm) or less at the top. Arrange the pots as shown. Place a cork in the bottom hole of the small pot and keep the small pot filled with water. Seeds can be planted and grown in the soil in the large pot. **03 Gc**

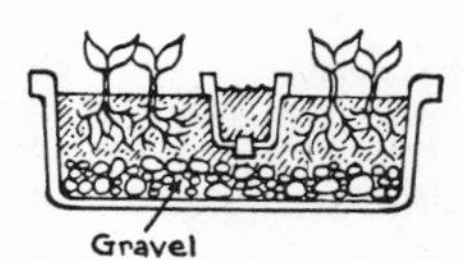

FIGURE 441.03

Preparing outdoor planting areas. The care of outdoor school or home gardens can provide interesting activities for students, but careful planning is necessary to make the activities fruitful learning experiences. Through such experiences, students can carry out their plans, work independently with various phases of gardening, and implement their knowledge about soils, plants, tools, etc. Before planting an outdoor garden, they should find out how much work is involved, then decide whether or not they wish to undertake the job. Outdoor gardening requires time and effort when the climate and other conditions are suitable, and not just when the gardener is ready. It should be noted that many problems concerning soil, climate, altitude, and seed selection can arise because of geographic locations. Because of the great variety of considerations involved in outdoor gardening, the following suggestions are general in nature. **04 Gc**

a. *Locale.* A planting area might be found on the school grounds, on a nearby lot, or on any land area that can be cleared and protected from vandals. The best location is one that receives sunlight all day, is level so that topsoil and seeds will not wash away, has proper drainage even after heavy rain, and has sandy or loamy soil. After a decision is made about the size of the area to be planted and the kinds of seeds to be grown, a plan should be made for the garden. The choice of seeds will determine the spacing and arrangement needed.

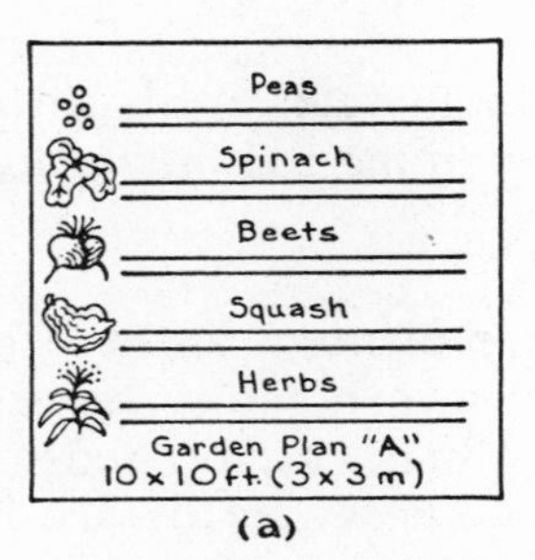

FIGURE 441.04

b. *Soil.* Proper preparation of the soil for an outdoor garden is important. When it is dry enough to crumble easily, the soil should be broken up with a spade to a depth of at least 6 inches (15 cm). Turn over each spadeful and break up the largest lumps as they are dug. A hoe and rake should be used to break up smaller lumps to make a fine, loose soil. Unbreakable lumps, stones, sticks, and other rubbish should be removed. The condition of the soil should be examined. Some soils become acid through time because leaves accumulate in them. Some become alkaline, the opposite of acid soils. The kind of soil needed for proper growth depends upon the kinds of plants to be grown. It is best to obtain information and advice from a local nursery. Generally, soils can be made acid by adding aluminum sulfate or can be rendered neutral by applying lime. Fertilizer is also important, but the kinds of fertilizer the soil needs depend upon what is already in the soil. Again local nurseries can help. Any fertilizer that is added to a soil should be applied according to the directions that accompany the fertilizer. Too much fertilizer can damage plants. Usually it is added to the soil in the spring when it can be worked into the ground.

c. *Planting.* The locale and choice of seeds determine the planting time. For most seeds, planting begins after the soil is prepared and when the weather is warm and not too wet. Because of differences in seeds, not all can be planted at the same time. Follow the directions for planting on seed packages.

d. *Watering.* Do not sprinkle the surface of the soil lightly. This will cause the roots of plants to grow too close to the surface. When the soil is dry beneath the surface, soak the garden slowly and thoroughly. Thorough soaking penetrates the soil deeply and keeps the roots at a proper depth. Such waterings might be done only once a week, depending upon the weather. It is usually best not to water during the heat of the day because much of the water evaporates.

e. *Weeding.* Weeds in an outdoor garden take food from the soil and moisture needed by the growing plants. If weeds are pulled out as soon as they appear, it is easier to keep the plot weed-free. The general appearance of the garden will also be improved.

f. *Transplanting.* At times it might be necessary to move a plant from a smaller container to a larger container or area. Transplanting should be done before plants are very large. Large plants have extensive root systems and more roots are injured when they are transplanted. Smaller plants suffer less damage than larger plants and begin renewed growth much sooner. It is preferable to either transplant in the evening or on an overcast day. If the plant to be moved is in a pot, water the pot thoroughly. Place a hand across the rim with fingers spread and turn the pot over. Tap the pot against something hard. The contents should drop out. If several plants are grown in a flat container, gently separate them. Roots tend to bind the contents of the soil together, so keep as much soil as possible with each plant. Have the new position for the plant prepared ahead of time. The soil should be moist, but not wet. Prepare a hole in the damp soil deep and wide enough to take all the roots. Place the seedling so that the roots are just below the ground level. Firm the soil by gently pressing inward and downward against the roots. Make a saucer-shaped depression around each seedling and give it a good watering. It might help to shade the seedlings for a few days if the weather is hot and sunny. Leafy branches, stiff cardboard, or shingles can be used for this purpose. Water frequently for the first few days until the plants are growing strongly. Frost-susceptible plants should be covered at night if the danger of frost is present.

442 VIVARIUMS

Generalization. Vivariums can be used to control factors that influence environmental conditions.

Contributing Idea A. The aquarium is a simulated environment designed for observing marine plants and animals.

01 Gc

Building aquariums. Almost any clean glass container works well for housing aquatic life. As a general rule, the surface of the water in the container should allow for at least 20 square inches (50 cm^2) of air surface for each inch (.5 cm) of body length of fish. Set the container in a strong light, but not in direct sunlight. Cover the bottom with 1 or 2 inches (3 cm) of clean sand mixed with gravel. Goldfish, guppies, and a variety of plants can be kept in these simple aquariums. They can be fed commercial fish food obtained from a pet store. Tropical or saltwater aquariums require special equipment (see *241.01*).

Contributing Idea B. The terrarium is a simulated environment designed for observing land plants and animals.

02 Gc

Building terrariums. Terrariums can be built from sheets of glass taped together and set in a baking pan filled with plaster of paris for permanent support. A glass top should be used. An aquarium tank or wide-mouth jar can also serve as a terrarium. If a jar is used, place it on its side to provide a larger land surface, and make a plaster, sand, or wood base to keep it from rolling. Various other containers can be used to create different environments (see *241.02*).

450 equipment for collecting

451 COLLECTIONS

Generalization. Collections show ways by which objects can be organized.

Contributing Idea A. Living objects can be collected, preserved, and organized on the basis of similarities and differences.

Collecting and preserving animals. Because of possible damage to habitats and the depletion of species, it is not recommended that students pick up samples of animals for collections. If seashore or other animals are temporarily studied, be sure they are replaced into the ecological niche from which they were taken. A possible exception to this recommendation might be made for insects that are abundant and that generally reproduce in quantity.

01
Dc
Gd

Collecting and preserving plants. Because of possible damage to habitats and the depletion of species, it is not recommended that samples of whole plants be collected. Parts of plants, however, can be collected without endangering an area if students are cautioned against damaging plants and property.

02
Dc
Gd

a. *Seeds.* A great variety of seeds can be obtained from a nursery. Others can be found scattered outdoors. Collected seeds can be kept in small baby food jars. Egg cartons also make fine containers because of their separate compartments. Hosiery boxes and shallow department store boxes make good containers when they are lined with cotton. Students can develop their own rationales for collections, and labels should reflect the rationale.

b. *Leaves.* Leaf specimens should be preserved as soon as possible after picking because they become dry and brittle very quickly. Until they can be mounted, keep them between sheets of moist newspapers in a flat, but natural position. Place a few sheets of another newspaper on top of them. These papers will absorb the moisture as the specimens dry. Place other leaves on top of the second layer of newspapers. Continue this arrangement for several layers, then place flat boards on the last layer of newspapers. These boards can be weighted down by stacks of heavy books. Make sure the weight is heavy enough to keep the specimens from wrinkling. If the specimens are very moist, the procedure might need to be repeated with fresh newspaper. An alternative method is to place the layers of newspapers and leaves between two flat boards. The boards can then be clamped or strapped tightly together, pressing the leaves and papers between them. After twenty-four hours, the specimens can be mounted to fairly heavy sheets of paper with tape, glue, or staples. The sheets can be prepared for scrapbooks or individually framed. Specimens can be labeled and organized into classifications determined by the students.

c. *Flowers.* Flower specimens can be pressed between sheets of waxed paper until they are dry, then mounted and arranged in some orderly way. Similarly, they can be placed immediately on tag paper and covered with transparent shelf paper. Either method preserves the coloration of the flowers for long periods of time.

Contributing Idea B. Nonliving objects can be collected and organized on the basis of similarities and differences.

Collecting rocks. Students can take a nature walk or a field trip to an excavation or rock quarry to collect samples of rocks. Collections can be housed in a variety of boxes (e.g., shoe boxes, cigar boxes, egg cartons). Each can be properly titled (e.g., smooth rocks, red rocks, sedimentary rocks). Individual labels (e.g., shale, quartz) might be added later after checking reference books or identification keys. A local rock collector might be invited to speak to the class and to share some of his or her collection. A list of students' questions prepared beforehand will help the collector with the talk.

03
Dc
Gd

INDEX GUIDE

SYNOPSIS OF SCIENCE-PROCESS CATEGORIES

A OBSERVING

- **Aa** Seeing
- **Ab** Feeling
- **Ac** Hearing
- **Ad** Smelling
- **Ae** Tasting
- **Af** Using Several Senses

B COMMUNICATING

- **Ba** Describing, Speaking, Sounding
- **Bb** Formulating Operational Definitions, Naming
- **Bc** Recording, Tabling, Writing
- **Bd** Researching the Literature, Reading, Referencing
- **Be** Picturing, Drawing, Illustrating
- **Bf** Graphing

C COMPARING

- **Ca** Making General Comparisons or Comparisons from Different Perspectives
- **Cb** Estimating
- **Cc** Making Numerical Comparisons, Counting
- **Cd** Measuring Lengths, Angles
- **Ce** Measuring Temperatures
- **Cf** Weighing
- **Cg** Measuring Areas, Volumes, Pressures
- **Ch** Making Time Comparisons, Measuring Rates

D ORGANIZING

- **Da** Seriating, Sequencing, Ordering
- **Db** Sorting, Matching, Grouping
- **Dc** Classifying

E EXPERIMENTING

- **Ea** Identifying a Problem, Formulating Questions
- **Eb** Hypothesizing
- **Ec** Controlling and Manipulating Variables, Testing

F INFERRING

- **Fa** Generalizing, Synthesizing, Evaluating
- **Fb** Using Indicators, Predicting
- **Fc** Using Explanatory Models, Theorizing

G APPLYING

- **Ga** Using Knowledge or Instruments, Identifying Examples
- **Gb** Inventing, Creating
- **Gc** Constructing
- **Gd** Growing, Raising
- **Ge** Collecting